BRITISH NUCLEAR WEAPONS AND THE TEST BAN 1954–1973

For my dad - John MacKinnon Muir Walker 1927–2010

British Nuclear Weapons and the Test Ban 1954–1973

Britain, the United States, Weapons Policies and Nuclear Testing: Tensions and Contradictions

JOHN R. WALKER
Foreign and Commonwealth Office, UK

Routledge
Taylor & Francis Group
LONDON AND NEW YORK

First published 2010 by Ashgate Publishing

2 Park Square, Milton Park, Abingdon, Oxfordshire OX14 4RN
711 Third Avenue, New York, NY 10017

Routledge is an imprint of the Taylor & Francis Group, an informa business

First issued in paperback 2018

British Library Cataloguing in Publication Data
Walker, John R.
British nuclear weapons and the test ban 1954–1973 : Britain, the United States, weapons policies and nuclear testing : tensions and contradictions. 1. Nuclear weapons–Government policy–Great Britain– History. 2. Nuclear weapons–Government policy–United States–History. 3. Nuclear weapons–Testing–Law and legislation–History. 4. Nuclear arms control–History. 5. Deterrence (Strategy)–History–20th century.
I. Title
327.1'747'0941-dc22

Library of Congress Cataloging-in-Publication Data
Walker, John R., 1960-
British nuclear weapons and the test ban 1954—1973 : Britain, the United States, weapons policies and nuclear testing : tensions and contradictions / John R. Walker.
p. cm.
Includes bibliographical references and index.
ISBN 978-1-4094-1112-3 (hardcover : alk. paper)
(ebook) 1. Nuclear weapons—Government policy—Great Britain—History. 2. Nuclear arms control—Great Britain—History. 3. Nucelar weapons—Testing—Law and legislation—History. 4. Nuclear Test Ban Treaty (1963)—History. 5. Great Britain—Foreign relations—United States. 6. United States—Foreign relations—Great Britain. I. Title.
U264.5.G7W35 2010
355.02'17094109045—dc22

2010024208

ISBN 978-1-4094-1112-3 (hbk)
ISBN 978-1-138-38397-5 (pbk)

Contents

List of Abbreviations

ABM (T)	Anti-Ballistic Missile (Treaty)
ACDA	Arms Control and Disarmament Agency
ACSA (N)	Assistant Chief Scientific Adviser (Nuclear)
AWRE	Atomic Weapons Research Establishment
CCD	Conference of the Committee on Disarmament
CDS	Chief of the Defence Staff
CTBT	Comprehensive Nuclear Test Ban Treaty
DARPA	Defence Advanced Research Projects Agency
DGAW	Director General Atomic Weapons
DOD	Department of Defense
DOP (O) C	Defence and Overseas Policy (Official) Committee
DRPC (AES)	Defence Research Policy Committee (Atomic Energy Subcommittee)
ENDC	Eighteen Nation Disarmament Committee
FCO	Foreign and Commonwealth Office
FO	Foreign Office
HE	High Explosive
HEU	Highly Enriched Uranium
HMA	Her Majesty's Ambassador
HMG	Her Majesty's Government
IAEA	International Atomic Energy Agency
JCCAE	Joint Congressional Committee on Atomic Energy
JCS	Joint Chiefs of Staff
JOWOG	Joint Working Group
MIRV	Multiple Independently-targetable Re-entry Vehicles
MOD	Ministry of Defence
MOU	Memorandum of Understanding
MRC	Medical Research Council
NATO	North Atlantic Treaty Organisation

NPT	Non-Proliferation Treaty
NRDC	Nuclear Requirements for Defence Committee
NTAC	Nuclear Tests Advisory Committee
NTPC	Nuclear Tests Policy Committee
PNE	Peaceful Nuclear Explosions
PTBT	Partial Test Ban Treaty
RAE	Royal Aircraft Establishment
REB	Re-entry Body
ROF	Royal Ordnance Factory
SAGW	Surface to Air Guided Weapon
SALT	Strategic Arms Limitation Talks/Treaty
SLBM	Submarine Launched Ballistic Missile
SPS	Single Point Safety
SRPAG	Seismic Research Programme Advisory Group
UKAEA	United Kingdom Atomic Energy Authority
ULMS	Underwater Long-range Missile System
UNDC	United Nations Disarmament Commission
UNGA	United Nations General Assembly
USAEC	United States Atomic Energy Commission

Chronology of UK Nuclear Tests, Testing Moratoria and Test-Ban Treaty Negotiations 1952–1974

1952

3 October	UK nuclear test (25 kilotons), Hurricane, Montebello, Australia.

1953

15 October	UK nuclear test Totem 1 (10 kilotons), EmuField, Australia.
27 October	UK nuclear test Totem 2 (8 kilotons), Emu field.

1956

16 May	UK nuclear test Mosaic (15 kilotons), Montebello.
19 June	UK nuclear test Mosaic (60 kilotons), Montebello.
27 September	UK nuclear test Buffalo One Tree (17 kilotons), Maralinga, Australia.
4 October	UK nuclear test Buffalo Marcoo (> 2 kilotons), Maralinga.
11 October	UK nuclear test Buffalo Kite (3 kilotons), Maralinga.
22 October	UK nuclear test Buffalo Breakaway (16 kilotons), Maralinga.

1957

15 May	UK nuclear test Grapple Short Granite (300 kilotons), Malden Island.
31 May	UK nuclear test Grapple Orange Herald (c800 kilotons), Malden Island.
19 June	UK nuclear test Grapple Purple Granite (200 kilotons), Malden Island.
2 July	UK proposes technical studies on control
14 September	UK nuclear test Antler Tadje (1 kiloton), Maralinga.
25 September	UK nuclear test Antler Biak (6 kilotons), Maralinga.
9 October	UK nuclear test Antler Taranaki (25 kilotons), Maralinga.
8 November	UK nuclear test Grapple X (1.8 megatons), Christmas Island.

1958

31 March	Soviet Union announces suspension of its test programme.
28 April	UK nuclear test Grapple Y (3 megatons), Christmas Island.
1 July– 22 August	Conference of Experts to Study the Possibility of Detecting Violations of a Possible Agreement on Suspension of Nuclear Tests.
22 August	UK nuclear test Grapple Z Pennant 2 (24 kilotons) Christmas Island.
22 August	UK and US announce test suspension of one year if Soviet Union follows suit and progress is made towards setting up an effective control system. Eisenhower proposes negotiation on a test ban treaty.
2 September	UK nuclear test Grapple Z Flagpole (1.21 megatons), Christmas Island.
11 September	UK nuclear test Grapple Z Halliard (800 kilotons), Christmas Island.
23 September	UK nuclear test Grapple Z Burgee (25 kilotons), Christmas Island
31 October	Conference on the Discontinuance of Nuclear Weapons Tests opens in Geneva.

1959

23 June–15 July	Technical Working Group 1 meets in Geneva as part of the Conference on the Discontinuance of Nuclear Weapons Tests to discuss high altitude nuclear tests.
20 July	UK announces that it will not resume testing for as long as negotiations on test ban treaty remain useful.
19 November– 19 December	Technical Working Group 2 meets in Geneva as part of the Conference on the Discontinuance of Nuclear Weapons Tests to review underground nuclear explosion identification issues.
29 December	Eisenhower announces that US would consider itself free to resume testing after 31 December, but would not do so without giving notice.

1960

29 March	Eisenhower and Macmillan propose technical research programme by three nuclear powers to address problems of detecting underground tests and that once treaty banning

	atmospheric test concluded. As soon as a test ban treaty was signed and research programme put in hand, they would be ready to begin a voluntary undertaking to refrain from holding any undergrounds tests below seismic level 4.75 mb for an agreed period.
3 May	Soviet Union proposes period for voluntary ban on small yield tests should be four to five years.
11–30 May	Seismic Research Programme Advisory Group meets in Geneva as part of the Conference on the Discontinuance of Nuclear Weapons Tests.

1961

21 March	Conference on the Discontinuance of Nuclear Weapon Tests resumes and UK and US table new proposals in Geneva on outstanding differences over the control system.
18 April	UK and US table complete draft treaty prohibiting tests in the atmosphere, under water and in outer space together with all underground tests above 4.75mb. It provided for a quota of 20 on-site inspections per year in the Soviet Union.
29 May	UK and US offered to substitute a variable quota of between 12 and 20 on-site inspections annually on the basis that within these limits each side would have the right to inspect 20 per cent of unidentified events above 4.75 mb.
30 August	Soviet Union announces end to its test moratorium
1 September	Soviet Union resumes testing.
3 September	UK and US propose that the three nuclear powers agree not to conduct tests in the atmosphere.
9 September	Conference on the Discontinuance of Nuclear Weapon Tests goes into recess.
15 September	US conducts underground test in Nevada
27 November	Soviet Union proposes treaty to ban all tests in atmosphere, outer space and under water. Verification would be based on national control only.
28 November	Conference on the Discontinuance of Nuclear Weapon Tests resumes.

1962

16 January	At the Geneva Conference the UK and US reject Soviet 27 November proposals and instead suggest resumption of negotiations on an internationally verifiable test ban treaty or that the Conference should adjourn and remit the test

	ban question to the new Eighteen Nation Disarmament Committee (ENDC).
29 January	Conference on the Discontinuance of Nuclear Weapons Tests holds final session in Geneva.
1 March	UK nuclear test Pampas, Nevada Test Site (about 10 kilotons).
2 March	US announces intention to resume testing.
14 March	ENDC holds first meeting in Geneva.
21 March	ENDC sets up nuclear test ban sub-committee consisting of UK, US and Soviet Union.
25 April	US begins atmospheric tests at Christmas Island.
27 August	UK and US table two alternative treaties in Geneva: a comprehensive test ban with minimum verification and a limited version covering tests other than those conducted underground. Soviet Union rejects both.
26 November	ENDC reconvenes. UK and US offer to reduce on-site inspections quota in proposed treaty from 12 to 20 to eight to ten.
7 December	UK nuclear test Tendrac, Nevada Test Site

1963

14–31 January	UK, US and Soviet Union hold private talks on test ban issues in Washington and New York.
12 February	ENDC reconvenes. Soviet Union makes no advance on its offer of two to three on-site inspections.
1 April	UK and US table Memorandum of Position summarising their proposals on on-site inspections.
10 June	UK and US announce that Lord Hailsham and Averill Harriman would visit Moscow for talks on test ban issues on 15 July.
18 June	UK announces that it would not conduct further atmospheric tests unless other countries did so first.
21 June	ENDC goes into recess.
25 July	UK, US and Soviet Union agree Partial Test Ban Treaty in Moscow.
5 August	Partial Test Ban Treaty formally signed in Moscow.
10 October	Partial Test Ban Treaty enters into force.

1964

17 July	UK nuclear test Cormorant, Nevada Test Site.
25 September	UK nuclear test Courser, Nevada Test Site.

1965

10 September UK nuclear test Charcoal, Nevada Test Site.

1974

23 May UK nuclear test Fallon, Nevada Test Site.

Acknowledgements

This book originated in my contribution to the Arts and Humanities Research Council funded project on the history of the UK nuclear weapons programme from 1953 to 1973. This project was based at the University of Southampton's Mountbatten Centre under the direction of Professor John Simpson. The original plan was that whilst on a part-time exchange from the Foreign and Commonwealth Office's Arms Control and Disarmament Research Unit between 2005 and 2007, I would contribute material for the then envisaged three volumes covering the periods 1953–58, 1959–64 and 1965–73 on the UK nuclear weapons programme and arms control. My original intention was to look at the Comprehensive Nuclear Test Ban Treaty (the test ban), the cut-off in the production of fissile material for weapons purposes, the Nuclear Non-Proliferation Treaty and the Anti-Ballistic Missile Treaty and the Strategic Arms Limitation Talks. Once I started looking into papers available at The National Archives it quickly became apparent that there was more than enough original documentary evidence to devote an entire book just to the interrelationships between the UK's nuclear weapons programme and the test ban – hence this volume.

Although this book is almost entirely based on primary sources several former defence scientists offered their comments and advice, both on the draft and in oral history seminars on more general points. I am extremely grateful to Frank Panton, Peter Marshall, Alan Douglas, Eric Carpenter, Frank Whiteway, Ken Johnston, Alwyn Davies, Roy Dommett, and Tom Lukeman. I benefited from discussion with, and advice from, scholars of British and Soviet nuclear history: John Simpson, Richard Moore, Robin Woolven, Kristan Stoddart, Brian Jamison, David Holloway, Lorna Arnold (a model and inspiration for any nuclear historian) and Kate Pyne. I would also like to thank all those who participated in the project advisory board and study groups and in particular to Angela Murphy and Marion Swinerd for their part in helping with this project. It goes without saying that any errors of fact or interpretation are my sole responsibility.

Although this book is based on research at The National Archives, I did have access to a few retained papers at AWE Aldermaston and Harwell in order to check a few points of detail on UK positions at the 1958 Geneva Experts' meeting and the reasons for the self-imposed testing moratorium between 1965 and 1974. I am grateful to the archivists at both establishments for facilitating my access. The information in these documents, however, was not used, paraphrased or cited in any of the pages that follow. This book thus in no way constitutes any sort of official history. I should also add that the views expressed in here are my own and not those of the Foreign and Commonwealth Office.

I would like to thank in particular Jez Littlewood for agreeing to participate in the exchange arrangement between Southampton University and FCO that enabled my participation in the UK nuclear weapons history project. I am extremely grateful too to all those at Ashgate for their help and commitment: Beth Dixon, Tom Gray, Nicole Norman, Claire Percy and Albert Stewart. And finally thanks to Lorna Miller for some very timely editorial advice – I am such a hopeless proof reader.

Chapter 1
Introduction

Historians have thus far ignored the impact of arms control on the UK nuclear weapons programme. Yet given the wealth of material now available in the National Archives, it is possible to undertake a detailed review of the one arms control issue that could have had the most impact on the nascent British weapons development programme, a test ban. There was also the possibility of a freeze on fissile material production, but it is arguable that a comprehensive test-ban treaty, had it come about, would have brought an end to UK weapon development in the 1950s. Parallel developments were occurring in the arcane world of warhead design and UK strategic and tactical nuclear plans and the rather more public arena of international negotiations for an end to all nuclear weapons testing. There were times when UK nuclear weapon objectives were contradictory and this story is as much about the Eden and Macmillan Conservative governments' attempts to square this circle as it is about warhead R&D. This story also involves an exploration of the nature of defence planning, Anglo-American relationships, the efficacy of British diplomacy and UK contributions to arms control and disarmament, both at the general and detailed technical levels. A key question for this study is to see just how the UK managed to balance the conflicting pressures created by its determination to become and remain a credible nuclear power whilst wanting to pursue disarmament objectives. The relative weight of these pressures changed over the period in question – 1954 to 1973 in response to both domestic and international imperatives. Charting these changes and their relationships is another key question to address in this study.

Whilst all this was going on at the diplomatic and political level, Britain struggled to build and sustain an effective nuclear weapons stockpile throughout the late 1950s, 1960s and early 1970s. This book will deal with the interrelationships and contradictions between the test ban and weapons development by providing a narrative of the key events that shaped this aspect of British nuclear weapons history. The Conservative governments of Harold Macmillan were stalwart supporters of the ban and played a key role the bringing about the Partial Test-Ban Treaty in 1963.[1] Between 1958 and 1962 the UK, along with the US and the USSR, participated in the Tripartite Conferences on the Discontinuance of Nuclear Weapon Tests, which sought to produce a comprehensive ban on all types of nuclear tests save for peaceful nuclear explosions in specific circumstances. The UK's involvement in these efforts at some of the key stages is highly significant,

---

1 Kendrick Oliver, *Kennedy, Macmillan and the Nuclear Test-Ban Debate, 1961–63* (Basingstoke, 1998), p. 209.

in particular how UK Ministers and officials pursued or reacted to arms control and disarmament initiatives. What is being offered here is not a detailed blow-by-blow account of the long and painful negotiations that took place in the Geneva-based Conference on the Discontinuance of Nuclear Weapon Tests. Rather it will highlight some of the key events and issues that help illuminate UK thinking and attitudes to working with the US; to verification; to wider nuclear arms control negotiations; and to the relationship between its own weapons programme requirements and arms control imperatives. Inevitably there will be some detailed narrative describing the chronology of some of the events in this period, which is necessary to help us understand the nature of UK thinking and rationales at particular moments in our story. The extensive footnoting is quite deliberate. A key objective of the Arts and Humanities Research Council (AHRC) funded UK Nuclear Weapons History project at the Mountbatten Centre was to provide a reference for other scholars who wish to explore some of the issues affecting the nuclear weapons programme, as well as its relationship to arms control, in much greater detail. A good deal of the available archival material has only been summarised in some places here. More detailed accounts of many aspects of this story could be produced, such as the procurement and deployment histories of individual nuclear weapons, or British attitudes to the institutional aspects of future test-ban treaty implementation. This book is not the final word.

Chapter 2 will look at the early pressures between 1954 and 1958; the UN Disarmament Committee and the Grapple tests; pre-1954 UK-US relations; AWRE's first attempts at thermonuclear designs and tests; and the growing public and international concerns. As well as the key Antler tests in Australia, Chapter 3 will examine the decisive year of 1958, which witnessed the successful planning and conduct of the Grapple Y and Z tests and which coincided with pressures for a moratorium and the resumption of the UK-US nuclear relationship with signing of the 1958 and 1959 Mutual Defence Agreements. Chapter 4 will look at the testing moratorium and how it impacted on the development of the first UK megaton range weapons: Yellow Sun; the US-designed Mark 28 warhead and its UK counterpart Red Snow; and the critical role that the Assessment Trials in Australia played in warhead design and adoption. Future UK test requirements and options will also be reviewed. Chapter 5 covers the testing moratorium and British responses to US policies. The Tripartite Conference on the Discontinuance of Nuclear Tests opened in 1958 and the latter part of this chapter will chronicle UK objectives and role in the Conference between 1958 and 1961 as well as Anglo-American differences. British efforts to solve the test-ban verification problems are reviewed in Chapter 6, which also provides the first detailed examination of the emergence of forensic seismology in the UK. This looks at the extent to which seismic research in the UK underpinned British approaches to test-ban verification. Chapter 7 begins with the end of the testing moratorium in 1961 and UK responses – a test of the new implosion system Super Octopus and other nuclear tests. The US request to use Christmas Island and the resumption of atmospheric testing presented significant problems and opportunities for the UK. Also, the Discontinuance Conference came to an end in 1962 and the shift of test-ban negotiations moving to the Eighteen Nation Disarmament Committee (ENDC). Chapter 8 deals with the lead up to, and participation by the UK in the Partial Test-Ban Treaty (PTBT) 1962 to 1963. Chapter 9 looks at the testing plans in the 1963–65 period including the Polaris warhead and WE177; UKAEA's proposed research test programme in 1964; the Polaris Economy tests; the arrival of Labour; and the impact on the testing and weapons programme of its 1965 defence review. British policy towards a CTBT from the mid 1960s until the early 1970s is reviewed in Chapter 10; this looks at the decline in superpower interest in the test ban and UK efforts to address the verification issues. Chapter 11 deals in detail with AWRE's future and the Kings Norton review; the Wilson government's nuclear policy review (no new strategic weapons); first steps in the Polaris Modernisation debate leading to Antelope, Super Antelope, STAG, Hybrid, Option M and Chevaline choices being addressed in 1966–73; and the decline and revival of Anglo-American nuclear relationship. As this modernisation work progressed the British return to testing with the UK effects tests and the decision to resume underground testing and the impact of this on CTBT policy is also examined. Finally, the concluding chapter will discuss how priorities were reconciled; when a test ban was a subordinate objective to the weapon programme requirements and when these priorities were reversed; the nature and impact of the UK-US relationship on UK test-ban policies; and the problems in acquiring and sustaining an effective and credible nuclear deterrent. Finally, the lessons of the British experience generally and for arms control and non-proliferation are reviewed in Chapter 12.

We should not forget that writing nuclear weapons history faces certain problems. These include very high levels of security and compartmented information systems that existed during the Cold War period – a strict need to know policy where only a handful of individuals were privy to the information; complex, diverse and changing bureaucratic structures responsible for or related to nuclear policy and implementation; changing understandings over time of the basic science and engineering of nuclear weapons; continued closure of archival material beyond the 30-year rule; sometimes contradictory and misleading evidence in open archives; and an ever diminishing band of original participants. None of these are new or novel. Given such obstacles historians might be forgiven if they decided to abandon any attempt at writing a comprehensive account of the UK nuclear weapons programme. However, the picture is not all bad; there is a surprisingly large amount of archival material available in The National Archives at Kew and in other places that enable a reconstruction of the detailed broad themes and directions of the UK programme. We can supplement archives with oral histories. The key is to be aware of the partial nature of the picture that might be painted and avoid extravagant claims based on limited evidence. Much can then be said about the UK programme, certainly in terms of main thrusts of policy, policies, rationales and a good deal too on the weapons systems themselves.

The threat (or opportunity) of a test-ban treaty hung over the UK nuclear weapons programme for much of the period covered by this book; the interrelationships changed over time as programme and international pressures varied. Exactly how these interrelationships operated at particular moments in time will, hopefully, become clearer in the following chapters.

Chapter 2
Early Pressures: 1954–1958

Introduction

The way in which the UK managed to cope with two diametrically opposed policy requirements in this period – being seen to favour limitations on testing, whilst resisting them at the same time as the national weapons programme strived to meet its demanding goals – has not been examined in detail before with the benefit of official papers. What ideas did the UK have on possible constraints on nuclear testing and to what extent were these driven by the need to be seen to respond to domestic and international concerns over atmospheric testing of megaton warheads. How did the UK's own nuclear testing programme, planned weapons deployments and test-ban pressures interact in this formative period? Could the UK demonstrate that it knew how to design, build and test successfully an H-Bomb before a test ban came into force?

Opening Gambits: The Thermonuclear Test Programme and Pressure for Cessation

International pressure for an end to atmospheric nuclear testing in the mid and late 1950s coincided with a key moment in the development of the British nuclear weapons programme. The Cabinet Defence Committee decided in 1954 that the UK should possess thermonuclear weapons and Aldermaston embarked on a crash programme to meet this requirement.[1] Politically the UK faced what appeared to be irreconcilable objectives: the need for thermonuclear weapons and the need to be seen to address domestic and international public concerns over the health, environmental and international security effects of nuclear tests. Sir Michael Wright, leader of the UK delegation to the 1958–62 negotiations for much of their existence, remarked in 1961 that it was not until the amendment of the 1954 US Atomic Energy Act that enabled the UK to access US know-how that the UK became committed supporters of a stand-alone test-ban treaty.[2] A key feature of this period was that AWRE was, as Arnold and Smith observe, working under pressure

---

1 Lorna Arnold, *Britain and the H-Bomb* (Basingstoke, 2001), pp. 110–11.

2 The National Archives (TNA) FO 371/157105, IAD 22/426/G, Geneva Nuclear Tests Conference: Review of Events to September 9, 1961 Sir Michael Wright to Lord Home (Received September 27) p. 3.

to complete its R&D programme as quickly as possible to beat the inevitable ban on nuclear testing.[3]

A joint MOD and UKAEA report noted in August 1955 that since international prestige was, at that time, so closely associated with prowess on the nuclear field, it was essential to avoid any agreement on prohibition of tests until the UK had conducted a successful test of a megaton weapon. Abolition of full-scale trials would seriously impede Britain's ability to make reliable megaton weapons, and it would therefore never possess any valid nuclear deterrent in the future when, if nuclear disarmament negotiations succeeded, the UK would have to live with the possibility of the clandestine Soviet retention of such weapons. In such circumstances, the only policy open to the UK was to let others make the suggestions on test bans and to try to influence US policy if Washington showed signs of agreeing to a complete abolition of tests.[4] Given the rising tide of public opinion against the further testing of thermonuclear weapons, it was imperative that the first UK trial take place as early as possible. For this reason, as Lt Gen Frederick Morgan, Controller of Atomic Weapons, Ministry of Supply, told the Atomic Weapons Trials Executive, the date for them was provisionally fixed as April/May 1957.[5]

The Defence Research Policy Committee (Atomic Energy Sub-committee) (DRPC (AES)) informed the Chiefs of Staff in mid December 1955 that political pressure in the UK and throughout the world against further megaton trials was increasing, despite the firm stand taken by HMG in parliament. It was thus possible that the UK's first megaton trials would have to be abandoned on political grounds. It was therefore essential that the series should be planned in such a way as to safeguard the future by obtaining the greatest possible amount of scientific knowledge and weapon design experience from them, to create a sound foundation for the UK's megaton development programme.[6] MOD opposition to even a one-year limitation on testing was considerable.[7] Eden himself informed the House of Commons that the UK would not accept arrangements that would put itself in a position of decisive inferiority to other great powers. That said, the UK was prepared to discuss methods of regulating and limiting test explosions that took

---

3 Lorna Arnold and Mark Smith, *Britain, Australia and the Bomb, The Nuclear Tests and Their Aftermath* (Basingstoke, 2006), p. 86; see also pp. 82–5 for the context in which test-ban pressures grew.

4 TNA AB 38/515, Operation Circus: a plutonium safeguards exercise 1955–58, Control of Atomic Weapons Prepared by B.T. Price and V.H.B. Macklen Part I.

5 TNA AVIA 65/827, Operation Gazette minutes of a meeting held on 1 September 1955.

6 TNA AIR 8/2468, COS Committee The British Megaton Warhead Trial Series Report by the Atomic Energy Sub-Committee of the DRPC COS (55) 336, 13 December 1955.

7 TNA PREM 11/1676, F.J. Stephens, MOD to P.F. de Zulueta, 10 Downing Street, 5 December 1955.

account of the UK's position as well as that of other powers.[8] At that time the limitations of British knowledge meant that the most certain way of producing a trial megaton explosion in 1957 was to use a large fission assembly in a Blue Danube ballistic case. Such a device would be big, heavy and extravagant in fissile material and could only be used as a free falling bomb. Sir William Penney (Director, AWRE) told the Chiefs in December 1955 when they met to discuss the DRPC (AES) paper that he would have preferred that the trial of Orange Herald (a small light megaton warhead for the Blue Streak intermediate range ballistic missile) should not be carried out in 1957 as this missile would not enter production until 1965. However, the danger that trials might be banned at some time in the near future made it essential that the UK should take every opportunity to test when it could.[9]

There were doubts at that time as to whether a further round of megaton trials would be possible by 1960 given the international opposition to atmospheric testing, so there seemed little choice but to test Orange Herald in 1957.[10] The Foreign Office representative at the same Chiefs of Staff meeting emphasised the growing political constraints arising from political opposition to atmospheric testing. In particular, the increasing emphasis on disarmament and the Medical Research Council's report on the dangers of radioactive pollution, which would be published before the 1957 thermonuclear tests, now code-named Grapple, might seize the public imagination. Thus the 1957 tests would need careful public presentation. Further tests might prove to be out of the question and the UK should therefore make the most of the opportunity.[11] In the meantime Aldermaston's task was to produce megaton range weapons for the RAF and only once that had been achieved was the UK ready to support a nuclear test ban.

Successful results from the Grapple tests scheduled for the spring of 1957 would certainly do a lot to change US attitudes to the sharing of information. However, there were concerns that this test might constitute the one and only

---

8 Hansard, House of Commons, Oral Answers, Column 1956, 6 December 1955.

9 TNA AVIA 65/1193. The initial estimate of the weight of the Orange Herald warhead was 4500 lb, but this would have required the MRBM to be a twin motor missile. A warhead of only 1800 lb could be produced but this would require a larger quantity of fissile material – nearly double the quantity needed for the 4500 lb warhead. C.G.W.L. to D.C.A.S and C.A.W, Medium Range ballistic Missile Warhead Weight – Effect on Range, 8 August 1955. With the knowledge then available at AWRE and having looked again at the weight limitations, likely longitudinal accelerations and vibrations in all directions, Penney could not guarantee to make a satisfactory warhead with a 1800 lb weight. W.G. Penney to Lt Gen Sir Frederick Morgan, Ministry of Supply, 20 September 1955.

10 TNA AIR 8/2468, Chiefs of Staff Committee The British Megaton Warhead Trial Series, Report by the Atomic Energy Sub-Committee of the DRPC, COS (55) 336, 13 December 1955.

11 TNA AIR 8/2468, Chiefs of Staff Committee Confidential Annex to COS (55) 105th Meeting, 20 December 1955.

British trial of this nature that would be permitted.[12] It was vital for political reasons that the target date of April or May 1957 be met.[13] There were certainly concerns in the MOD, and recognised too by the Ministry of Supply that the UK might only get one shot at testing megaton weapons.[14] For this reason therefore Penney planned to include in the trial four separate high yield explosions, although this would be very expensive in fissile material and raised opportunity costs through loss of service weapons as a result. The amount of fissile material required might reduce by half the number of Blue Danube weapons delivered to the RAF in 1956 and 1957, with the possibility of additional effects on deliveries scheduled for 1957 and 1958.[15]

Early Thoughts on Control

Domestic opposition to nuclear testing was an important factor in shaping British policy in the mid and late 1950s.[16] A moratorium on H-Bomb tests was first proposed by the Indian Prime Minister, Nehru, in April 1954 and was discussed by Sir Anthony Eden with John Foster Dulles in April, May and June of that year.[17] The UK view at that point, in the light of the relative state of progress in the

---

12 TNA AVIA 65/824, C.A.W. to C.A. et al, 3 October 1955. There was also another consideration at work here; the amount of fissile material needed for tests might reduce by half the numbers of weapons of the present type that could be delivered in 1956–57. Untitled minute, October 1955.

13 TNA AVIA 65/824, Note of a Meeting on Tuesday, 11 October 1955

14 TNA AVIA 65/824, L.J. Dunnett, Ministry of Supply to Sir Frederick Brundrett, Ministry of Defence, 6 December 1955.

15 TNA AVIA 65/824, October 1955. In fact the number of weapons planned for the Air Ministry in 1956/57 was 28, but the fissile material demands would reduce this to 15. G.A.C.Witherridge A.S./A.W. to Dep.Sec. (C) 20 October 1955. In May 1955 the Atomic Energy (Official) Committee had agreed that the military allocation of HEU would be as follows: 435 kg, 1956/57, 530 kg in 1957/58 and 750 kg for 1958/59. We can therefore see clearly why there would be concerns over weapons that were 'expensive' in their use of HEU; the Orange Herald designs used up to 120 kg of fissile material. TNA AVIA 65/775, Future Programme of Atomic Weapon Trials Appendix A and TNA CAB 134/756, Cabinet Official Committee on Atomic Energy A.E. (O) (55) 127, Fissile Material for Civil Purposes, Note by the Atomic Energy Authority, Tables I and IV, 9 November 1955.

16 J.P.G. Freeman, *Britain's Nuclear Arms Control Policy in the Context of Anglo-American Relations, 1957–68,* (London, 1986), pp. 95–6; see also DEFE 7/922, Note on Limitation of Nuclear Tests, 12 April 1957. This noted that the testing issue had become a highly political question because of health concerns and it was necessary to formulate a plan for the limitation of tests provided that this could be done without serious interference to military objectives.

17 TNA PREM 11/1676, Proposals for Regulation and Limitation of Tests of Nuclear Weapons, Ministry of Defence brief for the Prime Minister's visit to Washington, Walter Monkton to PM, 19 January 1956,

development of the H-Bomb reached by the US and USSR was that there might be considerable advantage in supporting the proposals.[18] A moratorium would give strategic advantage to the Western powers as the USSR did not at that point have a 'true' thermonuclear device; that is a two-stage device (or double-bomb in UK parlance) in which radiation from an exploding fission device is used to implode thermonuclear fuel in an adjacent spherical or cylindrical container. This is the so-called Teller-Ulam configuration.[19] The US was at first attracted by the idea for the same reason, but subsequently decided not to support it in view of the difficulties of delineating a precise limit to the size of permissible explosions and of proving that the limit had been overstepped. There was also the risk that the Russians might acquire all the information they needed from one successful megaton explosion that they would justify as an error in calculation. For this reason the UK decided, on reflection, to oppose the call for a moratorium, which in any case became unacceptable once the UK had decided in the summer of 1954 to embark on a programme to build thermonuclear weapons.[20] Despite this, Ministers and officials recognised that political pressure to ban tests might be easier to hold if the UK showed a readiness to consider plans for limiting them. Provided that the UK's position was safeguarded, there were thought to be considerable propaganda advantages for the West in not leaving the initiative to the Russians and Indians.[21] Moreover, such willingness might in fact ease rather than complicate HMG's plans to conduct megaton tests in 1957.[22]

There were further ideas on test limitations in 1955 that would limit H-bomb tests to one a year, but this was totally unacceptable to the UK. One of the main reasons for this was that it would be essential to have a successful result in the one and only permitted test. If limited to one explosion, this would have to be so arranged to guarantee a success; the efficient use of fissile material and development of an operational weapon would have to take second place. In order to ensure certainty, and to get the maximum value from the tests in the development of operational weapons the UK had to plan for both, which meant more than one

---

[18] TNA CAB 130/101, Brief for the Foreign Secretary, Proposed Moratorium on Experimental Hydrogen Bomb Explosions, 21 April 1954.

[19] See Chuck Hansen, *US Nuclear Weapons, The Secret History* (New York, 1988), p. 24; Lorna Arnold, *Britain and the H-Bomb* (2001), pp. 7, 15; David Holloway, *Stalin and the Bomb* (Yale, 1994), p. 307 and p. 314. A Soviet two-Stage device was not ready for testing until November 1955.

[20] TNA DEFE 13/353, Proposals for Regulation and Limitation of Tests of Nuclear Weapons, 1955.

[21] TNA PREM 11/1676, Telegram no. 9 from Secretary of State aboard RMS Queen Elizabeth to Foreign Office, 26 January 1956.

[22] TNA PREM 11/1676, Telegram no. 24 from Secretary of State aboard RMS Queen Elizabeth to Foreign Office, 27 January 1956.

explosion. Moreover, the economics of mounting such trials meant that it did not make any sense to incur such expense for only one shot.[23]

In considering the need for tests and a test ban, the Prime Minister was advised in mid December 1955 that while a distinction could be made between two types of kiloton weapon (pure fission weapons, all-up weight one to five tons, yield ten to 100 kilotons and tactical weapons, all up weight 300 to 1000 lb, yield one to 15 kilotons), megaton weapons were in a different category. It was conceivable that a clever technical country could make a pure fission weapon successfully without a single large-scale test. The yield would not be known within a factor of two either way unless the design was very conservative and fissile material was used extravagantly. Tactical weapons were much more problematic. Full-scale tests were almost essential. At this stage, the UKAEA believed that two or three tests would bring tremendous improvements in design, enabling production of three or four weapons instead of one. Several successful tests were thus necessary to develop a megaton weapon – a large yield weapon equivalent to one million tons of TNT and which need not necessarily be produced by fusion. It was at about this time, autumn 1955, that Penney and AWRE were developing the basis of what would become the UK's two-stage radiation implosion device.[24] At this point, the size of the yield was a key factor, not necessarily how it was achieved – large fission device, boosted or two-stage. As would become clear later, demonstrating an ability to produce a two-stage device was the key to unlocking the door to US nuclear secrets. Of these tests, at least two would have to be in the megaton range.

It was already certain, or so it seemed at that particular time to AWRE, that many versions of megaton weapons were possible: size, weights, yields and fissile contents could all be varied over a considerable range. Making them was another matter. If kiloton tests were prohibited the UK would be limited to fission weapons and to a crude design of tactical weapons, which would be very extravagant in fissile material and too unreliable to satisfy UK military requirements. If megaton weapons tests were prohibited, then the UK would be unable to produce and stockpile such weapons. Supply of both efficient kiloton tactical (the Red Beard weapon which was to replace the RAF's first nuclear weapon Blue Danube) and megaton strategic weapons to the services was thus dependent on the ability to test.[25]

Given this advice it is thus hardly surprising that Sir Anthony Eden, having succeeded Churchill as Prime Minister, told the House of Commons on 6 December 1955 that, 'while HMG will at all times welcome arrangements which contribute to world security, they are not prepared to accept arrangements which would put the UK in a position of decisive inferiority to other great powers, a position which is not justified by the state of our scientific knowledge and resources. HMG are

---

23 DEFE 13/353, A Minute to Mr Hanna, 5 December 1955.

24 Arnold, *Britain and the H-Bomb*, pp. 86–90.

25 TNA PREM 11/1676, Lord President to Prime Minister, The Status of Kiloton and Megaton Weapons in UK Strategy, 13 December 1955.

however prepared to discuss methods of regulating and limiting test explosions that take account of their own position as well as that of other powers'.[26] This in a nutshell was the UK position on a test ban or testing limitations until autumn 1958.

The Eden government had thus no intention of departing from its decision to manufacture and possess the 'hydrogen bomb as a deterrent to war'. It was an overriding concern to UK defence and foreign policy that the Grapple tests should be carried out as soon as possible. The tests would demonstrate that the UK was in a position to achieve its policy objectives of creating 'the major deterrent to war'. From the early Grapple planning phases at Air Ministry level, it had been made clear to the officials that it was of paramount importance, particularly politically, that the first Grapple trial as planned took place in April/May 1957.[27] For this reason the RAF was pressed by AWRE to take all possible action to meet a target date of April 1957. In the event the target date was put back to 1 May because of the inability of the technical control ship Narvik to complete her tasks in the 1956 Australian tests at Monte Bello, refit and position off the target island in the Pacific in the time required. The navy was unable to provide any alternative ship.[28] In order to do this planning had to start in February 1956.[29] The logistical planning aspects were considerable; there was no time to lose if everything was to be ready; if not, there was a risk that the tests might never take place for political reasons.[30] The Prime Minister made this clear to the US Secretary of State John Foster Dulles at the start of February 1956 that HMG would have to stand out

---

26 Hansard, House of Commons, Oral Answers, Columns 195 and 156, 6 December 1955.

27 TNA AIR 2/18184, F. Morgan C.A.W. to C.A.N., 1 November 1955. At that stage the device to be tested first was Green Bamboo. Orange Herald might also be included and to meet the date of April, it was going to be necessary for both of these warheads to be tested in 'lash-up' ballistic casings, probably that used the Blue Danube. Bomber Command was told that the trial should take place as soon as possible and that in order to prove the design of the firing circuits and the associated arming and safety devices, a comprehensive flying programme was required. Rabbles, ACAS (Operations) to AOC Bomber Command, 30 November 1955. By February 1956 the RAF were told to expect that they were to test four high yield weapons in the megaton range. R.B. Lees, ACAS (Ops) to ACAS (P) et al, 2 February 1956; the provisional dropping programme saw a Boosted Red Beard as the first live drop, Green Granite as the second, Green Bamboo as the third and Orange Herald as the last. R.B. Lees, ACAS (Ops) to CAS et al, Operation 'Grapple' Dropping programme, 4 October 1956. Later still the plan was to use Purple Granite for the fourth drop. This was believed to be 'an experimental boosted fission assembly'. A. Hughes PS to ACSA to PS Secretary of State, 5 June 1957.

28 TNA AIR 2/18184, R.B. Lees ACAS (Ops) to DCAS Operation 'Grapple' – Postponement of Target Date for First Live Drop, 3 May 1956.

29 TNA AIR 2/18184, R.B. Lees ACAS (Ops) to ACAS (P) et al, 2 February 1956.

30 TNA AIR 8/2468, Chairman Chiefs of Staff to ACAS Operation 'Grapple', 2 February 1956.

against any suggestion that tests should be suspended.[31] He emphasised that after the forthcoming US tests there was likely to be an upsurge of popular feeing that testing should stop. HMG would have to stand out against any such suggestion and hoped that the US would do the same.[32] This was not surprising, given that Eden had already agreed that Grapple should consist of four explosions.[33] If it were possible to make some proposals on this issue that would help to allay public apprehension about the health and safety aspects, the UK would be responsive. In a later meeting with Eisenhower, however, Eden again asked whether there could not be some mention of the regulation of tests in their communiqué after the meeting. The President was most reluctant to do this; he and his advisers left Eden believing that they thought regulation was impracticable and that continued testing had the highest value for military, scientific and research purposes. They agreed, however, that if the UK was asked about this in Parliament or by the press, Eden could say that the subject had been discussed and that the two governments held the firm conviction that if properly conducted, nuclear tests were not harmful to health.

Following on from this, Selwyn Lloyd, who had succeeded Eden as Foreign Secretary, told Dulles at the end of January 1956 that HMG did not intend to interrupt its own planned series of tests for 1956 and 1957.[34] Until the UK had completed these, its policy was to oppose any specific limitation that frustrated this objective. Until satisfactory disarmament arrangements were in place, which meant that effective verification measures were included in any arrangement, the UK refused to be placed at a disadvantage. Both the US and USSR had tested megaton weapons; the UK had not yet done so. Moreover, the UK certainly took the view privately and publicly that tests could be conducted safely if proper precautions were taken.[35] Yet in a January 1956 minute to the Prime Minister on the proposed tests, the Minister of Supply noted that, as opposition to tests was likely to grow, the UK could not rely on having further opportunities to test after the planned four Grapple shots.[36] And when the Chancellor of the Exchequer queried the need to test so many devices, the Minister of Defence repeated the point that the probability was that the UK would not be able to conduct tests after 1957, or indeed even ever again. This meant that maximum value had to be derived from the

31 TNA CAB 21/3090, Record of a Meeting at the White House, A Policy Regarding Tests of Nuclear Weapons, 1 February 1956.

32 TNA DEFE 7/2289, Record of a Meeting at the White House, 1 February 1956.

33 TNA AVIA 65/827, Atomic Weapons Executive, Operation Grapple Minutes of a Meeting, 1 February 1956.

34 TNA CAB 21/3090, Record of a Meeting at the White House on 31 January 1956.

35 TNA FO 371/123121, ZE 112/103, Banning of Nuclear Test Explosions, Draft brief for the Soviet leader's visit on the banning of nuclear test explosions, 5 April 1956.

36 TNA AVIA 65/824, Draft minute from Minister of Defence to Prime Minister under cover of letter Sir Richard Powell, Ministry of Defence to L.J. Dunnett, Ministry of Supply, 6 January 1956.

forthcoming trials.[37] As noted above, testing the Orange Herald device, the planned warhead for Blue Streak, should ideally have been postponed for three years on purely technical grounds until 1960. However, the combination of political factors (the test-ban pressures and the financial – the need to make maximum use of the logistics of conducting tests in the Pacific) made it unwise to assume that the UK could mount any further series of tests a few years after Grapple.[38] By May 1956 both Ministers and officials were worried that a growing weight of international opinion was developing against any continuance of major test explosions and thus reinforcing the belief that it would be politically impossible for the UK to hold any further major tests after those planned for the spring of 1957. At that time annual nuclear trials, both major and minor, were envisaged by the UK as being a technical necessity and an annual event from 1959 onwards.[39] Yet the UK might find itself compelled, by the force of public opinion, to accept an international prohibition of future tests.[40] In such circumstances, Ministers believed that it might be advantageous for the UK to take the initiative and propose discussions between the three weapon states on an agreement for the limitation, and then prohibition, of nuclear tests.[41]

Eden told the Cabinet in early June 1956 that he planned to announce the dates of the Pacific tests on 7 June, but that the US was concerned that any new initiative

---

37 TNA AVIA 65/824, Walter Monkton, Minister of Defence to Chancellor of the Exchequer, 8 February 1956.

38 TNA AVIA 65/824 Operation Grapple Policy 1955–56, Walter Monkton, Minister of Defence to Chancellor of the Exchequer, 8 February 1956. Note Trials of British Megaton Weapons, Ministry of Defence; TNA AVIA 65/1193, E.S. Jackson to Deputy Secretary noted on 24 August 1956 that 'this was one of the main reasons for testing Orange Herald at this Stage. Apart from the fact that in Green Bamboo we are trying to obtain a megaton yield economically, and in Orange Herald a megaton yield from the smallest possible warhead, the two warheads differ in the method of boosting. It is important to know which method will be the more effective'. Orange Herald was tested in 1957, but this put the warhead out of step with developments in missile engineering and re-entry body design. Not quite a chicken and egg problem, but the development was unable to proceed in a more coordinated and logical fashion because of the test-ban imperatives. TNA AVIA 65/1193, E.S. Jackson, DGAW to J.E. Serby, DG/GW, 16 June 1958.

39 TNA AVIA 65/870, F.B. Lloyd, DAW Trials to DGAW, Plan for the Future use of Maralinga, 11 May 1956.

40 The majority of Foreign Office papers at this time deal with domestic and international concerns and opposition to atmospheric testing: see for example, TNA FO 371/123131, 123137, 123138 and 123138 from 1956 and TNA FO 371/123131 from 1957. The National Council for Abolition of Nuclear Weapons Tests provided but one example of the sorts of groupings opposed to testing and it lobbied Ministers for the immediate cessation of all UK testing activities. TNA FO 371/129244, ZE 112/204, 27 June 1957; see also Arnold and Smith, *Britain, Australia and the Bomb*, pp. 76–8.

41 TNA CAB 130/115, GEN.529/1st Meeting, Nuclear Tests and Radiation, 28 May 1956.

by the UK on testing might make life difficult for Eisenhower in the forthcoming Presidential election. Another problem was the frosty relationship between Eden and John Foster Dulles, the US Secretary of State, occasioned by disagreement over handling French actions in Indochina in 1954 when Eden had been Foreign Secretary. Despite this, Eden thought that the UK should still make some sort of proactive statement on test limitations, but given US concerns, however, he proposed only to use the same language from his December 1955 statement to Parliament, namely that the UK would be prepared to join in any discussions on test limitations. The Cabinet agreed to this, but Parliamentary pressure was intense.[42] The Foreign Secretary briefed his Cabinet colleagues on 11 June that he and the Prime Minister were down to respond to several parliamentary questions on the testing issue.[43] He fully expected that the supplementaries would press HMG to state whether it would be take any initiatives on limiting testing. He asked for agreement to indicate that the UK would be ready to discuss international limitations without linking it to conclusion of a comprehensive agreement on disarmament. Although this line was agreed, it appears to have occurred in ignorance of the fact that the Foreign Office Minister of State had already announced to the UN Disarmament Commission meeting in London the day before that control should form an integral part of a disarmament programme. The UK could not make such an overnight change in policy, so Lloyd recommended that the UK stick to its general line, which was in accord with US preferences at that point. The bottom line was that any such scheme would have to take into account the need to safeguard essential UK testing requirements.[44]

As the international debate over this issue evolved, the Cabinet decided in early July that they would now take the line in Parliament that HMG would be ready to limit tests without insisting as before that discussion should be in the context of a comprehensive disarmament agreement.[45] This was a major departure from existing disarmament policy. However, it still seemed desirable to say nothing about the different possible methods of limiting test explosions, primarily because the exact details had not been fully analysed.[46] Lloyd, feeling the pressure from

---

42 TNA CAB 128/30 Part I, C.M. (56) 39th Conclusions, Item 9 Nuclear Weapons, 5 June 1956.

43 TNA CAB 128/30 Part II, C.M. (56) 48th Conclusions, Item 2 Nuclear Weapons Tests, 11 July 1956.

44 TNA CAB 128/30 Part II, C.M. (56) 49th Conclusions, Item 8 Nuclear Weapons Tests, 12 July 1956.

45 TNA FO 371/123702, UN 1192/527, Foreign Office telegram no. 631 to New York, 12 July 1956.

46 TNA FO 371/123702, UN 1192/527, Foreign Office telegram no. 633 to New York, 12 July 1956.

Parliament, thought above all that the UK should adopt a more forthcoming line instead of the negative attitudes taken to date.[47]

The Medical Research Council's (MRC) report on 'The Hazards to Man of Nuclear and Allied Radiations', which appeared in June 1956, was more reassuring about the risk than the public might have expected.[48] Its conclusions certainly ensured that the Foreign Secretary was able to convince his colleagues to maintain their 22 February decision to carry out a series of thermonuclear tests in the spring of 1957 and to announce it to Parliament before publication of the MRC report on 12 June.[49] Back in February, the Cabinet had thought that it was premature to make any announcement on the dates for the Christmas Island tests as this might only stimulate political controversy. Ministers decided that when the UK announced the forthcoming tests, HMG would also emphasise its willingness to discuss methods of regulating and limiting test explosions. This decision was directly attributable to the content of the MRC's report, which noted that if the contemporary rate of testing increased, and especially if greater numbers of thermonuclear weapons were exploded, there might be a risk of radiation levels increasing to a level where ill effects might be produced in a small number of the population.[50] The Prime Minister told his Cabinet colleagues in July 1956 that he would like to put some practical proposals to the US for an international system setting out some limits on testing and that a study of options for this should be undertaken as soon as possible.[51] Lloyd strongly agreed given his concerns over public reactions to the tests and suggested the government would expect to take some fresh initiative before very long. The Cabinet shared this view and the Foreign Secretary was invited to arrange for officials (also including some from the MRC) to study the issue and report back. An underlying concern here was that the first thermonuclear series planned for 1957 might not be successful. US and Soviet experience had shown that development of thermonuclear weapons required more than one series of tests, which suggests that the UK was analysing US test debris.[52]

---

47 TNA FO 371/123703, UN 1192/551/G, J.D. Murray to A.R.K. Mackenzie, UK Delegation to the UN, 12 July 1956.

48 TNA CAB 128/30 Part I, C.M. (56) 39th Conclusions, Item 3, Nuclear Weapons Tests in 1957, 5 June 1956.

49 TNA FO 371/123125, ZE 112/149/G, Nuclear Tests Memo by the Foreign Secretary, 1 June 1956; TNA CAB 12830 Part 1 C.M. (56) 16th Conclusions Item 12, 22 February 1956.

50 TNA FO 371/123125, ZE 112/167, Submits brief for Cabinet Meeting to be held on 7 June 1956, 6 June 1956.

51 TNA DEFE 7/2289, Extract from Cabinet Conclusions C.M. (56) 51st Conclusions, 20 July 1956.

52 TNA CAB 131/16–20, D.C. (56) 6th Meeting, 10 July 1956. Michael Goodman notes that the proficient monitoring of foreign nuclear test bomb debris played an important role in supporting UK's own nuclear weapons programme. See Michael Goodman, *Spying*

By the autumn of 1956 AWRE had come to the view that there were really only two routes to a megaton airdrop weapon: a large fission weapon in which they had confidence and the more problematical double-bomb design which they had started to work on.[53] The scientists and engineers doubted that they would have sufficient time to develop this design before tests were stopped by international agreement.[54] Another factor at play here was the uncertainty on whether neutron 'boosting', in any form would work.[55] This in turn was not helping the RAF and Air Ministry settle the Operational Requirements for its future weapons, including both Blue Steak and the stand-off propelled bomb – Blue Steel scheduled for entry into service in 1963. It seemed impracticable from the warhead point of view to give a firm estimate of the size and weight and yield until such times as AWRE had had the chance to evaluate the results from the Grapple tests – autumn of 1957 at the earliest. Air Ministry personnel were concerned that if there was a delay until then, and the warhead dimensions proved to be different from those that had been advised were likely to be achieved, introduction of Blue Steel into service could be delayed.[56] In the circumstances, this left Air Ministry personnel no option but to use arbitrary figures for their then current missile design work.[57]

Something Must be Done: A Possible Compromise?

In October 1956 the Foreign Secretary and Minister of Defence presented a paper to the Cabinet emphasising the growing pressure in the UK and elsewhere for abolishing all tests of nuclear weapons. Although Ministers lamented that much of this pressure was probably based on loose thinking or ignorance, it was very real and was an important factor in international and domestic affairs.[58] The UK

---

*on the Nuclear Bear, Anglo-American Intelligence and the Soviet Bomb (*Stanford, 2007), pp. 87 and 111.

53 A double-bomb was one in which a fission bomb was used to detonate an adjacent bomb containing fusionable material.

54 Arnold, *Britain and the H-Bomb* Appendix 3, The History of British R&D on Atomic Weapons. Probably written by Dr John Corner, p. 236.

55 Boosting entails solid lithium compounds or tritium gas being present or injected into the core of a fission device. If successful, this would mean that more of the core would fission, and thus much less fissile material would be needed.

56 TNA AIR 2/13748, W.R. Brotherhood, D.O.R. (C) to DGAW and D Arm, Ministry of Supply, 4 October 1956.

57 TNA AIR 2/13748, E.S. Jackson, DGAW, Ministry of Supply to Air Commodore W.R. Brotherhood, D.O.R. (C), 10 October 1956. He suggested that they opt for 9 feet in length, 46 inches in diameter with a gross weight (including a safety device in the weapon) of 5700 lb.

58 TNA CAB 129/83, C.P. (56) 243, Nuclear Weapons Tests Memorandum by the Secretary of State for Foreign Affairs and the Minister of Defence, 24 October 1956.

had to be seen to be doing something. At that stage a complete ban was out of the question for the UK given the state of the UK nuclear weapons development programme. The MOD was particularly concerned to avoid any action that would interfere or could interfere with the imminent Grapple series.[59] As a compromise, the Cabinet contemplated interim proposals to limit the world annual average of explosions to 15 megatons of fission for the three weapon states until such times as the nuclear powers could devise better systems of restraint. However, the UK would not propose any initiative publicly until such times as the US had been consulted given the ongoing campaign for Eisenhower to be re-elected in the November Presidential elections.[60] AWRE had earlier advised that the sort of proposal that could be lived with if push came to shove would be that the UK, US and USSR should each agree to limit their total test yields to five megatons of fission energy per year. AWRE also recognised that a lower figure of three megatons would probably be sufficient to meet UK requirements.[61] The origin of this idea went back to December 1955 when the Prime Minister had suggested to Sir William Penney that the UK should look into the implications of limiting nuclear weapons tests to one a year. In response Penney thought that it would be much more sensible and more in the UK's interests to have a 'ration' of megatons a year rather than a 'ration' of explosions. He felt that the UK weapons programme would not be hindered if it could be agreed that each country conducting tests gave an undertaking that the total amount of fission that it would cause in weapons tests in any calendar year would not exceed five megatons. The UK at least still had to conduct kiloton trials for a number of years yet.[62]

However, the Cabinet Defence Committee took the view that a three megaton limit might unduly constrain the UK's ability to develop megaton weapons; it would be better to have an equal share between the three testing states.[63] This idea nevertheless was only meant as a basis for discussion in the first instance with the US. Its purpose was simply to deflect domestic and international pressure for a test ban and to ameliorate the uncomfortable position that UK was in given its own

---

59 TNA DEFE 7/2289, R.R. Powell to Minister, Nuclear Weapons Tests, 24 October 1956.

60 TNA CAB 129/83, C.P. (56) 243, Nuclear Weapons Tests Memorandum by the Secretary of State for Foreign and the Minister of Defence, paragraph 11, p. 4, 24 October 1956.

61 DEFE 7/2289, A.M. Allen, UKAEA to Sir Richard Powell, Ministry of Defence, 9 July 1956.

62 TNA AB 16/1541, Sir Edwin Plowden to the Lord President, CH (55) Item 1, 8 December 1955 and Sir William Penney to Sir Edwin Plowden, UKAEA, 7 December 1955.

63 TNA CAB 131/16–20, D (56) 11th Meeting, Item 2 Nuclear Weapons Tests, 20 December 1956.

compelling testing requirements.[64] Indeed Antony Head, the Minister of Defence, made it clear to Cabinet that no system of limitation should restrict UK freedom to carry out its 1957 tests.[65]

Subsequent Cabinet discussion generated agreement that a UK initiative on limitations should be taken. Ministers recognised that if testing continued fall-out would increase to dangerous levels within the next generation. A maximum level of 12 megatons of fission per year would be more satisfactory than 15. There was little need for the UK to seek exact parity with the US or USSR, and in fact a four-four-three ratio might offer the prospect of an agreement which would limit on radioactive fallout to a more stringent 11 megatons.[66] Technical level discussions on these ideas had taken place with the Americans on 30 January 1956. Although US officials saw that some such scheme might be feasible, there was still a long way to go before the UK could convince all US agencies that the proposals were both safe and necessary from a technical or political point of view.[67] Views in Washington were split with the United States Atomic Energy Commission (USAEC) and Pentagon vociferously opposed to any test limitation in 1956.[68]

This presented problems for the Foreign Office. The Americans had convinced Sir William Penney that the whole basis on which it was proposed to control the limitation of tests was no longer valid because of their latest technical developments. However, it is not clear what these were though their existence could make a test ban even more harmful to the UK nuclear weapon programme. There was thus the prospect that the US would agree to any proposal to limit tests; and this view was unlikely to change in the very near future. Officials concluded therefore by February 1957 that it would be better to accept this view and make a bargain with the US whereby the Americans would make no moves towards a test ban unless agreed first with the UK.[69]

Uncertainty at the political level was matched at the engineering level where those responsible for designing the missiles that would carry the UK's megaton range weapons did not yet know the dimensions, centre of gravity and weight of the warhead to be carried. Much would depend on the results of the first thermonuclear

---

64 TNA DEFE 13/353, Telegram no. 9 from Secretary of State for Foreign Affairs aboard RMS Queen Elizabeth to Foreign Office, 26 January 1956.

65 TNA CAB 128/030 Part II, C.M. (56) 74th Conclusions, Item 6 Nuclear Weapon Tests, 25 October 1956.

66 TNA CAB 128/030 Part II, C.M. (56) 74th Conclusions, Item 6 Nuclear Weapon Tests, 25 October 1956.

67 TNA DEFE 13/353, Washington to Foreign Office, telegram no. 217, 31 January 1957.

68 Susanna Schrafstetter and Stephen Twigge, *Avoiding Armageddon. Europe, the United States, and the Struggle for Nuclear Nonproliferation, 1945–1970* (London, 2004), p. 88.

69 TNA FO 371/129802, UN 1192/169/G, I.T.M. Pink to J.D. Murray c/o UK Delegations to the UN, New York, 28 February 1957.

tests in Grapple. The Air Ministry noted in September 1956, in relation to the stand-off self-propelled bomb Blue Steel, that as far as the warhead was concerned they were still in the early experimental stages. It would be difficult, if not impossible, to guarantee the yield required within a given weight and size until AWRE had a chance of digesting some of the information acquired in Grapple. It was thus impracticable from the warhead point of view to give a firm estimate of the size and weight and yield until September 1957 at the earliest. If the Air Ministry waited until then and the dimensions proved to be different from those likely to be achieved, introduction of Blue Steel into service could be delayed. To avoid this delay, the Ministry thought that it might be necessary to obtain as good an estimate as it could of the warhead's size and weight and give these details to the missile designers. They could then subsequently accept a lower yield from the warhead if Grapple experience showed that the desired yield could not be achieved within the weight and dimensions of Blue Steel. Alternatively, there was always the possibility that the Grapple results might be better than expected and the warhead weight would be lower with a consequent increase in range.[70]

In the spring on 1957 the re-elected President Eisenhower and the new UK Prime Minister, Harold Macmillan, signed a private agreement at their Bermuda meeting in which they both agreed not to air test-ban limitation proposals publicly without first consulting the other. The US would also not air proposals on a cut-off without first checking with the UK.[71] In a separate and public statement on testing, Eisenhower and Macmillan noted *inter alia* that over the past months the UK and US governments had considered various methods of limiting tests. They had now jointly concluded that in the absence of more general nuclear control agreements of the kind they were seeking, a test limitation could not at that time be effectively enforced for technical reasons; nor could breaches of it be detected with confidence. They believed nevertheless that even before a general agreement was reached, self-imposed restraint could and should be exercised. The UK and US would continue to test only in such a manner as would keep world radiation from rising to more than a small fraction of the level that might be hazardous.[72]

---

70 TNA AIR 2/13748, W.R. Brotherhood, DOR (C) to DGAW and D Arm Ministry of Supply, 4 October 1956.

71 Harold Stassen, the President's disarmament special advisor, subsequently did just this on cut-off and caused heightened anxiety in London. Macmillan and his colleagues were not best pleased: '… Is this America's reply to our becoming nuclear power – to sell us down the river before we have a stockpile sufficient for our needs. Some of my colleagues suspect this.' Harold Macmillan, *Riding the Storm 1956–1959* (London, 1971), pp. 300–301. London feared that the cut-off date proposed, if accepted, would have serious effects on the UK military programme unless satisfactory arrangements were made in advance for the supply of fissile material.

72 TNA PREM 11/2166, Annex II Bermuda Meeting March 21–24, 1957; James C. Hagerty, Press Secretary to the President, C. Peter Hope, Foreign Office Spokesman, 24 March 1957.

This change in the UK policy position was embarrassing to explain away to Parliament, the UN Disarmament Commission (UNDC) Sub-Committee and domestic public opinion. A Foreign Affairs debate in the House of Commons on 23–24 July 1956 confirmed Ministerial concerns about domestic opinion. The opposition had made much more of disarmament and tests than in any debate for some time.[73] Although Ministers had ruled that possible methods of limiting and controlling tests should be examined, they accepted that in practice officials might not be able to suggest any suitable method, and HMG would be forced back on a simple expression of interest. And of course any new ideas would have to be discussed with the Americans first. A key point at this stage was that Ministers and officials were only thinking in terms of limitations: cessation was out of the question, and had to be coupled with the cessation of the production of nuclear weapons.

Moreover, it had become very clear to the UK at the end of 1956, following the conversations that Penney had in Washington on his way home from the Mosaic trials, that the key to access US nuclear secrets was that the British scientists should know how to make megaton weapons.[74] Arnold and Smith write that the growing likelihood of a global moratorium and/or ban on tests thus put a premium on finding dependable ways to produce megaton weapons quickly. Thus it was vital to be sure that UK calculations on boosting the yield of a fission weapon using light elements would not in fact lead to large yields. This was therefore the context of the Mosaic tests in 1956. AWRE wanted to examine the impact on yield of use of a layer of lithium deuteride surrounding the fissile core.[75] Mosaic was thus a trial to obtain information for future tests, especially Grapple and its results

---

[73] TNA FO 371/123705 UN 1192/593, P. Dean to Sir Roger Makins, Washington, 30 July 1956.

[74] The DRPC (AES) decided on 13 April 1955 that a test was required in April 1956 in order to get much needed information prior to the testing of the 'large bomb' in 1957. The reason for bringing this forward to April 1956 was that if AWRE waited until the Maralinga range was ready in October 1956, it would be too late for further investigation and changes for the 1957 tests in the Pacific. TNA AIR 2/18184, William Cook, Deputy Director AWRE to Air Marshall T.G. Pike, DCAS, Air Ministry, 15 April 1955. However, there were difficulties in producing lithium 6 for this trial and AWRE postponed it until May. TNA AIR 8/2469, DCAS to CAS and VCAS Mosaic, 26 October 1955.

[75] Arnold and Smith, *Britain, Australia and the Bomb*, pp. 106–7; AWRE had work well in hand by the end of 1955 in designing lithium 6 deuteride into weapons components. In November 1955 Plowden anticipated that the UKAEA would be able to test these in atomic trials during the next two to three years – see AB 16/2631, Lithium Supply and Requirements 1954–59. However, after the Mosaic trials, AWRE was less confident in its need for large quantities of lithium, but the Grapple trials the following year might increase the requirement substantially; AB 16/2631, Sir William Cook, AWRE to Sir Donald Perrott, UKAEA, 2 November 1956. The Soviet test Joe-4, in August 1953 used solid lithium deuteride; this would have been ascertained from radiochemical analysis of bomb debris see Arnold, *Britain and the H-Bomb*, p. 27 and Holloway, *Stalin and the Bomb*, p. 307.

were much less certain than usual.[76] As anticipated, this route to megaton yields was a dead end, but the two tests did make essential contributions to AWRE's knowledge of fusion reactions and its results were crucial for the planning for Grapple.[77] However, as Grapple was the priority for AWRE, this meant that it could not finalise the choice of safety device for Yellow Sun – the service megaton weapon planned to enter into service in 1959.[78] So the testing schedule was having damaging consequences for the service weapons programme.

During Anglo-American discussions on 7 December 1956 on the latest US disarmament proposals, Harold Stassen (Special Assistant to the President for Disarmament Policy) expressed interest in the latest UK ideas on limitation. He had previously expressed the strongest misgivings about any limitation of tests in advance of a ban on the production of fissile materials for weapon purposes. This was one reason for Lloyd and Head considering that the prospects of securing US agreement to UK proposals were not very good.[79] Britain's ability to pursue a truly independent nuclear arms control initiative then, or in the future, which was at variance with fundamental US interests, was simply non-existent. The UK still had to prevaricate when it came to discussing any limitations on testing. Until UK programme requirements were met, the emphasis was very much on limitation (see for example, the UK-French proposals to the UNDC Sub-Committee): outright banning would come later.[80] This remained UK policy in June 1956.

Competing Goals: A Test Ban and the Deterrent: Grapple May–June 1957

As Ian Clark observes, the hallmark of British policy in 1957 was its great sense of urgency, designed to achieve as much as possible before any constraints on atmospheric nuclear tests could be agreed or were imposed.[81] In fact the thermonuclear programme was conducted against the clock: the dates for the Christmas Island tests were set for political rather than technical reasons. By way of example of the sorts of extraneous political pressures HMG was under, the Japanese had called upon the UK several times to refrain from testing at Christmas

---

76 TNA PREM 11/1677, Minister of Supply to Minister of Defence, 13 March 1956.

77 Arnold and Smith, *Britain, Australia and the Bomb*, p. 131.

78 TNA AIR 2/13680, P.B. Wood to D.D.O.R.2, 2 May 1956.

79 TNA CAB 124/84 C.P. (56) 285, Nuclear Weapon Tests Memorandum by the Secretary of State for Foreign Affairs and the Minister of Defence paragraph 5 p. 2, 19 December 1956.

80 Cmd. 9770, *Report of the Proceedings of the Sub-Committee of the United Nations Disarmament Committee held at Lancaster House, London March 19–May 4, 1956*, London, 1956, p. 26.

81 Ian Clark, *Nuclear Diplomacy and the Special Relationship Britain's Deterrent and America 1957–1962* (Oxford, 1994), p. 191.

Island.[82] Such was the anxiety created by Japanese opposition that the UK went as far as to withdraw temporarily its 1955 declaration accepting compulsory jurisdiction of the International Court of Justice; this was done to forestall any Japanese attempt to take the UK to court in attempt to stop the nuclear testing programme at Christmas Island.[83] Embassy staff in Tokyo continued to receive a steady stream of delegations coming in to protest, as well as petitions against the forthcoming UK tests from a wide range of civil Japanese organisations.[84] And after news came through of the first test, there were two days of protests outside the Embassy in Tokyo.

The dates for the tests were so far in advance of what AWRE and MOD would have wished from the experimental and resources point of view that there were risks of failure, and officials knew that the Americans thought the same.[85] AWRE was preparing five possible assemblies; in the worst case four devices would be exploded, in the best only two would be needed. It was not even possible to determine the order of the shots well in advance because of the extremely tight programme on supplies of fissile material and the order in which the test assemblies would be ready. However, it does seem that in late January 1957, the firing sequence was to start with the most speculative design Green Granite 2 (two-stage radiation implosion device) and the one whose capacity to make a megaton explosion was the least in doubt, Orange Herald, would be fired later.[86] Meanwhile, even before the Grapple firing plan had been finalised, Sir William Cook, Deputy Director AWRE, was suggesting that authority be sought to plan on the basis of a further series of megaton trials in 1959.[87] He gave E.S. Jackson, Director General Atomic

---

82 TNA PREM 11/2166, Foreign Office telegram no. 96 to Tokyo, 6 March 1957; TNA PREM 11/2163, Sir Esler Dening to Mr Selwyn Lloyd, Japanese Reactions to United Kingdom Tests, 26 March 1957.

83 TNA PREM 11/2163, The Hague telegram no. 65 to Foreign Office, 12 April 1957; Foreign Office telegrams nos. 1377 and 1378 to New York, 12 April 1957; TNA FO 371/129266, ZE 212/245/G, Foreign Office telegram no. 1377 to New York, 12 April 1957 and Foreign Office telegram no. 1378 to New York, 12 April 1957.

84 TNA FO 371/129266, ZE 212/259, R.W. Selby, British Embassy Tokyo to H.C. Hainworth, Permanent Under Secretary's Department, 29 March 1957; ZE 212/262, R.W. Selby, British Embassy Tokyo to H.C. Hainworth, Permanent Under Secretary's Department, 5 April 1957; ZE 212/264, R.W. Selby, British Embassy Tokyo to H.C. Hainworth, Permanent Under Secretary's Department, 12 April 1957; and ZE 212/269, R.W. Selby, British Embassy Tokyo to H.C. Hainworth, Permanent Under Secretary's Department, 19 April 1957.

85 TNA DEFE 24/291, Note to Minister of Defence, 17 January 1957.

86 TNA DEFE 12/1012, Note of a meeting with the Minister on Thursday 24 January 1957, Grapple paragraph 1; the proposal to test the lighter version of Green Granite appears to have been put forward in December 1956. TNA AVIA 65/825, E.S. Jackson, DGAW, Ministry of Supply to Sir Frederick Brundrett, Ministry of Defence, 2 January 1957.

87 TNA AIR 2/18184, L.J. Dunnett to Sir Frederick Brundrett, Ministry of Defence, 15 April 1957.

Weapons in the Ministry of Supply, his best guess for likely testing requirements between 1958 and 1960 on 1 April.[88] This was, according to Cook, no more than a glance into the cloudy crystal ball. AWRE's wish list included an ambitious programme of eighteen tests:

1958 Maralinga
Test of techniques for small Green Granite for Blue Streak – 2 rounds
Test of strong source technique for two-stage trigger – 2 rounds
Test of artillery shell technique (in principle) – 1 round
Test of Seaslug (an anti-ship missile) warhead (in principle) – 1 round
Test of a mixed core (plutonium and HEU) boosted warhead (Red Beard or Blue Fox) – 1 round

1959 Christmas Island
Test of series of small thermonuclear warheads – 4 rounds

1960 Maralinga
Test of strong source warheads – 2 rounds[89]
Test of prototype stage 1.5 air to surface missile – (a) minimum fissile material for 5 kilotons (b) maximum yield – 2 rounds
Test of prototype Seaslug – 1 round
Test of artillery shell – 2 rounds

Fissile material for these tests would cost some £7 million for the Maralinga trials and £10 million for Christmas Island.[90] However, no budget had been set aside for this and the assumption had been that once the spring 1957 trials at Christmas Island were over no attempt would be made to maintain the airstrip or keep construction troops deployed. On reflection Cook subsequently told Jackson that if it were possible to have further trials at Christmas Island in 1959, six rounds instead of four might be desirable.[91] This ambitious plan was never put into effect for technical (results of Grapple X) and political (the testing moratorium of October 1958) reasons. However, the Ministry of Supply and AWRE continued to take the view that, if knowledge was to increase and there was to be an orderly and progressive development of weapons to meet service requirements, it would

---

88 TNA ES 1/1137, W. Cook to E.S. Jackson, DGAW, Ministry of Supply, 1 April 1957.

89 It is not clear from open sources what is meant by a 'strong source warhead'.

90 TNA AVIA 65/1217, Draft paper from DGAW for DRPC (AES) on Future Atomic Trials, 12 April 1957.

91 TNA AVIA 65/1217, E.S. Jackson, DGAW, Ministry of Supply to US (SAW), 12 April 1957.

be necessary to carry out major trials annually.[92] Cook's wish list formed the basis of a draft paper for DRPC (AES).

The final plan for the Grapple tests envisaged the firing of four shots. If Green Granite (Small) were successful, this would lead to a megaton warhead for the power guided bomb and it would not be necessary to test Green Bamboo or the larger Green Granite. Green Bamboo was the warhead being developed for use in the megaton and power guided bomb for the V-bombers, whilst Green Granite was an experimental round based on true thermonuclear principles, which offered the best prospect of economical use of material.[93] Orange Herald would then be fired. The programme was extremely tight and any unforeseen contingency would have set it back. AWRE held out no hope of improving on the schedule, although the scientists were prepared to bring forward any of the rounds by a few days if it proved possible to do so. Following a review of the progress it had made in preparing these rounds, AWRE changed the order of firing for Orange Herald and Green Bamboo. The reason for firing Orange Herald before Green Bamboo was to complete the programme on 8 June, which would be feasible if it proved possible to delete Green Bamboo and the large Green Granite from the programme after the firing of the first round. The firing order that Cook eventually decided upon was to fire Orange Herald (small) second as it did not depend on the results of the first shot so preparation could go ahead for firing it without waiting for the results of the first. Furthermore, if the first shot were completely successful the trial would finish three weeks earlier than formerly scheduled.[94]

AWRE and the services had made every effort to have these tests at the earliest possible date. The limiting factor was fabrication of the experimental test devices. Orange Herald was originally placed last in the programme, as it was the intention to fire a large version of this round, which was sure to explode, if all three earlier rounds had failed. Whilst this would have given a megaton explosion, it would not have enabled production of a warhead small enough for Blue Streak. When it was decided to introduce a second Green Granite into the programme (Green Granite (Large)), Orange Herald became the third round, as the second Green Granite could not be made ready before the beginning of July. AWRE now planned to abandon any thought of firing a large Orange Herald.[95] If the tests were successful, then production of a megaton warhead could begin in August 1957.[96]

---

[92] TNA AVIA 65/1217, draft paper from DGAW for DRPC (AES) on Future Atomic Trials, 12 April 1957.

[93] TNA DEFE 13/1012, Operation Grapple, DGAW, Ministry of Supply, 15 March 1957.

[94] TNA AVIA 65/1211, Copy of a signal sent by DD/AWRE to Air Vice Marshall Oulton, 13 April 1957.

[95] TNA DEFE 13/1012, DGAW, E.S. Jackson to PS/Minister, 15 April 1957.

[96] TNA AIR 2/13680, Note on DGAW's Suggestion for producing an Interim Red Beard and an Interim Megaton Bomb, 2 November 1956.

On the day before the first test, Brundrett advised Duncan Sandys, the Minister of Defence, of the firing plan for Grapple.[97] Test number I Green Granite II was a 30-inch diameter weapon weighing two tons. This weapon was the only one based on true thermonuclear principles.[98] Until recently this was expected to be the best approach the UK could possibly make towards a free falling bomb, though new ideas on the design of defensive weapons and the RI effect had undermined this judgement. Test number II, Orange Herald (Small), was a 45 inch diameter weapon weighing 2500 lb and was intended to be the prototype for warhead for Blue Streak. It was extremely expensive in fissile material; it was a wholly U-235 weapon, although it did have fusion material in the core to test this method of boosting.[99] There were two alternatives for the third test depending on whether or not test number one was successful. If Green Granite II was successful, the third test was going to be a device known as Blue Granite, which was 20-inch in diameter weighing 2500 lb. This was an extremely radical design only agreed late in the day. Its principles rested on the success of Green Granite II and the use of a trigger developed for defensive weapons. If it were successful, it would provide an extremely light and cheap solution for the free fall and propelled bombs as well as the Blue Streak warhead. If Green Granite II were not a success, then test three would be Green Granite I. This was the first approach to a true thermonuclear design intended for the free-fall and propelled bombs. It was 45 inches in diameter and weighed 10,000 lb, but AWRE hoped that it would not be needed.

Meanwhile, political opposition to atmospheric tests was growing, as too was Cabinet anxiety since, as the Foreign Secretary told his Cabinet colleagues, if the Grapple test was frustrated, the effect on UK R&D would be serious. For

---

[97] TNA DEFE 7/920, Brundrett to Minister, 14 May 1957. This operation was originally code named Green Bamboo and was initially to have been a ground (sea) burst of a large yield weapon – a layered fission design. However, by August 1955 AWRE thought that it was probable that it could produce an air droppable weapon. TNA AIR 8/2469, Nuclear Weapon Trials, N.C. Hyde A/ACAS (Ops) to VCAS, 8 August 1955. The name was changed to Gazette in September 1955, with Green Bamboo remaining as the code word for the warhead. R.B. Lees ACAS (Ops) to VCAS Nuclear Weapons Trials 1956/57, Gazette changed into Grapple by early November 1955. R.B. Lees ACAS (Ops) to DCAS Planning Progress Report – Atomic Weapons Trials 1956/57, 11 November 1955. Having started with only one shot, the firing plan had extended to three and possibly four weapons by end January 1956. R.B. Lees ACAS (Ops) to DCAS Planning Progress Report – Atomic Weapons Trials 1956/57, 27 January 1956.

[98] TNA DEFE 7/919, Operation Grapple, Note by the Chairman of the Atomic Weapons Trial Executive, 16 March 1957.

[99] TNA DEFE 7/919, Minute to Minister, 27 February 1957. This also noted that it was impossible to reach a firm decision on the number of megaton weapons that would be built until Grapple had been completed. On the old form of bomb, which was highly susceptible to the new defensive theory (R-I), the UK could make if it wished about 40 by 1961, thereby using up all the U-235 and relying only on plutonium for tactical and defensive weapons. If Granite II were a success, this would be reconsidered.

these reasons it was desirable not only to accelerate the tests as much as possible, but also to make further improvements in the safety precautions to be observed and to give the Japanese government sufficient information to allay its anxiety.[100] There was no prospect of any further acceleration, nor was there any chance of an effective deception plan to give the impression that the dates for the tests were later than actually planned.

Grapple: Results and Follow-up

In the event three rounds were fired in the Grapple trials on 15 May (Short Granite – 300 kilotons), 31 May (Orange Herald Small: 700 to 800 kilotons) and 19 June (Purple Granite – 300 kilotons).[101] Penney noted in June 1957 that one of the most important points that AWRE had understood from the Grapple tests was how to make a thermonuclear bomb. For the first time they saw the principles on which to focus all further development work and that they were only an extension of the Granite idea. For some weeks in June AWRE had been desperately trying to design an attachment to one of the Antler rounds to help them to understand how to develop clean bombs, namely those where most of the yield came from fusion rather than fission. Penney regarded the cost as small and they could work well within the 80 kiloton yield limit set by the Australians for a balloon test. However, there was a high probability that the attachment would not go off because the scientists would have to guess distances rather than optimise these by a full computer calculation. They also dared not put too much 'stuff' in the device in case it was more efficient than expected.[102] Penney was evidently talking about thermonuclear fuel.

The Royal Aircraft Establishment's (RAE) Armament Department remarked in early July 1957 that the most pressing need, if Yellow Sun development was to continue unchecked, was a decision on the type of warhead required. Much of the work done to date had assumed that Green Bamboo would be the warhead of choice. If, following Grapple, Short Granite was selected, it was even more important that an early decision was taken so that necessary changes in design and manufacture of development models should be made in time to meet the dates for planned trials and subsequent approval of the weapon.[103]

---

[100] TNA CAB 128/31, C.C. (57) 17 Conclusions, Item 4 Nuclear Weapons Tests, 12 March 1957.

[101] Arnold, *Britain and the H-Bomb*, pp. 143–7. It seems that a Canberra aircraft carrying the samples from the 15 May test back from the Pacific was lost off Newfoundland. AIR 8/2468, W.G. Penney to Air Chief Marshall Sir Dermot A. Boyle, CAS, 21 May 1957.

[102] TNA DEFE 7/919, Penney to Brundrett, 25 June 1957.

[103] TNA AIR 2/13681, Second Progress Statement on Yellow Sun, Royal Aircraft Establishment, Armament Department, Bomber Armament Division, 5 July 1957.

Unfortunately, the results had not provided sufficient data to enable a firm decision. On the evidence of the trials, the Air Staff eventually chose a Green Bamboo-type warhead to be eventually used in an interim megaton weapon – Violet Club. Air Staff officials accepted that any decision regarding the suitability of the Granite type warhead for a service megaton weapon could not be made for some months yet.[104] This uncertainty would have to be accepted and Yellow Sun designs should cater for both types (Green Bamboo or Granite type). One result of the Grapple trials was that the Air Staff were offered a choice of three yields for Yellow Sun, a bomb casing for the service megaton weapon; Green Grass a modified, although very similar to Green Bamboo with a yield of 500 kilotons – was selected because this gave most bombs for a reasonable yield.[105] If further AWRE trials proved a Granite warhead successful, the estimated time to produce this warhead for service use was six to 18 months longer than that required to reach the corresponding level with Green Grass depending mainly on size and safety device problems. The Air Staff delayed a decision for some time on which warhead to use for the Yellow Sun ballistic casing.

With certain modifications, AWRE planed to test the Short Granite design again at Grapple X. This experimental device was highly successful, but it did not represent a service warhead and the plan was to give priority to developing it into a safe, economical and reliable warhead for Blue Steel. To achieve this, considerable scientific and engineering design progress was necessary in order to reduce its size and weight and to provide adequate safety and operational reliability. Because of the need for further trials, the time factor suggested that Blue Steel would be its first application. In the meantime development and production of Green Grass for Yellow Sun and Blue Steel would continue and a decision to change over could be made when it was possible to better assess how well the development and weaponisation of the Grapple X device was proceeding.[106]

Further megaton tests were now needed since the yields from the Grapple series had been disappointing. Consequently, AWRE had to change its plans for warhead development, as it could no longer meet key elements of the operational

---

104 TNA AIR 2/13681, Minutes of Yellow Sun Progress Meeting, 16 July 1957. The Long Granite version however could not be accommodated inside the casing.

105 TNA AIR 2/13733, Air Council Progress Report on New Weapons Quarter Ended 30th September 1957, Note by DCAS, 16 October 1957 Appendix to A.C.(57) 81. At the same time the number of planned interim weapons (Violet Club) was reduced to 13 from 15 because of the additional demand for fissile materials required for the Grapple X trial. TNA AIR 2/13718, E.S. Jackson DGAW, to DCAS, Interim Megaton Weapons, 19 September 1957. In addition, original planning assumed seven Yellow Suns to be deployed in 1958/59, but it was possible that the number available in 1958/59 could be increased if certain trials were not agreed for 1959/60 (in other words saving on fissile materials). TNA AIR 2/13718, D.G.A.W, MOS to DCAS, 11 September 1957.

106 TNA AIR 2/13733, Appendix to A.C. (58) Yellow Sun RAF Bomb, Progress Report on Nuclear Weapons as at 31 December 1957.

requirements for Yellow Sun and Blue Steel.[107] Air Ministry staff thought it prudent to re-examine its requirement for a one megaton yield, and to see how low a yield it could accept.[108] AWRE meanwhile had six reasons for wanting another megaton test immediately. First, Short Granite and Purple Granite gave low yields, and although they proved the principle of the double bomb method, the thermonuclear reactions were very small. Second, the only megaton range weapon that the UK could make was a boosted or pure fission weapon, which was very expensive in fissile material – Orange Herald (Small). Officials calculated that the UK could only make 16 of these by April 1959, but if the double bomb method were perfected, numbers could be doubled, possibly trebled. Third, the boosted or pure fission weapon was very vulnerable to R-I.[109]

The double bomb in contrast was much less vulnerable and AWRE hoped that further tests would provide a design immune to R-I. Fourth, the Foreign Office view was that there was a real risk of a United Nations General Assembly (UNGA) resolution banning tests being passed in December, possibly even in November. The US attitude to this was unknown. Fifth, AWRE could be ready to fire two rounds in early November 1957, but a crash programme of two rounds by early October would be possible, but these would be less well designed. AWRE was therefore looking to make all four test devices in case the political situation required the earliest possible firing date. Duncan Sandys had written to the Prime Minister in mid July 1957 to emphasise that developments in the disarmament field made it essential for the UK to conduct its next megaton trial as quickly as possible.[110]

---

107 TNA AIR 2/13748, D.A. Arm to DGAW, Nuclear Warhead Policy, 6 August 1957. Penney had apparently indicated that Yellow Sun would not accommodate the service version of the megaton warhead and now had plans for warhead development subsequent to the Green Bamboo type in the Interim Megaton Weapon. By March 1958 AWRE could only offer a Green Grass warhead for Blue Steel. Minutes of the second meeting held at AV Roe & Co Limited Harrow to discuss the Warhead Installation on Blue Steel, 14 March 1958. AWRE were however working on a new one ton megaton warhead which would have several advantages, notably economical use of fissile material and easier safety arrangements; the difficulty was that no design details were available for use in the Blue Steel weapon design; it was hoped to have this available in 1960/61. Notes on a meeting at AVRO Woodford on 4 February 1958 to discuss the Warhead Installation for the Blue Steel Project.

108 TNA AIR 2/13748, P.B. Wood A.DOR (C) to D/SAAM, Yield of Deterrent Weapons, 29 August 1957.

109 The Radiation Interaction – R-I – was apparently an effect, a project and an access caveat, depending on whose records are accessed. The UK had considered many possible adverse effects of nuclear explosions on its own weapons in the mid 1950s. One was the capture of neutrons by fissile material and the way this would change the isotopes involved leading to short-term radioactive decays releasing more neutrons. Thus if there was an attempt to detonate the device within minutes of an explosion, it could have an excess of neutrons and pre-detonate, giving a much lower yield that it was designed for.

110 TNA PREM 11/2859, Sandys to Prime Minister, 14 July 1957. This resulted from an oral recommendation from of a small group of officials chaired by Sir Frederick Brundrett

Aldermaston was reasonably confident of the November rounds, but less so for the October ones. Therefore, November was the most desirable date scientifically.[111] In addition, the experience at Grapple had shown that it was not satisfactory to make hurried modifications during a short trial. A much better system would be to hold 'step-by-step' tests, with explosions at six to eight week intervals that could be adjusted at will. If disarmament negotiations made it likely that tests would soon be brought to an end, it would be desirable to go back to Christmas Island as soon as possible, possibly November.[112] The UK fear was that in the event of a UN resolution proposing a suspension of tests, HMG would find itself in great difficulty with public opinion and might be obliged to acquiesce. In order to forestall this risk, Macmillan decided to proceed with a test programme in the autumn.[113]

Even at this late stage, Cook thought that if feasible it might be better to cancel the Antler trials in Australia and move two of its shots to Christmas Island.[114] The MOD too wanted to know in light of international pressure on testing just how soon AWRE could go back to the Pacific. Brundrett told Jackson that in view of advice from the Foreign Office that there was a very serious danger indeed that nuclear testing would not be allowed after the end of 1957, it was vital that the UK had another large scale explosion, or indeed more than one if possible.[115] Aubrey Jones, Minister of Supply, wrote to Macmillan on 13 June 1957 to note that the recent US disarmament proposals might conceivably lead to a moratorium on nuclear tests as early as 1 July 1958, with the possibility that they might never be resumed. The Minister of Defence suggested that in these circumstances the UK ought to maintain the Christmas Island facilities so that it could conduct tests in the first half of 1958. Decisions were needed to suspend action to dismantle the facilities. Macmillan agreed.

– others involved were Powell (Permanent Under Secretary, MOD) Dean (Permanent Under Secretary Foreign Office), Plowden, Chairman (UKAEA), Musgrave, Permanent Under Secretary, Ministry of Supply. Macmillan did not like departments sending a lot of paper around on this issue, as some of it was sure to leak. Macmillan to Mr Bishop, 20 July 1957.

[111] TNA DEFE 7/919, Reasons For wanting another Megaton Test immediately, undated minute.

[112] TNA AVIA 65/1432, Note of a meeting held on 4 July to discuss possible future megaton trials.

[113] TNA PREM 11/2859, Commonwealth Relations Office (CRO) telegram to UK High Commissioner in New Zealand, Prime Minister's Personal Telegram, Serial No. T272/57, 27 July 1957.

[114] Antler was the code name given to as series of fission device trials at Maralinga in September 1956.

[115] TNA AVIA 65/1431, Brundrett to Jackson, DGAW, Ministry of Supply, 9 July 1957.

Logistical considerations were such that it would not have been practicable to hold tests in the summer of 1958 if such a decision had not been reached.[116] Officials had originally assumed that if the UK did not test again for two or three years it would have been cheaper to remove everything from the island at the end of the Grapple series. The international situation changed the basis of calculations. Despite the pressure the earliest that the Ministry of Supply thought that a test could be mounted, taking into account AWRE and service inputs, would be early December and that could only be done by cancelling Antler and dropping bombs off Christmas Island rather than Malden Island. If Antler proceeded, then the date moved to early spring 1958, possibly February. These dates assumed an immediate decision to test again.[117] Interestingly enough in light of what was to transpire in 1958, one of the assumptions under which the Ministry of Supply had to operate was that there would be no tests in the 18 months after May 1958.[118] This pause was the policy of HMG and to which all concerned had to work.[119] It is not clear whether this was largely for economic, political, technical or logistical reasons – or a combination of all four.

Macmillan was told at the end of June that if the UK was unable to carry out any further tests after the autumn of 1957, the UK would, in comparison with both the US and USSR, be able to make only inefficient (and expensive) use of fissile material. Indeed the possibility of a test ban had led directly to AWRE having to test a device (Orange Herald) earlier than was necessary or desirable in relation to the weapon programme.[120] Plowden (Chairman, UKAEA) had explained the limitations in knowledge obtained from the first megaton test to Macmillan and emphasised the need to test again in view of the pressure for a suspension of tests.[121] There was thus, in Aldermaston's view, a very real need for continuing tests in both megaton and kiloton ranges for a period of at least two years, but officials felt at that time that it was very much more important to be able to continue the kiloton rather than megaton tests. Presumably, this was because even then scientists recognised the critical importance of the fission primaries in the design of multi-stage weapons,

116 TNA AVIA 65/1431, Aubrey Jones to Prime Minister, 13 June 1957 and Macmillan to Minister of Supply, 14 June 1957.

117 TNA AVIA 65/1431, E.S. Jackson to Sir Frederick Brundrett, Ministry of Defence, 10 July 1957.

118 TNA AVIA 65/1431, E.S. Jackson, DGAW to Air Vice Marshall Oulton, Task Force Commander Grapple, 30 August 1957. Cook thought however that there was a case for trials at Christmas Island for at least two years to come at intervals of about three months. The Ministry of Supply thought that Cook was beginning a 'softening up' process to get the recent decision about future trials altered. US/SAW to AS/AW, 16 October 1957.

119 TNA AVIA 65/1440, E.S. Jackson to Major-General G.E.R. Bastin, Director of Weapons and Development, The War Office, 12 November 1957.

120 TNA AVIA 65/1248, Nuclear Tests Policy Committee, Brief for meeting on 18 November, Appendix A, E.S. Jackson, DGAW, 15 November 1957.

121 TNA PREM 11/2859, J. Bishop to Prime Minister, 23 July 1957.

especially when economy in the use of fissile material and the need for lighter designs with smaller diameters were so important. The possibility of providing any form of defence against attack by megaton weapons was entirely dependent upon the ability to produce small defensive kiloton weapons, a capability that the UK lacked in mid 1957.[122]

Macmillan informed the Ministerial Committee on Atomic Energy on 24 July that further tests were needed to enable the UK to design nuclear weapons with a higher proportion of fusion to fission in their yields, and which would make more economical use of fissile material. The Committee noted too that the longer the UK delayed such tests, the greater were likely to be the political difficulties in view of the growing pressures of world opinion in favour of a ban.[123] It does seem that the concerns over pressure for a test suspension in the UNGA played a major role in the decision to proceed with a further megaton test in the autumn of 1957; Macmillan said as much in a letter to Australian Prime Minister Robert Menzies at the end of July.[124]

AWRE's Testing Plans: Antler and the Test Ban

The UK had to juggle conflicting requirements when it came to deciding what had to be tested: service pressures for target response (exposure of military equipment to the effects of a nuclear explosion) and indoctrination tests, validation of service weapons, new concepts and service requirements, economy in the use of fissile material as well as more general economic pressures and public demands for a test ban.[125] The Minister of Defence had told Macmillan in January 1957 that it was essential that the UK conduct a further series of kiloton trials as soon as possible. We can see this clearly from DRPC (AES) deliberations on the firing programme for the Antler trials scheduled for late summer 1957.[126] Overall it seems that

---

[122] TNA DEFE 7/922, Sir F. Brundrett, MOD to Sir Edwin Plowden, UKAEA, 21 June 1957.

[123] TNA CAB 134/1328, Ministerial Committee on Atomic Energy Confidential Annex to A. (57) 1st meeting, Minute 1, 24 July 1957.

[124] TNA PREM 11/2859, Macmillan to R.G. Menzies, 26 July 1957 A similar message went to the New Zealand and the Canadian Prime Ministers – CRO telegram no. 493 to UK High Commissioner in New Zealand, 27 July 1957 and CRO telegram to UK High Commissioner in Canada 30 July 1957; 'Pressure may … grow for the suspension of tests .. and a resolution prosing this may well be introduced in the General Assembly … in that event we should be in great difficulty with or public opinion and might be obliged to acquiesce.' Their assistance was needed to help with logistical aspects of the tests.

[125] TNA WO 291/1526, The Value of Live Indoctrination at a Nuclear Weapon Trial (Operation Buffalo), 20 November 1957.

[126] TNA AVIA 65/1116, D.R.P. (A.E.S)/M (57) 5 Meeting 30 July 1957; TNA AIR 6/115 Air Council Memoranda 1957 Further Trials of Atomic Weapons Operation 'Volcano',

AWRE's test plans were in a constant state of flux in this period. Aldermaston did not decide exactly which shots should be set off until the test programme was underway. Admiral Brooking (AWRE), for instance, told an Atomic Weapons Trials Executive meeting on 14 August 1957 that an informal AWRE view on the future programme would be for two more tests after Grapple X, the latter one possibly in April 1958, but it was too soon to be definite.[127] There had been enormous pressures on the thermonuclear development programme, but what of the other warhead requirements? There were pressures here too and when combined with those from the thermonuclear programme it was clear to the UK that a good deal of detailed technical calculation and engineering design work remained.

The armed forces were worried that pressure for a test ban might prevent the UK from acquiring sufficient target response data of its own; they emphasised the need to make maximum use for these purposes of those tests that were conducted.[128] The DPRC (AES) considered a joint note from the Director General, Atomic Weapons at the Ministry of Supply (DGAW) and Director AWRE at the end of 1956. This emphasised that if the weapons programme, especially in the field of small warheads, was to proceed with all speed, further trials were needed at Maralinga in 1957.[129] Unfortunately, it was not then possible to say what rounds it would be possible or desirable to test. However, five objectives ought to be covered as early as possible: to check yield against fissile compression, including the low compressions applicable to low-yield warheads; developments of core boosting using tritium; development of an atomic six inch gun shell (however AWRE would not be ready to test in connection with this weapon); development of Red Beard (the new tactical weapon to replace the first generation device Blue Danube) with small loading tube and investigations in aid of development (reduction in size of the fission triggers) of thermonuclear warheads.[130] These Red Beard development rounds were separately described as 'two special rounds on techniques in connection with smaller Green Granite'.[131]

September/October 1957 DCAS, 18 January 1957.

[127] TNA AB 16/2371, Atomic Weapons Trials Executive Operation Grapple X Meeting held at St Giles Court on 14 August 1957.

[128] TNA AVIA 65/1116, D.R.P. (A.E.S)/M (57) 1 Minutes of Meeting 12 February 1957. Target response data was concerned with the effects of nuclear explosions on a wide range of military equipment.

[129] TNA AVIA 65/1114, D.R.P./(A.E.S) P (56) 16, DRPC Sub-Committee on Atomic Energy Nuclear Weapons Trials at Maralinga in 1957, Note by DGAW and DAWRE, 20 November 1956.

[130] TNA DEFE 7/911, Mancroft to Prime Minister, 31 January 1957.

[131] TNA DEFE 7/911, DRP (AES) P (56) 16; and F. Brundrett to Minister, Nuclear Weapons Trials in Australia in 1957, 8 January 1957; TNA AVIA 65/854, R.G. Elkington, Ministry of Supply I. de L. Radice, H.M. Treasury, 10 January 1957. This noted that six tests could be conducted, but that as of January 1957 AWRE were not in a position to test a gun shell.

The Committee's approval was sought and there was also a need to obtain Chiefs of Staff authorisation for expenditure of fissile material for these tests equivalent to 25 kg of plutonium.[132] This was required because in mid 1956 the planned weapons production programme envisaged an eventual stockpile total of 170 megaton and 840 kiloton weapons, so any impact on fissile material supplies had to be considered carefully given the shortages and expense of both plutonium and U-235 at that time.[133] Depending on the firing programme adopted, the trials could consume up to 30 kg of plutonium and 40 kg of HEU.[134] The Chiefs of Staff agreed that the trials were necessary and accepted that it would mean diversion of fissile material from weapons production.[135] And the Foreign Secretary in giving his agreement pointed out that the latest US disarmament proposals and growing public opposition to testing made it particularly important that UK should go ahead with essential nuclear tests without delay, as it might make it more difficult to test in 1958.[136]

The Prime Minister had initially questioned the need for these trials.[137] In response, the MOD and Foreign Secretary had advised that they were of critical importance.[138] Lloyd noted that the latest US disarmament proposals, growing public opposition to testing and possibly future reports of the United Nations Scientific Committee on Effects of Atomic Radiation could make it more difficult to carry out tests in 1958. Macmillan agreed that the tests should proceed as rapidly and quietly as possible, but noted that the government had to recognise that out of all the international discussion on disarmament something might emerge that would affect these plans.[139]

---

[132] TNA DEFE 7/911, DRPC Sub-Committee on Atomic Energy Nuclear Weapons trials at Maralinga in 1957, Report to the Chiefs of Staff, F. Brundrett 12 December 1956; HEU was also used in the trials – TNA DEFE 7/911, Atomic Weapons Trials Executive, Operation Antler 12 June 1957 (ii) Transport of Fissile Material. All but one load of HEU went by sea; the other went by air; TNA AVIA 65/854, J.M. Gibson, MOD to G.S. Downey, 18 January 1957.

[133] TNA AVIA 65/1114, DRPC (AES), The Nuclear Weapons Programme, 6 June 1956.

[134] TNA AVIA 65/854, U.S./S.A.W to Dep Sec (A), 18 February 1957.

[135] TNA DEFE 16/603, Minister of Defence to Prime Minister, 31 January 1957; see also TNA DEFE 7/911, C.W. Wright to Sir F. Brundrett, 1 January 1957.

[136] TNA DEFE 7/911, D.S. Laskey, Foreign Office to J.M. Gibbon, Ministry of Defence, 24 January 1957; Selwyn Lloyd to Prime Minister, 30 January 1957; D.S. Laskey to J.M. Gibbon, 24 January 1957.

[137] TNA PREM 11/2844, Manuscript comments on D. Sandys to Prime Minister, Nuclear Trials in Australia, 25 January 1957.

[138] TNA AVIA 65/854, Lord Mancroft, Parliamentary Secretary MOD to Prime Minister, 31 January 1957.

[139] TNA DEFE 7/911, Macmillan to Parliamentary Secretary, MOD 1 February 1957.

The original Antler firing plan was for six rounds, depending on how development work progressed between April and October.[140] However, the results of Grapple caused AWRE to reconsider the nature and number of rounds.[141] Aldermaston changed its plans to fire two Red Beards, Blue Fox (the prototype for stage 1 and a half Surface to Air Guided Weapon (SAGW) and Pixie, which had some relevance to Seaslug.[142] There was a fifth round, described separately as a 'radiation round'.[143] Brundrett (Chief Scientific Adviser, MOD) thought that the Red Beard tests could be delayed, but wondered what AWRE thought. (Red Beard was seen as of great importance in giving RAF tactical air forces in Germany a nuclear capability, but the RAF did not get nuclear weapons in Germany until the 1970s.[144]) Cook commented that if no account were taken of the possibility of tests being suspended, then it would be perfectly possible to delay the firings until 1958, given the planned in-service dates for Red Beard, which was planned for the end of that year. However, in view of the political situation on testing it would be better to fire these rounds during Antler. Macmillan felt that the politics of testing were such that there would be enormous advantage in planning to complete Antler

---

140 TNA AVIA 65/1201, W.G. Penney to E.S. Jackson D.G.A.W, Ministry of Supply, 15 April 1957; maximum yields ranged from three to 80 kilotons. Although by June the weapon known as C2 (possibly Pixie) would have a yield of one to two kilotons. J.T. Tombling, Superintendent Planning Trials, AWRE Antler Firing Programme: Second Revision 5 June 1957. U-235 was apparently used in some or all of the devices given the need to transport it to Australia. ES 1/747 Antex – committee meetings minutes – Antler 1957 Atomic Weapons Trials Executive Operation Antler, Minutes of meeting 13 March 1957; see also TNA ES 1/759, H.C. Hainworth, Foreign Office to Admiral P.W.B. Brooking, AWRE, 29 July 1957, but only 'a very small amount of U-235' went by air, Brooking to Hainworth, 19 July 1957.

141 TNA AVIA 65/1201, Operation 'Antler' Firing Programme 1957 to Wheeler, Melbourne from Lloyd, 16 July 1957.

142 Apparently, one problem with nuclear Seaslug was that it was intended to be used against Soviet cruisers whose guns had a longer ranger than the missile. Its extravagance in use of fissile material and the shortage of such material were also factors. Personal correspondence and see Richard Moore, *The Royal Navy and Nuclear Weapons* (London, 2001), pp. 111–12.

143 TNA ES 1/758, Pilgrim to Director Antler – order of firing, 5 June 1957. This listed round one as Blue Fox, round two as Blue Fox Boosted, round three as Red Beard (Production), round 4 as Tom (production) and round five radiation round and if possible round six Pixie. This firing programme was agreed in Penney's office on 5 June. Pixie is described as weapon C6 with an estimated yield of 80 kilotons. Antler Firing programme: Second revision, 5 June 1957. Blue Fox seems to have been based on or connected to Orange Herald see J. Corner to SSTD, Yields Expected at Antler, 18 June 1957. The explosive used in Pixie had not been settled and even with better explosives, it was unlikely to exceed two kilotons. Corner did not know anything about round five.

144 TNA AIR 2/18210, Nuclear Weapons Annex B to ACAS (Ops)/98 A, 1 October 1970; and J.G. Matthews, DD Ops (S) (RAF) to DD Air Plans, RAF Germany Buccaneer Strike Capability, 15 October 1970.

before parliament re-assembled towards the end of October.[145] Penney told him that AWRE would do its best, weather permitting.[146]

This meeting seemed to endorse AWRE's plans to fire Blue Fox and Red Beard, but to omit Pixie. This latter device was not a prototype service weapon, but an experimental device weighing 250 lb with a diameter of 16 inches, which might be of value if there was a need for such a small weapon.[147] AWRE were unsure whether their current theoretical calculations for such a small device were accurate hence the need for a test. Penney reminded the Committee that there was shortage of fissile material and that there was a danger that the tests might use up an undue proportion of the available military stock. In his view, it was essential to conserve it. Pixie, although it would only have a yield of one kiloton, would require an amount of plutonium that would be almost sufficient for two Red Beards. This latter point was especially telling given the current proposals circulating internationally for a cut-off in the production of fissile material for weapons purposes, which suggests that there was a compelling need to reduce the Antler programme to conserve this material.

In the end, the Committee decided to fire Pixie, one Red Beard (a mixed core using HEU and plutonium) and Blue Fox.[148] As Arnold and Smith observe, Antler had gone through more changes in plans than any other tests in Australia.[149] Such last minute changes reflected the pace of technological research at Aldermaston. The final decision on Pixie and Red Beard depended on minor trials results planned for completion by 4 September for the former and 21 September for the latter.[150] It was possible AWRE could accept deleting the standard Red Beard from the programme as they were confident that it would work and while it was important to carry out such a test, in view of the possibilities of a test ban, the risk was worth taking. This was then changed and the mixed Red Beard (uranium and plutonium

145 TNA ES 1/758, Sir F. Brundrett, MOD to Sir William Penney, 21 June 1957.

146 TNA ES 1/758, W.G. Penney to Mr C.A. Adams, Chief of Research, 26 June 1957.

147 TNA AVIA 65/906, Minutes of a meeting to discuss Nuclear Warheads for Stage 1 and 1b 6 September 1957. AWRE noted at this meeting that the minimum possible weight was 260 lb, length 28 inches and diameter 15 inches. This include proviso for the 'ball' and its mounting and the firing circuit. It was suggested that civil plutonium might be used, but AWRE felt that it was extremely unlikely that this could in fact be employed, especially for Pixie, because the yield might be reduced to an unacceptably small value. Nevertheless, it was agreed that information on the possible use of civil plutonium in small warheads should be acquired.

148 Pixie was originally a model scale HE implosion assembly produced as a quicker method of testing implosion theory than a large assembly like Blue Danube. It later been suggested that it might be useful to try the effect of including a plutonium ball.

149 Arnold and Smith, *Britain, Australia and the Bomb*, p. 187.

150 TNA AVIA 65/1201, Charles Adams, AWRE to E.S. Jackson, DGAW, Ministry of Supply, 21 July 1957.

cores) was cancelled and replaced with a device known as Tom (a fission trigger) because it would be of value to the thermonuclear programme.[151]

Penney however was not convinced that all the arguments on Pixie had been properly aired.[152] He wrote to the Chiefs of Staff to make these clear. As result, a further meeting took place to review the Antler firing programme. Penney told this meeting that even if Pixie proved successful, he could still not guarantee to develop a weapon without another test. He thought that this should be considered on the basis of two alternative hypotheses: first, that nuclear tests would not be stopped after Antler; and secondly, that they would. On the first hypothesis, the best course would be to work out precise service requirements in terms of weight and size, preferably one requirement for all weapons in this class and to develop a warhead specifically to meet it. If the second hypothesis came to pass, it had to be borne in mind not only that the firing of Pixie might not be sufficient for development purposes, but also that Western disarmament proposals for a cessation of tests were conditional upon agreement that no fissile material produced after a certain date could be used for weapons. For these reasons it was questionable whether material should be used for a round that was as expensive and profligate as Pixie and also whether material would be available for the production of the weapon ultimately developed.

Sir Richard Powell (Permanent Under-Secretary, Ministry of Defence) who chaired this meeting said that it could not be assumed that if tests were suspended, a cut-off would take effect shortly thereafter. However, it was extremely difficult to forecast disarmament developments, but he thought that test-ban pressures might be so heavy in the autumn that the UK and US would not be able to insist on making a halt in testing conditional on a cut-off. In any case, the US had assured the UK of its intention to supply fissile material if the condition should in fact be adopted. Given all this, would it be worth firing Pixie? Penney thought that it would be given Powell's comments, but if Pixie were to be fired a number of design changes would be needed before a properly engineered small warhead could be produced; a further test might still be needed. In light of this, the meeting agreed that Pixie should be fired at Antler. It was, but in the following year after the exchange of warhead information with the US, the DRPC (AES) agreed in mid November 1958 to drop further development of Pixie.[153] In fact the writing was on the wall for Pixie well before November 1958. A Ministry of Supply study of nuclear warheads in SAGW concluded in April 1958 that the use of Pixie instead

[151] TNA AVIA 65/1116, MISC/M (57) 101, Meeting 26 August 1957 to discuss the firing programme for ANTLER 1. PIXIE. AWRE anticipated that the maximum yield from this would be 30 kilotons. TNA AVIA 65/1201, AWRE Aldermaston to UKMOSS (a) Melbourne, 27 August 1957.

[152] TNA AVIA 65/1116, 1960 MISC/M (57) 101, Meeting to discuss the firing programme for Antler 2, Red Beard, 26 August 1957.

[153] TNA AVIA 65/1116, Confidential Annex to D.R.P. (A.E.S)/M (58) 2, Development of US Warheads for British Weapons, 11 November, 1958.

of the six and a half kiloton Indigo Hammer (formerly Blue Fox) would give no significant gain in performance in any of the missiles under consideration – Red Duster and Yellow Temple – as its yield was smaller, about one kiloton, which would reduce the altitude and range at which acceptable lethality could be achieved.[154] Moreover, it would use also one third more fissile material and be less safe since its core was nearer criticality.[155]

Testing Constraints: Further Moves, 1957

Even though the UK nuclear weapons programme was moving towards its first megaton range tests in the spring of 1957, HMG had, as we have seen, already indicated its readiness to contemplate constraints on nuclear testing. The Bermuda Declaration of 24 March 1957 committed the UK and US to conduct their nuclear tests in such a way as to avoid any significant increase in world levels of radioactivity. This also proposed that the UK and US would announce their tests well in advance; in addition, advance registration with the UN and limited international observation would be acceptable if the USSR would do the same.[156]

Harold Stassen (Eisenhower's disarmament representative) suggested to Commander Allan Noble (Minister of State, Foreign Office) in mid April 1957 that when the US and UK had completed their tests scheduled for summer 1957, they should seriously consider whether they would gain much from continuing with further tests. To do so would only encourage others to follow suit, and the UK might then be faced by a resolution in the UNGA that autumn calling for the immediate ending of all tests. He suggested that, although this problem was not immediate, it should be considered seriously.[157] However, Washington advised that Stassen was apt to run some length ahead of the field. To halt all US tests after the current series would be to reverse present US policy as specified by the National Security Council.[158] In a subsequent meeting between the two, Stassen recalled that at Bermuda in the spring the UK and US had agreed that action on testing would take place after the cut-off in the production of fissile material for military purposes. The US view seems to have been that nuclear physics had become such a precise science its scientists

---

[154] TNA AVIA 65/2086, Extract from DRP (AES)/M (58)1 Nuclear Warheads in S.A.G.W., 31 July 1958.

[155] TNA AVIA 65/207, DRP (AES)/P (58) 4, Nuclear Warheads in S.A.G.W. Note by the Ministry of Supply, 18 April 1958. Cook told the Committee that further tests or US information might enable AWRE to obtain this yield with smaller dimensions.

[156] Cmnd.126, *Final Communiqué from the Bermuda Conference (with annexes)* (London, 1957).

[157] TNA FO 371/129812, UN 1192/299, Foreign Office telegram no. 1878 to Washington, 18 April 1957.

[158] TNA FO 371/129820, UN 1192/455/G, J.E. Coulsen, Washington to I.T.M. Pink, Foreign Office, 23 May 1956.

believed it possible to design a weapon that did not require nuclear explosive testing. Acceptance of Soviet proposals to halt testing would thus achieve nothing; the priority now was to reach agreement on stopping fissile material production for weapons purposes. The US was now also prepared to move a little on the testing issue. If the UK agreed, they might agree to action concerning tests before 1 April 1959, but in the context of a partial agreement. Stassen wanted to know what the UK ideas were.[159] Washington was also thinking about a moratorium on tests between the entry into force of an agreement and the date of the cut-off. The UK felt that its position was clear, but it would await any new American ideas.

On 6 May 1957 at the UNDC Sub-Committee the UK tabled proposals for advance registration of tests, a group of experts to work out methods for limitations and controls and that cessation of tests should follow a cut-off in the production of fissile material for weapons purposes.[160] An earlier version of the paper had been held back in early April in deference to strong objections from France and Canada.[161] In particular, the UK proposed that the US, UK and USSR should draft an agreement on the advance registration of nuclear test explosions with the UN and work on this should begin immediately; this had originally been suggested by Canada, Norway and Japan in January 1957. The UK was prepared to include a provision for limited international observation of such tests. Furthermore, it suggested creation of a committee of technical experts within the framework of the UNDC Sub-Committee to consider possible methods of limiting tests and to investigate the requirements for effective supervision of any agreement to limit such explosions. The results of this might be coordinated with the work of the group of experts that the US had suggested should prepare a general nuclear control system.

The UK made clear however, that its position was that cessation of all nuclear tests should follow the prohibition of the production of fissile material for weapons purposes and as part of a general disarmament agreement.[162] (This latter element was specifically sought by the MOD.[163]) Macmillan told the Cabinet that in addition to this, the UK was also contemplating making a statement that in the meantime the UK would voluntarily undertake to conduct tests only in such a manner as would ensure that world radiation did not rise to more than a small fraction of the levels that might be hazardous. The voluntary offer would only be

---

159 TNA FO 371/129820 UN 1192/470/G, Record of Conversation between Commander Noble and Mr Harold Stassen on May 28 1957.

160 TNA FO 371/157110, IAD 22/515, Chronology of the Nuclear Test Negotiations Part I – before the Geneva Conference, 1961.

161 TNA FO 371/129820, UN 1192/466, manuscript minute, 31 May 1957.

162 United Nations Disarmament Commission, Sub-Committee of the Disarmament Commission, United Kingdom: Memorandum on Nuclear Test Explosions, DC/SC.1/56, 6 May 1957.

163 TNA FO 371/129813, UN 1192/345, H.C. Hainworth to Mr Mackenzie, Nuclear Tests Explosions, 29 April 1957.

made orally when the main proposal was tabled.[164] The MOD meanwhile made clear to the Foreign Office that it could not support any disarmament proposals that suggested that the UK would be willing to accept cessation of testing and a cut-off in isolation from a general disarmament agreement. To do so would put the UK in a hopelessly inferior position in relation to the USSR and US.[165]

Further work was undertaken in the Foreign Office in June 1957 on testing policy. One suggestion was that HMG should advocate an immediate moratorium for two years on tests of offensive weapons with appropriate controls and a moratorium on the testing of defensive weapons from the date of entry into force of a disarmament agreement.[166] Now that the USSR had accepted international control at the UNDC Sub-Committee on 14 June 1957 over the cessation of tests, it was unlikely that the West would be able to resist for very much longer the line that a cessation of tests was only possible if linked to a cessation of fissile material production.[167] It would certainly be difficult to defend this position in Parliament or the UNGA. A two-year period would allow the UK to continue its own testing programme for the 'smaller types of weapons'. Meanwhile, the Defence Secretary informed his colleagues on the Cabinet Defence Committee that the possibility that the UK might require further tests could not be excluded.

The Russians rejected this proposal in the UNDC Sub-Committee, but when they accepted the notion that there should be international control on 7 June, the position on testing in the UNDC Sub-Committee completely altered. Given the US proposal for a ten-month moratorium to start from July 1958 and the then state of public opinion on this subject at home and abroad, Lloyd told the Cabinet that it would now be very difficult for the UK to continue to stand out against any moratorium. Once tests were suspended even on a temporary basis, he thought it would in practice be very difficult if not impossible to resume even though the US did not share this view.[168] Certainly, AWRE wanted to work on the assumption of megaton range trials in 1958 if disarmament considerations were ignored.[169] Test

---

164 TNA CAB 128/31 Part I, C.C. (57) 37th Conclusions, Item 1 Nuclear Weapons Tests, 2 May 1957.

165 TNA FO 371/129816, UN 1192/380, Discussions on disarmament at UN, V.H.B. Macklen, MOD to H.C. Hainworth, Foreign Office, 1 May 1957.

166 TNA FO 371/129825, UN 1192/573/G, A.R.K. Mackenzie to I.T.M. Pink, Policy on Nuclear Tests, 15 June 1957.

167 Soviet proposal Introduced in the Disarmament Subcommittee: Cessation of Atomic and Hydrogen Weapons tests, 14 June 1957 in Department of State, *Documents on Disarmament 1945–1959, Volume II – 1957–1959* (Washington, 1960), p. 791.

168 TNA CAB 129/87 C. (57) 146, Disarmament Memorandum by the Secretary of State for Foreign Affairs, 21 June 1957.

169 TNA AVIA 65/870, E.S. Jackson, DGAW to Private Secretary, Minister, 25 June 1957.

limitations also featured in the four Western powers' disarmament proposals on 2 July and 2 August 1957 in the UNDC Sub-Committee.[170]

Lloyd was aware of the dangers to the UK nuclear programme that would result from a moratorium. It would mean that the UK would be unable to complete its development of kiloton range weapons and, although the UK could make megaton range weapons, the opportunity of developing efficient methods of producing them with less fissile material would be lost. Macmillan had agreed, given the disappointing yields of the Grapple devices, to test again in the autumn. Although the UK would still possess powerful weapons, it would be handicapped unless it could persuade the US to pass on its latest know-how. From the UK perspective, it would suit the national programme very much better if there were to be no question of a tests suspension for at least two years.

Duncan Sandys (Minister of Defence) weighed in with his own memorandum for the Cabinet; he emphasised that the proposal for the early cessation of nuclear tests and of the production of fissile material for military purposes would virtually knock Britain out as a nuclear power. Neither the US nor the USSR would be seriously affected by such a proposal.[171] Sandys accepted that it would be extremely awkward for the UK to stand out and to lay itself open to the charge of wrecking the first hopeful step towards disarmament. At the same time, he did not think that the US should design its plan in such a way that the first stage, which might prove to be the last, should be achieved almost wholly at the UK's expense. Such an outcome would leave Britain naked without the means of retaliation or defence. Any UK support for the US would be contingent on some key conditions that would not leave the UK at such a disadvantage, amongst which included US provision of its nuclear know-how and fissile material supplies to help the UK build up a reasonable stock of nuclear weapons, primarily for defensive purposes such as low-yield anti-aircraft missiles.

Selwyn Lloyd stressed to Stassen that the UK could not afford to be the odd one out if the USSR and US accepted a moratorium. Stassen had suggested that the UK should accept a moratorium, but link it to a first stage disarmament agreement, including among other points provisions for a cut-off. The UK would still strive for a comprehensive disarmament agreement, but it could not escape the fact that, unless it were confident about achieving such an agreement, an early cut-off would completely disrupt the defence programme, deprive the UK of a credible deterrent capability, and upset the whole basis on which its present defence planning was based. Test-ban pressure would also increase in the UNGA in the autumn. Indeed it did. The Soviet Union proposed an additional agenda item on the 'discontinuance under international control of tests of atomic and hydrogen

---

[170] Cmnd. 333, *Report of the Proceedings of the Sub-Committee of the United Nations Disarmament Commission held at Lancaster House, London, March 18–September 6, 1957* (London, 1957), pp. 75 and 96.

[171] TNA CAB 129/88, C. (57) 151, Disarmament Note by the Minister of Defence, 23 June 1957.

weapons', and submitted a draft resolution that repeated the proposals that it had made in the UNDC Sub-Committee on 14 June. Draft resolutions submitted by Japan, India and Yugoslavia called for an urgent agreement on the suspension of all nuclear weapon tests.[172] The resolution that passed in the end – UNGA 1148 (XI) – on 14 November called for, 'the immediate suspension of testing of nuclear weapons with prompt installation of effective international control'. Language on control had come from the US and UK.

On balance Lloyd advised his colleagues that the UK should back the Stassen proposal, suitably amended to cover UK concerns and that it might form the basis for a Four Power draft on nuclear tests that could be presented to the Russians. This issue was the sole item on the Cabinet agenda on Monday 24 June 1957.[173] Macmillan said that the proposals in the draft US statement represented the Administration's considered views. This meant that the UK faced difficult decisions. If the UK abandoned its own test programme, much of the effort that had been expended to date would be wasted and the further improvements in efficiency and economy in design and use of fissile material, that were key objectives, would be frustrated. Moreover, AWRE had indicated that it would have liked to continue with megaton trials if the UK ignored disarmament considerations.[174] A cut-off after 1960 would also undermine defence policy. One of the main objectives of that policy was to maintain UK political independence and this would be frustrated if the UK were prevented from producing its own weapons and the US and USSR refused to destroy their stocks by 1960.

In the subsequent discussion the Cabinet realised that it would be important to discuss with the US what technical help (fissile material or weapons) they would be prepared to provide to meet British requirements if they were to endorse current US proposals on disarmament. It would be equally important to ensure that the US provided technical information too. Although the US had undertaken to look into the supply of material, this was without commitment given the need to convince Congress. In any event they would not move on this until the Russians indicated some readiness to accept US proposals. Thus the UK would have to be very clear in any understandings about what the US would need to offer and when they would do so if London were to support Stassen's proposals. Macmillan would therefore send a message to the President setting out the understandings on which the UK would be prepared to support US proposals for partial disarmament. In addition, a new committee of senior officials would be appointed to keep under review further developments in disarmament. At that point the Russians rejected the Four Power proposal tabled on 2 July 1957.

---

[172] United Nations, *The United Nations and Disarmament 1945–1965* (New York, 1967), p. 142.

[173] TNA CAB 128/31, Part II C.C. (57) 47th Conclusions, 24 June 1957.

[174] TNA AVIA 65/870, E.S. Jackson, DGAW, Ministry of Supply to Private Secretary, Minister, 25 June 1957.

Lloyd sought Cabinet approval on 11 July 1957 to announce that afternoon that the UK supported the US proposal for an initial suspension of nuclear tests for a period of ten months; Canada and France had already done so. Lloyd's colleagues approved.[175] In August the US made a new proposal for a first step disarmament agreement to include a suspension of testing for a period of two years under certain conditions and safeguards. These included Soviet acceptance of the US position that within that period, there would be a permanent cessation of fissile material production for weapons purposes and installation of inspection systems to ensure performance. Until such a first step arms control agreement came into force the US would conduct such nuclear tests as security required.[176] The UK must have received the assurances it was seeking as it joined the US in the presentation of the new initiative.

The Four Power paper (US, UK, France and Canada) submitted to the UNDC Sub-Committee on 29 August 1957 on proposals for partial measures of disarmament included a separate section on measures for testing.[177] All parties to such an agreement would undertake to refrain from conducting tests for a period of 12 months from the date of entry into force, provided that all sides agreed on the installation and maintenance of the necessary controls. Experts would meet to design the inspection system. After the first 12 months, the parties would be free to test again unless there was agreement for a further period under effective control. If the inspection system was working effectively to each party's satisfaction and the sides were making satisfactory progress in the preparation of an inspection system for a cut-off, all parties would undertake to refrain from conducting nuclear explosions for a further period of 12 months. Such an extension came with conditions. It would only be made with the understanding that testing could, at the discretion of each party, be conducted 24 months after the entry into force of the partial disarmament agreement and if the installation of the cut-off verification system had not been installed to the satisfaction of each party concerned before the end of the 24 months and if the cessation of the production of fissile material had not been put onto effect. If tests were resumed, then they would be announced and registered in advance along with their dates and the range of total energy to be released; observation of the tests would be allowed; and would be limited in the amount of radioactive material to be released into the atmosphere.[178] Here we can see the considerable conditionality that was so important to the preservation of the

175 TNA CAB 128/31, Part II C.C. (57) 51st Conclusions, Item 4 Disarmament, 11 July 1957.

176 TNA FO 371/129840, UN 1192/916, Statement by the President, 21 August 1957.

177 Department of State, *Documents on Disarmament*, p. 871. The UN Disarmament Commission was the main international forum in which the major powers discussed possible disarmament measures. See United Nations, *The United Nations and Disarmament*, pp. 135–47.

178 TNA FO 371/129840, IAD 1192/929, Canada, France, the UK and the US, Working Paper: Proposals for Partial Measures of Disarmament, 29 August 1957.

UK's own nuclear activities in the near term. Such conditionality would make it virtually certain that the proposals would not come into effect.

Future Testing Plans

In light of the political pressures to stop and the scientific requirements to test, Macmillan proposed establishment of a committee of officials to formulate advice on policy questions relating to nuclear tests. This met under the auspices of the Cabinet Office and held its first meeting in 1957. The Nuclear Test Policy Committee (NTPC) would therefore be the main forum for developing recommendations on preferred options and requirements for the UK testing programme. As the prospects for international agreement on test limitations became more pressing, officials had to balance the external pressures and internal requirements when shaping their proposals. There were also uncertainties over how many further tests the UK needed. Much was unclear. Although the indications were that an agreed moratorium on tests would be unlikely before July 1958, the Minister of Supply felt that there was a case for the UK to proceed with further trials as rapidly as possible.[179]

The Foreign Secretary's main concern was to avoid any criticisms in the disarmament context. Lloyd thought that HMG should not say anything at that stage about another series of tests in 1958 nor suggest that the UK intended to press forward with tests more quickly than it had original planned.[180] AWRE's informal view in August 1957 had been that the future test programme might need two more tests after Grapple X, scheduled for late October or early November, but it was too soon to be definitive.[181] The dates for the next Pacific tests, Grapples Y and Z in 1958 depended on the outcome of Grapple X.[182] AWRE outlined its preferred testing programme proposed in mid November 1957 to the Committee. As a result of the November test (Grapple X) AWRE wanted a continuous programme of megaton test operations over the next two to three years to carry out two main projects. These were to consolidate and exploit the results of the Grapple X operation and to develop a warhead for Blue Streak. Cook informed Jackson that the objective was also to produce an immune warhead (namely one not susceptible to pre-detonation caused by neutrons flowing from a nearby exploding nuclear bomb), weighing one ton or less and suitable for Blue Streak. (The existing warhead, Orange Herald, was extravagant in use of fissile material and was very vulnerable.)[183] Brundrett

179 TNA AVIA 65/1217, US/SAW to DGAW, 24 June 1957.

180 TNA AVIA 65/1211, Foreign Secretary to The Rt Hon Aubrey Jones, 20 June 1957.

181 TNA ES 1/578, Meeting held at St Giles Court on 14 August 1957.

182 TNA AIR 8/2467, Cyril Musgrave, Ministry of Supply to Sir Maurice Dean, Air Ministry, 10 October 1957.

183 TNA AVIA 65/775, Future Programme of Atomic Weapon Trials Appendix A; the Orange Herald designs used up to 120 kg of fissile material. TNA AIR 2/13680, Note on

told the Chiefs of Staff that the original programme for Christmas Island had been to hold trials in February and May 1958 and then to have an 18-month pause, but the intention now was to have a test operation every four to six months.[184] In short, testing requirements and plans appear to have been in a permanent state of flux.

As a result of the outcome of Grapple X the matter was again reconsidered. Officials now proposed to hold a series of trials, starting in March 1958. Political considerations had caused previous trials to be rather hurriedly organised. It was the intention in the next series to proceed more methodically, changing only one variable at a time. The general aim of the megaton series would be to improve the ratio of nuclear yield to the weight of the weapon.[185] It was necessary to proceed with trials since no one knew how far the US would go in sharing its atomic secrets. However, AWRE and the MOD believed that US scientists had great respect for the progress made by the UK, and the more they knew of the UK programme, the greater would be their readiness to exchange knowledge. The Tests Committee concluded that in view of the political position concerning disarmament and nuclear testing at that time, the UK would be able to proceed with megaton tests for a little while longer. Therefore, they agreed to let Aldermaston proceed as it had proposed, but the programme would be re-appraised continually in light of such help as the US decided to provide. Even at this early stage, officials saw the semi-permanent establishment of support facilities at Christmas Island as an incentive for the Americans. Such an establishment there would make it more likely that the US would wish to use the island for its own tests.[186] This wish was to come true, but that was for the future.

DGAW's Suggestions for producing an Interim Red Beard and an Interim Megaton Bomb, 2 November 1956.

[184] TNA CAB 134/2274, Nuclear Tests Policy Committee N.T. (57) 4th meeting, 18 November 1957. The future programme, if approved by Ministers, was as for tests in: 1958 March and August at Christmas Island; 1959 January, Christmas Island; May, Maralinga; September, Christmas Island; 1960 February Christmas Island, Operation Grapple Finance Committee Future trials policy and the work of the Committee, TNA AB 16/2371, G.M.P. Myers, 2 December 1957.

[185] TNA AIR 8/2467, Chiefs of Staff Committee, COS (57) 90th meeting, Christmas Island (Pacific Ocean) Retention as a Base, 21 November 1957.

[186] TNA CAB 134/2274, Cabinet Nuclear Tests Policy Committee, N.T. (57) 4th meeting, 18 November 1957.

Conference of Experts to Study the Possibility of Detecting Violations of a Possible Agreement on Suspension of Nuclear Tests: Geneva 1 July to 21 August 1958[187]

A conference of scientific experts, originally proposed by the UK in July 1957 and in the August 1957 Four Power disarmament paper, opened in Geneva on 1 July and ended on 21 August 1958. Penney, who went on to play a key role, and Sir John Cockcroft represented the UK, to review techniques for detecting nuclear tests, although the main work was in fact done by Penney.[188] A secondary objective for the UK in these talks was to find out as much as possible about the Soviet monitoring system and its capabilities.[189] However, the immediate precursor for these important discussions was Eisenhower's proposal to Khrushchev of 28 April 1958 that technical experts meet to study the possibility of detecting violation of a possible agreement on suspension of nuclear tests. Khrushchev accepted.[190] British experts would participate on the basis that they would deal only with scientific facts and opinions. The ideal result would be a technical definition, agreed by both sides for submission to the governments represented, of a system offering the maximum possibility of detecting any test anywhere together with the minimum control structure able to achieve this. On the key seismic question – detection and identification – the experts concluded that the seismic method by itself would not identify all events, whether earthquakes or explosions, and that inspection would be necessary.[191] It adopted conclusions on the utility of various detection methods: registration of acoustic waves, seismic waves and radio signals as well as the collection of samples of radioactive debris; it also adopted conclusions on the detection of nuclear explosions at very high altitudes.[192] The number of control posts was perhaps the most difficult point to agree, but in the end it was sorted

[187] Cmnd.551, *Report of the Conference of Experts to Study the Methods of Detecting Violations of a possible Agreement on the Suspension of Nuclear Tests*, Geneva, 1 July to 21 August (London, 1958).

[188] TNA FO 371/135566, Technical conference on detection of nuclear tests 1958, H.T. Morgan to Mr O'Neill, 8 September 1958, Detection of Nuclear Tests; the convening of the conference 'gave particular satisfaction' to HMG since its origins lay in UK proposals. Cmnd.551, *Report of the Conference of Experts*, p. 3.

[189] Michael Goodman, *Spying on the Nuclear Bear, Anglo-American Intelligence and the Soviet Bomb* (Stanford, 2007), p. 115.

[190] TNA FO 371/135556, Technical conference on the detection of nuclear tests 1958, Geneva talks on nuclear tests political brief for United Kingdom Representatives, 26 June 1958.

[191] TNA FO 371/135562, Technical conference on detection of nuclear tests 1958, Nuclear tests conference: 19th meeting, Geneva telegram no. 385 to Foreign Office, 24 July 1958.

[192] TNA FO 371/135567, ZE 122/218, Geneva telegram no. 451 to Foreign Office, 14 August 1958.

out when the Russians finally went along with the numbers suggested by Penney: 160–70.[193] Penney in fact played a key role in finding a compromise overall as he suggested that the Conference set a threshold for a proposed detection system at five kilotons given the difficulties of achieving a one kiloton threshold.[194]

Crucially the experts judged that there was good probability of recording seismic signals from deep underground nuclear explosions in continents equivalent to one kiloton and above taking place on the surface of the earth and up to ten kilometres altitudes. There would be a good probability of recording seismic signals from deep underground nuclear explosions in continents equivalent to one kiloton and above. The capability of the control system to identify underground nuclear explosions of one to five kilotons depended on three factors. The first of these was the small fraction of earthquakes that could be identified from data obtained from the control posts alone. The second problem concerned the fraction of earthquakes identifiable with the aid of supplementary data obtained from existing seismic stations; and, the third issue was the fraction of events still left unidentified which might be nuclear explosions and for which the international control organisation would need to conduct on-site inspection.[195]

The report noted that, notwithstanding the great difficulty that might be experienced in obtaining positive identification of a carefully concealed deep underground nuclear explosion, there was always the possibility of detection by means of on-site inspection. As Sir Michael Wright noted, this conference reached the hopeful conclusion that given an adequate international control system – fixed control posts, overflights and on-site inspection – all tests could be adequately monitored in the atmosphere, underwater and underground.[196] Monitoring in space problems would be considered later.[197]

UK officials back in London were pleasantly surprised that the Conference revealed that the Russians were prepared to accept that there should be

---

[193] TNA FO 371/135567, Technical conference on the detection of nuclear tests 1958, Geneva telegram no. 466 to Foreign Office, 18 August 1958; Personal communication. Of these 160–70 posts 100–110 would be on continents, 20 on large oceanic islands and 40 on small oceanic islands.

[194] Carl Romney, *Detecting the Bomb The Role of Seismology in the Cold War* (Washington, 2009), pp. 96–7.

[195] Cmnd. 551, *Report of the Conference of Experts*, pp. 25–6. TNA FO 371/135568, ZE 122/233, Technical conference on the detection of nuclear tests 1958, Geneva telegram no. 472 to Foreign Office, 19 August 1958; TNA FO 371/135568, ZE 122/243, Technical conference on the detection of nuclear tests 1958, Report of the Conference of Experts to study the possibility of detecting violations of a possible agreement on suspension of nuclear tests, EXP/NUC/28, 20 August 1958; UNGA A/3897, 28 August 1958.

[196] Sir Michael Wright, *Disarm and Verify An Explanation of the Central Difficulties and of National Policies* (London, 1964), p 24.

[197] High altitude tests were dealt with by a Technical Working Party of Experts of the UK, US and USSR. This reported to the Discontinuance Conference and set out agreed recommendations on ways of detecting such tests. Wright, *Disarm and Verify*, p. 53.

internationally manned control stations on their territory.[198] London suspected, however, that the Russians were confident that political pressures would build up on the Western powers ahead of the UNGA in the autumn 1958 that would make it difficult for them to avoid a suspension of nuclear tests. Although there were still strong defence objections to a suspension of tests, UK-US discussions over the preceding months had left the UK in a position where London was contemplating agreeing with the US on a fairly early suspension. For the UK, the successful report represented a new and helpful development.[199] Officials regarded the outcome as a clear justification of their technical approach to the problems of control and inspection in the field of disarmament, which Lloyd had first proposed on 17 July 1957 to the UNDC.[200]

This report provided the essential technological basis on which to build further progress. The next task was to find solutions to the important practical problems relating to the organisation, installation and functioning of a control organisation. In parallel work began at the AWRE's Weapons Group on seismological techniques of verification.[201] For Macmillan, the report's outcome was certainly evidence that an effective control system was possible in theory, but many practical and political problems needed to be solved first before the report could be translated into reality. London needed time to think about this. Nevertheless, the UK was prepared to enter into negotiations with other governments that had tested nuclear weapons with a view to solving these problems and concluding an agreement for the suspension of tests under effective international control. The UK would be ready to begin such negotiations by 31 October 1958. It is clear from Cabinet records that the UK felt itself under considerable pressure to complete its own testing activities and meet national and international demands for test limitations.[202] The pressure of public opinion against nuclear testing had by this time become such a serious factor that no British government could have afforded to be seen dragging its feet.[203] There was no let up in public disquiet on atmospheric testing, which put considerable pressure on the Macmillan government.

---

[198] TNA FO 371/132699, IAD 123/40, Test Suspension Proposals, Geneva Talks on Nuclear Tests, C.D.W. O'Neill to Private Secretary, 26 July 1958.

[199] TNA FO 371/132698, IAD 123/29, Attachment to Advice to the Prime Minister for a Parliamentary Question, 22 August 1958.

[200] Atomic Weapons Research Establishment, *The Detection and Recognition of Underground Explosions. A Special Report of the United Kingdom Atomic Energy Authority* (London, 1965).

[201] Atomic Weapons Research Establishment, *The Detection and Recognition*, p. 9; see Chapter 6 for more detailed discussion of UK work in this area.

[202] Clark, *Nuclear Diplomacy*, pp. 196–7.

[203] Wright, *Disarm and Verify*, p. 135.

The Next Megaton Test: Grapple X

The situation on future megaton tests became much clearer once the scientific data from the Grapple series had been analysed. Orange Herald gave a yield of about 750 kilotons and was highly successful. A large fission weapon of approximately one megaton in a form suitable for a ballistic missile was therefore now possible; such a warhead would weigh about 2400 lb and could also be used as an interim weapon in a free falling bomb in a Blue Danube case. Green Granite was the first attempt at the hydrogen weapon, but the yield was somewhere between 300 and 350 kilotons. The third shot was a modification of this design using slightly more fissile material, but it was unsuccessful in that it was intended to give a larger yield but in fact only managed slightly over 200 kilotons. The primaries for both of these rounds operated satisfactorily, but the secondary did not work according to theory. All this meant that it was necessary to reconsider urgently current trial plans for the Antler series in Australia.[204] Grapple meant that the only megaton bomb that the UK could produce immediately was a fission weapon using U-235 only. To provide a small stockpile of these weapons in the next two years would use all the available U-235 and there would thus be no material to spare for Red Beard. There was therefore no point in a mixed core Red Beard test at Antler. The two shots connected to the H-bomb would be of no purpose, as they would not provide data to help elucidate the present design troubles.[205] It seems that if the megaton weapons adopted used only U-235, then Red Beard using only plutonium would go into production, but if a mixed plutonium-U-235 megaton weapon were adopted mixed plutonium/U-235 Red Beards would be made. One might infer that this was because they would use the same primaries, and thereby economise on production costs.

Improvements to the UK's thermonuclear weapon design were imperative and Aldermaston was strongly of the view that progressive testing rather than a single series offered the best way of doing this. They would be ready in November for a further test taking full account of the Grapple lessons. One recurring factor that seriously affected the timetable was the likelihood of the stoppage of tests arising

---

[204] This programme was intended to test, provided that they were ready in time: Red Beard with either a pure plutonium or a mixed core; two shots designed to give further information about megaton bomb design; Blue Fox, which was to provide data leading to the development of a five kiloton warhead for defensive guided missiles; and finally Pixie, a small scale trial design to give data that might lead to the development of a shell weapon – Seaslug. Therefore, the Antler programme came down to three possible shots: the standard Red Beard, Blue Fox and Pixie. MOD thought that the Red Beard trial could be deferred and since Pixie was highly speculative and there were differences of opinion, there was a preference to postpone this one too. That said, the Admiralty was worried that a ban on testing would prevent testing and that a Red Beard proving round for production should therefore be fired.

[205] TNA AVIA 65/1432, Draft Conclusions of a meeting held in Room 358 St Giles Court, 10 July 1957.

from disarmament talks. Foreign Office officials' view was that the pressure for cessation was so great throughout the world that the chances of being able to conduct any more tests after the end of 1957 were very small. Brundrett therefore advised his Minister that he regarded it as absolutely essential to make arrangements to carry out such further trials of megaton devices as were possible before the end of the year. Whether cancellation of Antler and incorporation of its tests could be carried out at Christmas Island would simplify logistics was also examined, but later rejected.[206] Macmillan agreed and the Ministerial Committee on Atomic Energy reviewed the proposal on 24 July. Ministers agreed that further megaton tests take place as soon as AWRE was ready for them. The longer the delay, the greater the likely political difficulties in view of the growing pressure of world opinion in favour of a ban.[207] Preparations would therefore proceed expeditiously with a view to holding the tests in November. Political pressure to acquiesce in a test ban or moratorium was a determining factor in this decision.[208]

Against this technical background, the Foreign Office anticipated that the UNGA session starting in September was likely to pass by a two-thirds majority a resolution condemning the further holding of nuclear tests in late November/ early December. This was likely to arise in connection with the report of the UNDC Sub-Committee, which normally reported by the end of October. A special resolution might even be presented earlier. For this reason the Foreign Office advised Ministers that it was essential for the UK to plan for tests to take place not later than early November.[209] Pressure was mounting on HMG to acquiesce to a ban and the government decided to hold further megaton tests as soon as possible.[210] Macmillan told the Cabinet on 25 July that scientific reasons called for further tests and to establish the best method of making more economical use of fissile material. It had proved possible to advance the date of the text trials to November 1957 instead of the initial date of June 1958.[211] There were signs that Cook wanted to press for megaton trials to continue after May 1958.[212]

Unfortunately the Ministry of Supply, who funded the test programme had made no financial provision for a further test after Antler and it had no spare funds available either. In addition, there would have to be expenditure on fissile material that would have otherwise been paid for from funds allocated to weapon production. Treasury officials had to be persuaded to part with more cash; it was

---

206 TNA DEFE 7/919, Brundrett to Minister, 9 July 1957.

207 TNA DEFE 7/919, Cabinet Ministerial Committee on Atomic Energy, Confidential Annex to A. (57) 1st Meeting, Minute 1, 24 July 1957.

208 TNA DEFE 7/919, Foreign Office telegram no. 3052 to Washington, 30 July 1957.

209 TNA AVIA 65/1431, E.S. Jackson, DGAW to Secretary, 25 July 1957.

210 TNA AVIA 65/1431, Foreign Office telegram no. 3052 to Washington, 30 July 1957.

211 TNA CAB 128/31 Part II, C.C. 57 (57) Item 1 Nuclear Tests, 25 July 1957.

212 TNA AVIA 65/870, L.J. Dunnett to Secretary, Future Trials from 1956 – Tentative Programme, 7 November 1957.

found somehow.[213] In addition, Macmillan emphasised that no publicity should be given to this decision. The directive was given orally to Brundrett and Plowden. Macmillan attached special security to the 1957 test series results because of the domestic political pressures to halt further H-Bomb tests. He wanted to keep to as small a circle as possible the knowledge that the tests had failed to provide the UK with the megaton range striking power hoped for.[214] Yields had to be passed as far as possible verbally, rather than in writing to the minimum number of essential recipients.[215] He also gave instructions that knowledge of the Grapple results and the reason for conducting a further trial so soon should be limited to the very small circle of officials who had already been informed.[216] He regarded the further tests as necessary both for scientific reasons and in order to establish the best method of making more economical use of fissile material.[217] The expense and scarcity of fissile material and threat of a cut-off were critical factors in this secrecy. AWRE and the MOD hoped that access to US design information would enable the UK to make the most economical use of its own fissile material.[218] In fact Ministers agreed that Brundrett could tell the Americans that whatever the final decision on the future UK testing programme might be, the UK certainly had plans for further tests at intervals over the next two or three years.[219]

Grapple X and Continuing Pressures

A new Cabinet Office Committee – the Working Group on Disarmament – was told in mid September 1957 that Macmillan, with a view to countering Soviet propaganda, wished to suggest to Eisenhower that the UK and US should revive the earlier UK proposal that the two countries should voluntarily undertake to limit the fission yield of their tests to a stated amount. Although the US had been

[213] TNA AVIA 65/1431, Meeting of Ministerial Committee on Atomic Energy Brief for Minister, Nuclear Tests.

[214] TNA CAB 21/4536, W. Geraghty to Sir Norman Brook, Christmas Island Tests, 17 February 1961.

[215] TNA PREM 11/3583, Harold Watkinson to the Prime Minister, 13 February 1961.

[216] TNA AVIA 65/1248, Richard Powell, MOD to Sir Cyril Musgrave, Ministry of Supply, 19 July 1957.

[217] TNA CAB 128/31 Part 2, C.C. (57) 57th Conclusions, Minute No.1 Nuclear Tests, 25 July 1957.

[218] TNA PREM 11/2329, Briefing for Prime Minister on Defence Matters, 22 October 1957. By November 1958 year Plowden was able to tell the Cabinet Defence Committee that US design information would greatly extend the UK's capacity to manufacture warheads from British stocks of fissile material. TNA CAB 131/19, Cabinet Defence Committee, D (58) 28th Meeting, Item 2 Fissile Material, 26 November 1958.

[219] TNA AB 16/2371, F. Bishop, Downing Street to D.W. Ward, MOD, 28 November 1957.

unable to accept this in January, there were now signs of a possible change in opinion.[220] Macmillan thought that this undertaking could be given during the autumn UNGA. From the UK perspective such a declaration would have to be a joint one with the US and specify an upper limit of fission yield from the future nuclear tests of the two countries taken together. Although the US had admitted that in the past its tests had generated high fission yields, it was now able to test devices where the total fission yield was still a very much smaller percentage of its new thermonuclear designs it had ready to test. In these circumstances, given the volume of fission yield which the UK at that time expected from its proposed test programme, Macmillan thought it should be possible to make a declaration about total fission yield from UK-US test over the next three years. These would meet the needs of the defence programmes of both countries and would simultaneously generate considerable public credit. It seems, therefore, that a significant driver for this proposal was the compelling need to defeat any Soviet attempts to derive propaganda advantage by painting the West in a poor light given global concerns over the perceived adverse health effects of radioactive fallout. If Eisenhower were prepared to buy into this, it would be possible for the UK and US experts to work out the fission yield and each country's share in it. The working group in the end endorsed the Prime Minister's proposal on the lines that he envisaged, but in the event it was not pursued energetically.

Grapple X preparations took place in the greatest secrecy to avoid public pressure that might put the testing programme in jeopardy.[221] These pressures meant that planners were forced into a situation where they were improvising in order to carry out a new trial in November 1957 when the initial thinking had been that the next round would not take place until June 1958.[222] However, the state of flux in thinking was evident at this time since an earlier estimate in mid August 1957 saw AWRE thinking in terms of two more tests after Grapple X, the latter possibly in April 1958.[223] Consequently, all preparations were rushed and service personnel posted overseas at short notice to an island where facilities for long term occupation were inadequate. The November trial was rushed forward to an extent that was undesirable scientifically. Jackson felt that it would have been better to have allowed more time to analyse and digest the results of the previous trials.[224] However, the new US disarmament proposals, which had been put to the Russians, threatened to lead to a moratorium on tests as early as 1 July 1958. In these circumstances, the UK decided that the best course was to keep the facilities

[220] TNA CAB 130/130, GEN.604/5th Meeting, Item 1 Proposed Declaration on Limitation of Fission Yield, 17 September 1957.

[221] TNA DEFE 7/920, Draft Colonial Office telegram to Commissioner for Western Pacific, 6 September 1957.

[222] TNA AVIA 65/1248, DGAW Correspondence 1957, Atomic Weapon Trials.

[223] TNA DEFE 7/920, Atomic Weapons Trials Executive Operation Grapple X Meeting at St Giles Court, Grapex (57) 2nd meeting, 14 August 1957.

[224] TNA AVIA 65/1440, D.G.A.W to Permanent Secretary, 27 September 1957.

at Christmas Island in being for tests in the first half of 1958.[225] That said, the Foreign Secretary was anxious to avoid criticism in the disarmament context of any decision since he did not think that the UK would want to suggest that it had in mind to press forward with tests more quickly than was originally intended.[226] Nevertheless, progress was being made on the design front with two options under development. Shortly before Grapple X, William Cook had told Jackson that:

> Our experimentation to date shows that there is good possibility of being able to fire a round to determine the functioning of lithium deuteride as a thermonuclear multiplying substance at the compressions we hope to achieve in a spherical arrangement, as the second element of a two bomb system. The lithium deuteride will contain a mixture if Li^6 and Li^7 isotopes and this system is an alternative to the $U238/Li^6D$ multiplying system that we are testing at Grapple X. The firing of Grapple X will confirm, or otherwise, the correctness of our calculations of compression as well as testing the U238/Li6D system. Many more calculations are required before we shall be satisfied with the system proposed for February.[227]

[225] TNA DEFE 7/919, Duncan Sandys to Minister of Supply, 11 June 1957.

[226] TNA DEFE 7/919, Selwyn Lloyd to Aubrey Jones, 20 June 1957.

[227] TNA DEFE 7/922, W.R. Cook, AWRE to E.S. Jackson, DGAW, Ministry of Supply, 29 October 1957. This letter is retained in TNA AVIA 65/1217. However, we see it in TNA AB 16/2371 and that this letter stated AWRE would be ready to fire at Grapple Y a trial round to determine the functioning of lithium deuteride as a thermonuclear multiplying substance at the compressions they hoped to achieve in a spherical arrangement as the second element in a 2-bomb system. Jackson concluded from this that the round was designed to promote AWRE's knowledge of thermonuclear principles rather than one designed to test a particular warhead developed for service use. G.A.C. Witheridge, MOS to H.C. Hudspith, UKAEA 1 November 1957. The trial aimed at providing knowledge to enable AWRE to carry out successful development of a Granite type of megaton weapon. P.W.B. Brooking, AWRE to H.C. Hudspith, Finance Branch, UKAEA, 5 November 1957. There is discussion in TNA ES 1/583 of bonding lithium deuteride powder into blocks for the Grapple Y rounds, D.T. Lewis, SSCD to K.T.B. Scott, 5 November 1957. The UK would have learned from debris analysis that the Soviet Union had been the first to use lithium 6 deuteride in a two Stage device in 1955. G.A. Goncharov, 'American and Soviet H-bomb development programmes; historical background', Physics – Uspekhi vol. 39 (10) 1033–44 (1996) Russian Academy of Sciences October 1996, p. 1043. There are closed Aldermaston files with open descriptions at the National Archives which refer to Operation Strategy, which was about monitoring Soviet tests in this period for instance TNA ES 1/844 Operation STRATEGY (monitoring Soviet tests); see also TNA ES 1/852 Soviet tests ELIGIBLE and RACKET; analysis 1953–57 and TNA ES 1/1114 Soviet atomic explosions; thermonuclear weapon design 1954. It is also clear from an explanatory note sent by the Minister of Defence to the Chancellor of the Exchequer that the UK had used debris analysis to obtain clues on design information referring to Granite type warheads. He said that, 'we believe that both the US and USSR have exploded assemblies of this type'. TNA AVIA 65/824, Walter Monkton, Minister of Defence to Chancellor of the Exchequer,

AWRE prepared three rounds for Grapple X, two megaton and one fission trigger. The first one was planned to be ready for flying out to the Pacific by 26 October.[228] If the first megaton shot proved successful, the other two would not be fired. However, November would not give Aldermaston the complete answer. Trials in 1958 formed an essential part of its plans.[229] The 8 November 1957 Grapple X shot was a turning point in that up to that point AWRE scientists had argued that they would need an 18-month gap between tests for theoretical studies.

Grapple X produced a yield of about 1.8 megatons – much bigger than anticipated.[230] It seems clear that the availability of improved computing power was a key factor. The Ferranti and English Electric computers, installed in 1955 and 1956, were not powerful enough for the necessary calculations for the thermonuclear reactions. However, an IBM 704, installed early in 1957, successfully evaluated alternative designs for the secondary due to be tested at Grapple X.[231] The impact on service planning appears to have been decisive. Although successful from a scientific point of view, the device did not represent a service warhead. However, the plan was now to develop it into a safe, reliable and economical warhead. To achieve this considerable scientific and engineering design progress would be necessary in order to reduce it in size and weight and to provide adequate safety and operational reliability. Further trials would be needed and the time factor suggested that Blue Steel would be its first application.[232] However, by the end of April 1958 the plan was changed and on AWRE advice Green Grass was to be the warhead for Blue Steel.[233] Testing plans and weapons engineering were thus clearly in a state of flux; the pressure on AWRE's engineering design capabilities must have been behind this decision.

AWRE did not have political or financial approval for further tests, but Cook thought that they would want to continue improving and testing until a

---

Trials of British Megaton Warheads, 8 February 1956; see also Arnold, *Britain and the H-Bomb*, p. 91, which notes that the AWRE radiochemists were very good and had been able to calculate the approximate masses of lithium 6 deuteride and U-238 in the Soviet test of November 1955.

228 TNA ES 1/642, Minutes of Third Planning Meeting, paragraph 12 Transport of Weapon Components, 7 August 1957.

229 TNA AVIA 65/1248, Cabinet Nuclear Test Policy Committee D.G.A.W. Correspondence 1957, E.S. Jackson DGAW to Deputy Secretary, 23rd October 1957.

230 Arnold, *Britain and the H-Bomb*, p. 161.

231 David J. Hawkings, *Keeping the Peace The Aldermaston Story* (AWE, 2000), pp. 38–9.

232 TNA AIR 2/13733, Progress Report on Nuclear Weapons as at 31 December, 1957 Appendix A.C. (58) 1. By February 1958 the advantages of the Granite warhead over Green Grass were so great that the Air Ministry asked that possible action should be taken to equip Blue Steel with the Granite warhead from the beginning. See also TNA AIR 2/13748, W.R. Brotherhood, D.O.R. (C) to D (RAF) B/M.O.S., 25 March 1958.

233 TNA AIR 2/13748, Minutes of 5th Blue Steel Progress Meeting, 24 April 1958.

test ban stopped them.[234] Cook made clear that AWRE would not wish to defer the testing programme until the UK had achieved an exchange of information with the Americans and that the next stage would be to plan to test a warhead for Blue Streak.[235] After Grapple X AWRE believed that the most effective way to the RAF operational requirement for megaton warheads for Blue Steak and powered guided bombs was to proceed with a trial programme consisting of a test at Christmas Island every four to six months for a period of two to three years.[236] AWRE had in mind a series of intermittent tests designed to perfect the Granite type of weapon (the two-stage radiation implosion device). Each test would check a limited number of variables, and each step forward would be dependent on the results of the preceding test. These tests would continue at roughly three monthly intervals until either success was achieved or a ban was placed on further testing.[237] Nevertheless, this test had been an experimental design to show whether the basic UK ideas on two stage boosted fission warheads were sound. Results were in line with pre-test calculations. AWRE scientists were therefore confident that a yield of 1.5 megatons could be obtained from a two stage boosted fission device weighing three and a half tons; and, moreover, with a design that could be made invulnerable to the R-I problem.[238]

Sandys minuted the Prime Minister on 28 November 1957 to tell him that in light of Grapple X the UK was now well on the way to evolving a device that was light, economical in fissile material and immune to counter-measures. In order to complete this development expeditiously AWRE felt it necessary to hold further tests at four to six month intervals. All of which would be subject to continuous review in the light of cooperation with the Americans.[239] The previous development programme had envisaged an interval of 18 months after May 1958 for concentrated theoretical studies, but the results of Grapple X had changed that; a more regular series would now be needed.[240] Tests were essential if the UK was to meet its requirement for an immune trigger for a warhead of one megaton and one ton in weight.

The course of development towards the final design was influenced by several factors, the most important of which were the functioning of different

---

[234] Air Vice Marshall Wilfrid E. Oulton, *Christmas Island Cracker* (London, 1986), p. 358.

[235] TNA AVIA 65/1248, E.S. Jackson DGAW to Secretary, 18 November 1957.

[236] TNA DEFE 7/922, Draft letter from Prime Minister to Prime Ministers of Canada, Australia and New Zealand, Duncan Sandys, Minister of Defence to Prime Minister, 6 January 1958; and Brief for Prime Minister's Visit to Australia and New Zealand.

[237] TNA AVIA 65/1248, Future Nuclear Tests, DGAW, 29 August 1957.

[238] TNA DEFE 7/922, Results of Operation Grapple and Grapple X, 28 November 1957.

[239] TNA DEFE 7/92, Duncan Sandys to Prime Minister, 28 November 1957.

[240] TNA AVIA 65/1440, Brief for Prime Minister's Visit to Australia and New Zealand, Megaton and Kiloton Weapon Trials.

thermonuclear mixtures, effects of reduction in size, changes in geometry, reduction in case weight and change in implosion pressures and the testing of trigger devices immune to countermeasures to marry up with progress on implosion pressures.[241] Technically only one variable at a time could be changed, so the next test in the spring of 1958 would test a change in the thermonuclear mixture of the basic Grapple X design. Another mixture would be tested in the late summer/autumn tests. Ministers agreed on 6 December 1957 that the March tests (Grapple Y) could proceed, but they would take a final decision near the time for the summer/autumn tests, although they did agree that the minimum necessary preparations could be made for these latter tests.[242] However, an accident at Aldermaston's lithium 6-deuteride production building set back the Grapple Y programme by four weeks until new safety equipment was installed.[243]

The testing schedule carried potentially severe opportunity costs for the wider weapons programme; it seems that the extensive programme planned by AWRE would consume a significant part of UK fissile material stocks.[244] The Antler trials for example had resulted in the diversion of fissile material allocated for weapons purposes.[245] However, both the DRPC (AES) and the Chiefs of Staff had strongly endorsed the need for these trials. Jackson felt obliged to remark in November 1957 that in striving for perfection the UK might end up expending so much material and money in tests that it would have been cheaper to have made the less efficient weapons and done without further tests. The argument that a more efficient warhead would enable the UK to complete its stockpile targets at an earlier date turned on how much fissile material or thermonuclear material would be expended in the process. DGAW even questioned the need for a warhead immune from the R-I problem. He felt that even if the USSR had effective defences, some non-immune warhead might still get through. The effect of the 1956 tests on the supply of fissile material made it important to decide whether the need for such a warhead

241 TNA DEFE 7/921, Further Information on Proposals for Megaton Tests, MOD, 2 December 1957.

242 TNA DEFE 7/921, Prime Minister's Personal Minute, Serial No. M.601/57, 6 December 1957; TNA AVIA 65/1217, Draft paper from DGAW for DRPC (AES) on future atomic trials, 12 April 1957; Sir Richard Powell, MOD to E.S. Jackson, Ministry of Supply, 16 December 1957. In 1958 a brief for the NTPC meeting on 12 February noted that Ministers would probably need to know the outcome of Grapple Y both scientifically and in its effects on opinion at home and abroad, before authorising Grapple Z. For political reasons, there was a chance that the actual decision to fire might be postponed until a very late Stage. See TNA AB 16/2371.

243 TNA AVIA 65/1440, William Cook, UKAEA to E.S. Jackson, DGAW, Ministry of Supply, 1 January 1958; TNA DEFE 7/2379, Cpt F.B. Lloyd, Ministry of Supply to V.H.B. Macklen, MOD, 2 January 1958.

244 TNA AVIA 65/1248, Nuclear Tests Policy Committee (Brief for meeting on 18 November, Appendix A).

245 TNA DEFE 16/375, J.M. Gibbon to G.S. Downey, Treasury, 18 January 1957.

justified the impact on the supply of atomic weapons for the services. Officials noted that the equivalent of five and a half weapons' worth of fissile material might be consumed if all the test shots were fired.[246]

This appears to be the only occasion, however, when the Ministry of Supply questioned AWRE's objectives. Whatever the merit of DGAW's concerns, they did not undermine the UK's progression towards small, efficient and immune warheads. All three were keys to help unlock the door to the treasure trove that would be access to US nuclear data on thermonuclear two-stage weapons. Penney's view was that, under conditions where at any time a cessation of trials could be forced upon the UK, further firings in February and May 1958 would still be needed. These were totally dependent upon the result of Grapple X. However, any progress would be such as to either allow an interval before any further tests, or sufficiently poor to compel a compete rethink of the design for the large weapon.[247]

Conclusions: Early Pressures 1954–1958

The UK had to run with the hare and hunt with the hounds on testing. It wanted the political benefits of being seen to respond to concerns over atmospheric tests – hence its ideas on fission limits, registration and proposals for experts to examine the options for a verifiable ban. Ministers, officials, senior service personnel and AWRE scientists also desperately needed to bring the UK's thermonuclear programme to the point where a successful warhead design was tested before any meaningful limitations or prohibitions took effect. Additional requirements on AWRE by the three services for kiloton range weapons also required testing. A key imperative was to be able to build weapons economical in their use of scare and expensive fissile materials, HEU especially. Such weapons also had to be immune to the R-I effect. By the end of 1958, the UK had nearly reached the point at which it would be able to meet these requirements, but considerable further work was still needed if AWRE was to master the art of thermonuclear warhead design, but that was only half the battle. Design and engineering of weapons that could be safely stored, handled and transported by the RAF was something else again in terms of effort and expense. Further testing of megaton range devices would still be needed as an essential component in the UK weapons programme, but test-ban pressures were not going to go away either. Running with the hare and the hounds would have to continue for some time to come.

---

[246] TNA DEFE 16/599, DRPC Sub-Committee on Atomic Energy D.R.P. (A.E.S)/P (55) 20, Overseas Atomic Weapon Test Programme – 1956 Note by C.A.W and DAWRE Appendix III, 5 July 1955.

[247] TNA DEFE 7/920, Minute from Sir Richard Powell to Prime Minister, 30 August 1957.

Chapter 3
1958: A Decisive Year

Introduction

There is a very clear sense from available National Archive papers as well as Lorna Arnold's book 'Britain and the H-Bomb', that time was running out for the UK's atmospheric nuclear tests: but just how severe were these time pressures and what impact did they have on the direction and content of the weapons programme? How did the UK manage to juggle its conflicting policy objectives: ensuring the development of an effective and economical UK deterrent whilst supporting constraints on atmospheric nuclear testing? Just what were the key benefits from continued testing that would allow the UK to satisfy its nuclear defence policy requirements? The key to the door of US nuclear secrets was demonstration of an independent capability to design, build and test successfully two-stage thermonuclear explosive devices. 1958 was to prove a decisive year in this regard and through this door would open the way for a more front-footed UK approach to a test ban.

UK Weapons Tests and Test Limitations

International pressure became more acute in the first half of 1958. AWRE were at that time still thinking in terms of a continuous series of tests – single megaton shots once every three or four months. Any delays in holding a trial would cause a corresponding delay in meeting the full weapons programme developmental requirements. In Victor Macklen's view, Ministers had to understand that the present trial programme was not designed as a crash programme to fit in everything that was humanly possible before a sudden stop to tests as a whole.[1] Macklen was on Brundrett's staff in the Chief Scientific Adviser's department, MOD as a key adviser on nuclear weapons matters. At the same time, the Chiefs of Staff felt that continued tests, especially in the kiloton range, were essential from the military point of view. Even if the US McMahon Act were amended, the control of information would still be in the hands of the US. If the UK was to develop improved British nuclear warheads, it had to complete its own tests before any suspension became effective. The Chiefs did not regard it as satisfactory to have

---

1 TNA DEFE 7/923, V.H.B. Macklen to Sir Richard Powell, 3 March 1957.

to rely on US information.[2] If a test moratorium took effect from January 1959 this would still mean that the UK would not be able to test the full range of weapons that it wanted to develop, although progress with megaton weapons should have gone far enough by that time.[3]

After Grapple X, the UK was able to design a one-megaton warhead for its Blue Streak missile, but it would be profligate in fissile material and vulnerable to counter measures with no prospect of decreasing the design's vulnerability.[4] The operational requirement for a Blue Streak warhead of one ton and immune to countermeasures required further trials involving several different variables, which at that stage of knowledge had to be tackled seriatim. These were the functioning of different thermonuclear mixtures, the effects of reduction in size, change in geometry of fissile and non-fissile components, reduction of case weight and changes in implosion pressures and the testing of triggers immune to countermeasures. A test ban would make such methodical progression impossible. As things stood, the first interim megaton weapon (Green Grass in Violet Club) was planned for delivery to RAF Wittering on 4 March 1958. However, given the safety concerns surrounding this weapon – it did not contain all the elaborate safety arrangements which were necessary to meet the full operational requirements – Penney made it clear to the Chief of the Air Staff that any movements or transport of these weapons would only be done in agreement and after consultation with AWRE.[5] The Green Grass warhead was not to be moved in an assembled condition by road.[6] It seems that the handling aspects of this warhead were a factor in the formation of the MOD's Nuclear Weapons Safety Committee. Violet Club – was very heavy at 7000 lb, expensive (£1.5 million), dirty (practically all fission and very little fusion for its 500 kiloton yield[7]), not very safe in flight and extremely vulnerable to countermeasures. It was an experimental weapon. In fact the original number estimated that could be produced, 16, was reduced to 13 because of the need for further megaton trials.[8] In practice planning intended that this number would

[2] TNA CAB 130/19, Cabinet Defence Committee D (58) 15, Views of the Chiefs of Staff on Disarmament, 18 March 1958.

[3] TNA CAB 130/130, Working Group on Disarmament GEN.604/3 Draft Submission to Ministers Nuclear Disarmament, Foreign Office, 31 March 1958.

[4] TNA DEFE 7/922, T.W. Ellner to F.A. Bishop, attaching Memorandum for Prime Minister on Megaton Trials, 2 December 1957.

[5] TNA DEFE 7/2294, W.G. Penney to Air Chief Marshall Sir Delmont A. Boyle, Chief of the Air Staff, 26 February 1958; and undated manuscript minutes and accompanying drafts from V.H.B. Macklen to Sir Frederick Brundrett.

[6] TNA DEFE 7/2294, J.T. Tomblin, ACT to Mr E.F. Newley, CWD, Transport of Assembled Green Grass, 15 October 1958.

[7] TNA AIR 2/13718, D.A.W.D.P to D.O.R. (C) Interim Megaton Bomb, 3 December 1957.

[8] TNA AIR 2/13718, E.S. Jackson DGAW to DCAS, Interim Megaton Weapons, 19 September 1957. The Admiralty was allocated some HEU for the propulsion programme

not be reached until March 1959, with only five being planned to be available by July 1958. However, weapon number five was running late and was not available until mid December 1958.[9] The first five of these were in a Blue Danube case, the remainder in a Yellow Sun Mark I case; 12 appeared to have been produced in total. This was the number that the RAF could 'tolerably' accept.[10] Manufacture of the weapon had been adversely influenced by the use of design and manufacturing capacity and material in the preparation of trial stores (devices for Grapple X[11]) and the bringing forward of the dates of Grapple Z.[12]

Macmillan told Eisenhower on 1 January 1958 that the UK could not accept abolition or suspension of tests at that stage given the then state of British knowledge. However, if the US had made as much progress as it had wanted and was prepared to make information available to the UK, the position would be different.[13] However, a test ban achieved under those circumstances would be important, as it would prevent further tests and further nuclear weapons states. Completion of the UK-US Agreement for Cooperation on the uses of Atomic Energy for Mutual Defence Purposes (Mutual Defence Agreement) in July was not enough. There still had to be the scientific exchanges on weapons design and engineering that the agreement permitted. There were plans for short series of tests – Grapple Z involved four shots in August and September 1958. It was critically important to complete these before the talks with the Americans began.[14] As Anglo-American nuclear collaboration was developing British scientists needed as much weapons design information as possible in order to ensure that UK-US nuclear collaboration could be resumed and sustained. For Penney at least, speaking in mid August 1958 in advising Macmillan, this meant that the UK would need to ensure that the bilateral talks went as far as possible on the US side in the expectation that this would remove the UK's need for tests in 1959.

---

that accounted for the other side of the reduction.

9 TNA AVIA 65/1218, DAWDP to SSWP/AWRE, 18 November 1958. Violet Club production implies that that the Air Ministry indicated at a DRPC (AES) meeting on 11 November 1958 that they wanted to stop Violet Club at the end of the calendar year, which meant putting the chop on after No. 5 or No 6. No. 5 was running late and not scheduled to appear until mid December 1958.

10 TNA AVIA 65/1116, D.R.P. (A.E.S)/M (57) 6, DRPC Sub-Committee on Atomic Energy meeting 17 December 1957 Item 3 Interim Report on Allocation of Military Fissile Material Stock and Weapons Programme; and see also *TNA AIR 2/13705, J. Tait A/Dof Ops (B & R) to ACAS (Ops) Violet Club and Yellow Sun Mark 1, 17 January 1959.*

11 TNA AIR 2/13718, G. Hampton for Deputy Director of Bomber Operations to AOC Bomber Command, 27 September 1957.

12 TNA AIR 2/13733, Progress Report on Nuclear Weapons Appendix to A.C. (58) 61, Progress Report on the Development of New Weapons up to 31 August 1958 Note by DCAS, 8 September 1958.

13 TNA PREM 11/2554, Extract from Telegram no. 4, from Foreign Office to Washington (Prime Minister's Personal Telegram T. 1/58), 1 January 1958.

14 Arnold, *Britain and the H-Bomb*, p. 176.

The Cabinet Defence Committee invited the Minister of Defence in March 1958 to consider the conditions on which the UK could accept a suspension of nuclear tests unrelated to the cut-off. The main objective was to establish the technical information that the UK would need from the US if it agreed to a suspension of nuclear tests.[15] In discussion in the Cabinet Office Working Group on Disarmament it was stated that from the military point of view a limitation on further tests might be acceptable, but a complete suspension of tests would raise greater difficulties that could differ according to the date on which the suspension would come into operation. The question of the conditions on which the UK would be prepared to accept a suspension might well arise at an early date. Accordingly, Sir Frederick Brundrett, MOD Scientific Adviser, was tasked to consult the UKAEA and Chiefs of Staff to see what these conditions might be from a military point of view. However, we cannot see what the officials came up with, as the relevant paper in The National Archives file – CAB 130/2288 – remains closed. The Minister of Defence instructed his officials during a meeting with the Chiefs of Staff on 12 March 1958 to examine whether it would be feasible to suspend megaton but not kiloton tests; and, as part of this exercise, find out what effect the agreement to suspend nuclear tests by the end of 1958 would have on the UK megaton test programme.[16] What did this mean in practice for operational weapons for the RAF? The Air Council was told in January 1958 that the plan was to develop the Grapple X device into a safe economical and reliable warhead for Blue Steel – the stand-off air delivered bomb. To do this considerable scientific and engineering design progress was necessary in order to reduce the weight and size and to provide adequate safety and operational reliability.[17] In March 1958 the warhead for Blue Steel was proceeding on the basis of Green Grass as then installed in Violet Club.[18] However, the status of the UK's megaton deterrent was far from satisfactory in spring 1958; more tests were needed to improve capabilities, reduce the use of fissile material and enhance safety.

The pressing design objective was to produce a megaton weapon of lightweight about 2000 lb that was safe and immune. To achieve this more tests were necessary with a megaton series of tests every three to four months as the quickest route to meet the requirements. Therefore, if successful over the next two trials (Grapple Y and Grapple Z) officials hoped that the UK should be in a position where it could design a warhead for Blue Steel and possibly Blue Streak. Grapple Y was aimed at providing knowledge to enable AWRE to carry out the successful development

[15] TNA CAB 130/130, GEN.604/6th Meeting, Item 2 Conditions for Suspension or Limitation of Nuclear Tests, 26 March 1958.

[16] TNA AIR 2/13753, Extract from minutes of meeting of Chiefs of Staff with Minister of Defence, 12 March 1958.

[17] TNA AIR 6/116, Development of New Weapons – up to 31 August 1958 Appendix Progress Report on Nuclear Weapons, Note by DCAS, 15 January 1958.

[18] TNA AIR 6/116, Note by DCAS Appendix Progress Report on Nuclear Weapons as at 31 March 1958, 9 April 1958.

of a Granite type megaton weapon.[19] However, it would not be immune to R-I and at least one more year of testing would be required to be certain of the design for Blue Streak. If megaton trials were stopped at that point in time – early March 1958, the UK would be unable to meet the Blue Steel or Blue Streak requirements and would also be unable to guarantee a safe and satisfactory free falling bomb. A further two to three years further testing would be needed.[20] Against this background, a moratorium would only be acceptable to the UK if it could be sure that the US would provide all the data it needed which would otherwise have to be acquired from testing. This would not be enough because there would still be a need to seek assurance by the occasional test that US information was sufficient and complete and that the UK had assimilated it properly. Such an occasional test would be contrary to the suspension and the UK would have to take all the US information on trust without experimental verification.[21] In this context we should not lose sight of Goodman's observation that atomic intelligence – the monitoring of US and Soviet tests – made an important contribution to the UK's thermonuclear programme.[22]

Approval for Grapple Z

Sandys minuted Macmillan on 14 March to seek formal approval for preparations for further megaton tests in the autumn. Sandys pointed out that the UK's megaton trials were planned on the assumption of continuous development. If all went well, these new tests would produce information that would lead to a lighter megaton warhead that was more economical in fissile material and as such suitable for use in Blue Steel and Blue Streak. Grapple Y in April would, if successful, provide further information that would be essential for the planned line of development, but would not in itself enable the UK to design a warhead satisfying the desired aforementioned criteria.[23] The object of any further megaton tests after September/

---

[19] ES 1/583, P.W.B. Brooking, Assistant Deputy Director AWRE to H.C. Hudspith, UKAEA Finance Branch, London, 5 November 1957.

[20] TNA DEFE 7/923, V.H.B. Macklen to Mr Ferguson, 3 March 1958.

[21] TNA CAB 130/130, GEN.604/3, Draft Submission to Ministers Nuclear Disarmament, Foreign Office, 31 March 1958.

[22] Goodman, *Spying on the Nuclear Bear*, p. 167.

[23] AWRE sent out 3.5 kg of plutonium for Grapple Y and 72 kg of U-235 plus 370 kg of U-238. Presumably, all required for the tests. TNA DEFE 7/923, W. Surtees, AWRE to V.H.B. Macklen, MOD, Operation Grapple Y – Air transport RA Material, 19 March 1958; AWRE calculated that the maximum weapon yield might be 5 megatons, with the expected yield at 2.5 megatons. TNA ES 1/583, W.H. Stephens, Deputy Director, RAE (E) to Air Vice Marshall John Grandy, Task Force Grapple, 21 December 1957. The actual yield was thought to be 2.9 megatons – see Operation Grapple Y – Interim Report, Trials Planning Branch AWRE July 1958 p. 4. Cook told Air Commodore Weir of Task Force Grapple on

October would be to provide knowledge of the means of reducing the vulnerability of bombs and warheads to nuclear countermeasures exploiting the R-I effect. Fissile material for the tests would come from UKAEA stocks.[24] Sandys urged Macmillan to agree to these tests in view of the possibility of a suspension of tests.[25] Macmillan agreed and instructed the date of Grapple Y to be advanced as much as possible.[26] This meant that the service personnel on the ground out in the Pacific had to work extra hard to ensure that the infrastructure was ready in time to support the test.[27] British thinking, as of end of March, was that if any further megaton tests were to be needed after those envisaged for the autumn of 1958, then it would be to provide knowledge of the means for reducing warhead vulnerability to nuclear countermeasures.[28] There was, however, still a fair degree of uncertainty as to future intentions in the minds of the Grapple Task force leadership. Air Vice Marshall John Grandy commented to the Deputy Chief of the Air Staff in early June 1958 that it was unlikely that they would get a firm long-term policy from the government on the test programme, financial authority for tests to date had been on a 'next test only' basis. Despite this they had still been able to obtain approval for limited expenditure on the base itself.[29]

To assist in this acceleration Sir William Cook had been loaned back to AWRE from UKAEA headquarters.[30] He informed the Atomic Trials Committee in mid May that if tests were not to be suspended, then it would be desirable to limit Grapple Z planning (scheduled for September) as originally planned. However, it is clear from the files that there were repeated requests to those involved with Grapple Z for acceleration of the programme, largely caused by the international politics of the test ban. Britain was uniquely vulnerable to such a ban.[31] Grapple

---

21 January that the maximum yield was possibly overestimated by at least one quarter. A 5 megaton yield was unlikely and if was, then the champagne would flow. After Grapple Y AWRE planned to reduce the yield right down and the likely yield for all trials after Y was 1 megaton. TNA ES 1/584, W.R. Cook to Air Commodore C.T. Weir, Force Grapple, 21 January 1958.

24 TNA DEFE 7/923, Duncan Sandys to Prime Minister, 14 March 1958.

25 TNA DEFE 7/923, Duncan Sandys to Prime Minister, 22 May 1958; TNA DEFE 7/2379, Minute to Minister, 16 May 1958.

26 TNA DEFE 7/923, Attached note to R.R. Powell to Sir Charles Cunningham, Home Office, 13 June 1958; TNA ES 1/578, Atomic Weapons Trials Executive Operation Grapple Meeting held at St Giles Court, 17 April 1958.

27 John Haggas, *Christmas Island The Wrong Place at the Wrong Time* (Leicestershire, 2002), pp. 52–3.

28 TNA DEFE 7/2379, Duncan Sandys to Prime Minister, 20 March 1958.

29 TNA AIR 20/9807, Christmas Island, John Grandy Commander, Task Force Grapple to DCAS, 4 June 1958.

30 TNA DEFE 7/905, Atomic Weapons Trials Executive Operation, Grapple Meeting Held at St Giles Court, 17 April 1958.

31 Arnold and Smith, *Britain, Australia and the Bomb*, p. 210.

Z was in the event accelerated to the maximum degree.[32] This impacted adversely the delivery rate of the interim megaton weapon – Violet Club; the RAF only had three of these by the end August 1958, with the next one scheduled for delivery in October. They had expected to have five by July; the falling off in delivery rate was a direct result of the bringing forward of Grapple Z.[33] At the planning stage in February 1958, for instance, the Trials Executive assumed that Grapple Z would involve balloon shots as well as air drops. The series would thus take up to six weeks and it was unlikely that the first firing could take place before the last week in September.[34] Test-ban pressures had a direct impact on the nature of what devices would be tested. Jackson told J.E. Serby, Director General of Guided Weapons at the Ministry of Supply, that AWRE had to test rounds that were not ready because there might never be another opportunity to test them. It was not just a question of whether AWRE would be able to develop a megaton warhead of the Granite type for Blue Streak and if so what its characteristics were likely to be, but whether they would be able to develop one before further testing ceased.[35]

Ministers finally agreed on 1 April 1958 that full preparations could go ahead for Grapple Z in the autumn of 1958.[36] The planned date of end September and mid October for Grapple Z was unacceptable given the political pressures seeking an end to atmospheric testing.[37] Instructions were given that the date of the operation was to be brought forward as much as practicable. Macklen went to Aldermaston to discuss the possibilities with Penney on 2 April. AWRE in response were able to state that they could be ready for one balloon firing on 15 August and an airdrop on 1 September 1958. The second airdrop appears to have been of a 'three bomb construction' of a type that the UK had not hitherto investigated.[38]

---

[32] TNA ES 1/1098, Item 2 Report on Grapple Z by AVM J. Grandy, Task Force Commander, Atomic Weapons Trial Executive Meeting held at St Giles Court, 16 October 1958.

[33] TNA AIR 6/116, Progress Report on the Development of New Weapons – up to 31 August 1958, Note by DCAS Appendix Progress Report on Nuclear Weapons, 15 January 1958.

[34] TNA ES 1/578, Atomic Trials Executive Operation Grapple Meeting held at St Giles Court, 19 February 1958.

[35] TNA AVIA 65/1193, E.S. Jackson, DGAW, Ministry of Supply to J.E. Serby, DG/GW, Ministry of Supply, 5 June 1958.

[36] TNA AVIA 65/1140, D.A.W. Trials to U.S. (S.A.W.), 2 April 1958.

[37] TNA ES 1/578, Atomic Weapons Trials Executive Operation Grapple Meeting held at St Giles Court, 17 April 1958.

[38] TNA AVIA 65/1249, William Penney UKAEA to E.S. Jackson, DGAW, Ministry of Supply, 14 July 1958. A three-stage device would include two thermonuclear stages instead of one.

Grapple Z Timetable Pressures

Brundrett stood by in case Macmillan gave any further instructions to take even more 'violent steps' to bring the date still further forward.[39] The Ministry of Supply and AWRE looked again at the dates and concluded that it was not possible to advance the date of the balloon test before 15 August or the airdrop before 1 September, but all would be done by the Ministry of Supply to ensure that these dates were met.[40] If there were to be any question of the suspension or abolition of tests then Macmillan felt that he would have to ask 'for a fresh confirmation of Eisenhower's promise to see us through'.[41] Lloyd was worried that it might leak out in Washington that officials had advised Dulles to accept a suspension of tests. If that became known in the UK it would make it extremely difficult for the UK to refuse to follow suit.[42] Looking to the longer term, Ministers felt that it was opportune, in view of the risk that the UK might not be able to conduct tests after October, to hold back on the programme of semi-permanent building at Christmas Island, but they did not want everything stopped.[43] Meanwhile, Grapple Y was facing further problems when there were last minute difficulties with the fission trigger known as a 'Tom' in the Aldermaston argot: AWRE staff at Christmas Island cryptically telegraphed home to report that Tom had suspected tummy trouble (separately described as five ulcers and nine gaps) and asked for his twin to be despatched as soon as possible. The twin was despatched from RAF Wittering on 20 April, expected arrival time on 22 April.[44]

---

39 TNA DEFE 7/2379, Sir Frederick Brundrett to L.J. Dunnett, Ministry of Supply, 15 April 1958; see also TNA AB 16/2371, Atomic Weapons Trials Executive Operation Grapple Meeting Held at St Giles Court, 17 April when the Committee was told that the planning date of end September and mid October for Grapple Z was unacceptable and must be advanced.

40 TNA AVIA 65/859, L.J. Dunnett, Deputy Secretary (A), Ministry of Supply to Sir Frederick Brundrett, MOD, 16 April 1958.

41 Harold Macmillan, *Riding the Storm 1956–1959* (London, 1971), p. 485.

42 TNA FO 371/135549, ZE 120/20, Foreign Office telegram no. 2312 to Washington, 28 April 1958.

43 TNA AVIA 65/859, R.R. Powell, MOD to F.A. Bishop, 2 April 1958.

44 TNA ES 1/587, AWRE Christmas Island to AWRE Aldermaston, 17 April 1958; AWRE Aldermaston to AWRE Christmas Island 17 April; AWRE Christmas Island to AWRE Aldermaston 19 April 1958. There were separate concerns about static build up on the perspex used in Tom and whether anti-static polishes or a wet rag should be used. AWRE Aldermaston to AWRE Christmas Island 22 April 1958. It also seems that some Red Beard Weapons and supercharges were left at Christmas Island after Grapple X.; some of the Red Beards were ear marked for the live trial. C.M. Bean, SXX to Mr Moyce SSXD, 21 March 1958. Christmas Island: Building B 4 alleged removal of HE during inter-trial period.

Penney asked on 8 April for the possibility of earlier firings still for Grapple Z to be investigated.[45] However, if there was a chance for an agreement on limitations effective from the end of 1958, then it would be prudent to extend the preparations for Grapple Z so additional tests could take place.[46] It was agreed that Grapple Z would now consist of four shots: two balloons and two air drops, starting on 15 August and ending on 1 October, but these dates were to change.[47] Ministers confirmed that it was essential that the first airdrop take place by the planning date of 1 September. There was discretion however on the timing of the balloon shot and this could be moved up to 22 August from 15 September provided the task force commander was content that wind and weather conditions would permit firing in time for an air drop on 1 September. Both rounds had to be fired and for political reasons it was most unlikely that, in the event of unsuitable weather, the Commander would be allowed to fire balloon rounds after the airdrop.[48] Cook was even toying with the idea of bringing forward parts of Grapple M, provisionally scheduled for the following year, by tacking it onto the end of Grapple Z in particular the possibility of including Grapple M on 8 September, or alternatively aiming for 1 December if practicable.[49] This would have entailed one airdrop and one balloon round, but test-ban politics and practicalities on the ground made this impossible.[50] Political considerations took precedence over the scientific imperatives when it came to determining the intervals between the shots at Grapple Z.[51] Delay was unacceptable politically, indeed the whole object of AWRE's efforts in late April was to accelerate the trial to 1 September for the main shot; and for the same reason there were planning pressures to complete Grapple M before the end of 1958 as this might become the date for test suspension.

Brundrett noted at the end of May 1958 that in view of the serious possibility of agreement being reached to prohibit megaton testing after the end of the year, the NTPC had considered an AWRE proposal to extend the Grapple Z trial. This would have the objective of doing the maximum to ensure that the UK gained knowledge that would enable it to render the megaton weapon immune to countermeasures. Given the outstanding success of Grapple Y, AWRE had little doubt that the test of a

---

45 TNA ES 1/609, AWRE Aldermaston to AWRE Christmas Island, Adams to Pilgrim, 8 April 1958.

46 TNA CAB 134/2274, Cabinet Nuclear Tests Policy Committee N.T. (58) 2nd meeting, 14 May 1958.

47 The actual dates were 2 and 11 September for the megaton shots and the balloon shots on 16 August and 22 September.

48 TNA AVIA 65/1140, Ministry of Supply, London to Headquarters Grapple, Christmas Island, 25 April 1958.

49 TNA ES 1/609, AWRE Aldermaston to TFC HQ Grapple Christmas Island for Task Force Commander from Sir William Cook, 21 April 1958.

50 TNA AVIA 65/1140, D.A.W. Trials to U.S. (S.A.W.), 28 April 1958.

51 TNA ES 1/609, Adams to Pilgrim, Aldermaston to Christmas Island, 11 April 1958.

one ton, one megaton warhead planned for 1 September 1958 would be successful, but it would not provide the data necessary to make it immune to countermeasures. This was the object of the kiloton test on 15 August for an immune trigger for a two-stage weapon; this one was based purely on theoretical considerations and it was not possible to be anything like so sanguine about its successful outcome. If it were successful, it could be applied as the trigger for the device tested on 1 September. Without this trial the UK would be entirely dependent upon what the US chose to impart.[52] However, in the event of failure, an alternative approach would be needed, but one that could not be applied to the megaton device tested on 1 September. A second megaton shot would therefore be required. Were there no question of an early cessation of tests, AWRE would have preferred to spend longer on the construction of these devices and carry out trials in 1959. However, in view of the very real possibility of an end to testing in 1958 and in view of the very great importance that the UK knew how to make a megaton warhead immune, it was most desirable to test these devices in 1958.[53] For this reason the NTPC agreed that Grapple Z should be extended, if necessary to provide a second line of approach in case the 15 August test failed. However, matters were made worse when the US stated at the end of April 1958 that it would halt its programme of tests for six months in order to see whether discussions could proceed with the Russians on some form of control mechanism. As a direct result of this development, Ministers tasked the UKAEA to see whether the Grapple Z tests scheduled for the autumn could be brought forward because of the urgency. Penney then investigated every possibility of advancing the date of the autumn tests. By re-arranging the work, dropping some of it and with additional help from William Cook, he felt that it was possible to bring the date forward from late October to the first week in September. (In the event Grapple Z was subject to 'extreme compression', and the reasons for which were not always apparent to the scientific staff on the ground.[54]) The plan in mid April called for two tests: a kiloton test and a lightweight megaton test, but in the end only one shot was fired.[55] Macmillan was informed in late July/early August that preparations were in hand to fire four shots at Christmas Island. These consisted of a first attempt at an immune trigger of kiloton yield, a one ton–one

---

[52] TNA DEFE 7/2379, Draft minute to the Prime Minister, 20 May 1958.

[53] TNA DEFE 7/923, Sir Frederick Brundrett to Minister, 20 May 1958 and draft minute to the Prime Minister, Megaton Test, September 1958.

[54] TNA ES 1/599, R. Pilgrim, Scientific Director to All Members of the Scientific Staff, 4 September 1958. When the trials were over there was some concern amongst the scientific staff that the pattern set at Grapple Z (namely the great rush occasioned by test-ban pressures) would be used as a precedent for the future. Pilgrim and Noah Pearce were agreed that this would have been a quite unacceptable principle upon which to plan further trials. Pilgrim could not visualise a repetition of the particular circumstances that had made compression of the programme necessary. R. Pilgrim to All Members of the Scientific Staff, 24 September 1958.

[55] TNA DEFE 13/150, E.N. Plowden to Prime Minister, 16 April 1958.

megaton shot based upon the April results from Grapple Y; an alternative design of an immune trigger device of kiloton yield and finally a megaton device of a more advanced design.[56] If the first two shots were completely successful, last two would not be needed as the UK would have learnt enough to engineer a warhead for Blue Streak and Blue Steel and to give it some immunity against countermeasures.

Further Trials: Grapples M, O and Lighthouse 1959

There were good scientific reasons for further trials after Grapple Z.[57] Sir Richard Powell, Committee Chairman, reminded NTPC members that future UK policy on tests would be reviewed in September 1958. Three facts would have to be taken into account: the results of the Geneva Experts' Conference on test-ban verification, the outcome of the technical discussions with the US on implementing the recently signed Mutual Defence Agreement and the effect of world opinion following the United Nations Scientific Committee on the Effects of Atomic Radiation final report. Cook reported that depending on the outcome of Grapple Z and the results of the exchanges with the US, it was still not possible to say whether in the event of the US giving the UK useful information, proved and safe service weapons could be produced without a further test. In response to some Foreign Office questions, the Official Committee on Disarmament's Working Group commented on the defence for further UK tests.

The Ministry of Supply asked AWRE in mid April 1958 for information on proposals that should be made for future Grapple trials in 1959.[58] In response, Cook told the Director General of Atomic Weapons (DGAW) that the proposal had been bound up with the possibility of extending Grapple Z, but now that the programme was clear, it was possible to give an indication of the future possibilities.[59] If Grapple Z showed that tritium boosting was successful, then AWRE would want to apply this knowledge to kiloton weapons either to increase yield or reduce the fissile material used. However, such tests ought to await the development of a system likely to meet with service acceptance. Cook noted that on the assumption that Grapple Z was fired as proposed, and assuming that megaton firings were still permissible politically, it would be wise to plan for Grapple M to be fired starting on 1 March 1959 on the scale of one to two air drop megaton tests together with two

---

[56] TNA DEFE 7/2380, Draft Minute to the Prime Minister, Megaton Tests at Christmas Island, 28 July 1958.

[57] TNA ES 1/578, Atomic Weapons Trials Executive Operation Grapple Meeting held at St Giles Court on 17 April 1958.

[58] TNA AVIA 65/1440, Capt F.B. Lloyd RN (retd) to Sir William Penney, 18 April 1958.

[59] TNA AVIA 65/1441, William Cook, UKAEA to DGAW, 19 May 1958.

balloon tests in the kiloton range.[60] The actual rounds could not be specified at that point in time as they depended on the results of Grapple Z. There were nevertheless a number of options: a further test of a two-stage one ton–one megaton warhead if the Grapple Z test failed; further test of a three-stage one ton–one megaton if the Grapple Z test failed; test of a two-stage one ton–one megaton warhead with reduced U-235 content if Grapple Z test succeeded; test of a one megaton warhead of still lighter weight; test of a one megaton warhead incorporating an immune trigger, possible inclusion of any safety measures required; test of an immune boosted trigger for two-stage or three-stage warhead to give the required yield, as estimated from the trigger firing at Grapple Z; further attempts at an immune trigger if Grapple Z failed; and a kiloton test of a two-stage trigger if Grapple Z failed.

Penney had told Aldermaston's Weapons Development Policy Committee on 25 June that divisional estimates for the next financial year should be based on the assumption that tests would not be suspended and the US-UK bilateral agreement would not reduce the necessity for tests and may in fact increase it. For the financial year 1959–60, therefore AWRE should estimate for three to four rounds on Lighthouse and two balloon firings and two air drops at a spring 1960 Grapple.[61] Lighthouse tentatively included plans for two boosted kiloton warheads, but these depended on the results of Grapple Z.[62]

By July AWRE had submitted proposals for agreement for minimum preparations needed to conduct a megaton test (Grapple M) in March 1959.[63] Penney circulated a note on trial preparation policy in mid September given the then uncertainty about the future of major trials. Preparations were to continue for Grapple M (February/March) and for Lighthouse (August/September), but as far as finance was concerned, no orders were permitted between 15 September and 30 October if they could be delayed without prejudice to the proposed trial dates. Orders that had to be placed by contract should, if possible, carry a break clause

60 Sufficient plutonium was available from stocks for trials such as Grapple M, but it would be necessary to supplement this with about £2.75 million worth of U-235, which could only be made available at the expense of weapons production. AB 16/2371, Brief for the Ministerial meeting on Thursday 29 May 1958 to discuss Atomic Weapons trials, V.H. Cole, Finance Branch, UKAEA to Mr Rawlinson – points for UKAEA Chairman, 28 May 1958.

61 TNA ES 1/1138, Extract from Minutes of 63rd WDPC Meeting on 25 June 1958. A significant target response programme was to be associated with Lighthouse along with an indoctrination exercise. TNA AVIA 65/1217, draft paper from DGAW for DRPC (AES) on future atomic trials, Maralinga 1959, 12 April 1957, DGAW, 30 January 1958. Lighthouse was the code name for a further series of fission device trials at Maralinga scheduled for 1959.

62 TNA AVIA 65/1242, 1958 E.S. Jackson, DGAW, Ministry of Supply to Deputy Secretary), 28 May 1958. Other activities would have included an experiment on the burning of plutonium without explosion, which was very important for AWRE's work on safety problems. U.S./S.A.W to Dep Sec 'A' 29 May 1958.

63 TNA CAB 134/2274, Cabinet Nuclear Tests Policy Committee, N.T. (58) 4th meeting, 16 July 1958.

in order to limit AWRE's commitments, should test suspension become effective. Even in early September, it was not entirely clear to AWRE what the future testing policy would be and this uncertainty had implications for the degree of packing for equipment to be left on the island and the amount of stores to be returned to the UK. This would depend on the date of the next trial; the Scientific Director, Roy Pilgrim, decided that to meet this uncertainty, the rundown should be based on the assumption that there would be a trial in the spring of 1959.[64] Plans were in hand for the redeployment of staff if trials were suspended.[65] By the end of October Penney was now telling Jackson that AWRE definitely did not want to carry out a Grapple M trial earlier than June 1959 and it might even be July. In fact Aldermaston could not see a need, or indeed a possibility of holding a useful megaton trial in 1959. This appears to have been a result of the flow of US design information.[66] However, since they were going to be producing a large number of Red Beards over the next few years, it would be advisable to have at least two operational drops of the stockpiled weapon.[67] Grapple M was thus moved to a target date of 15 June with four shots; two air dropped and two balloon, all in the kiloton range. These tests were dependent on assessment tests at Maralinga and were required whether major tests were suspended or not.[68]

Diplomatic Pressures, Test Ban Negotiations and Testing Requirements

The UK position in early 1958 was that there should still be a link between nuclear tests and a cut-off. UK acceptance of this in the summer of 1957 had been on the assumption that in the event of such a prohibition the UK would be able to rely on the US to meet its fissile material requirements. Officials knew in March 1958 that the US would not be able to provide the UK with the material that it most required.[69] In any case, a cut-off linked to a test suspension was seen as the third best solution for the UK.[70] The Cabinet Defence Committee had recently decided

[64] TNA ES 1/599, N. Pearce, Scientific Superintendent to all group leaders, rundown at Conclusion of Grapple Z, 4 September 1958.

[65] TNA ES 1/987, Director to All Banded Staff, 15 September 1958.

[66] TNA AVIA 65/1217, E.S. Jackson, Future trials Programme, Note by DGAW.

[67] TNA AVIA 65/1217, W.G. Penney to E.S. Jackson, DGAW, Ministry of Supply, 31 October 1958.

[68] TNA AVIA 65/1217, E.S. Jackson, Ministry of Supply to Sir Richard Powell, MOD, 5 November 1958.

[69] TNA CAB 130/130, GEN.604/6th Meeting, Item 5 Amendments to United States Atomic Energy Legislation, 26 March 1958. It is not clear what was meant here by 'material'; by 1959 however with the 1959 amendment to the 1958 Mutual Defence agreement, the US and UK agreed to exchange UK plutonium for US HEU and tritium.

[70] TNA FO 371/135549, ZE 120/21/G, H.C. Hainworth to Sir P. Dean, 30 April 1958.

that, as a matter of tactics, the UK should not propose any alteration in its formal position, but should leave it to the US to take the initiative. If it did so, the UK could then make its agreement conditional upon the US providing the UK with certain technical information.[71] Despite the low probability of Soviet acceptance of such a link, in view of the serious military implications of any ban in production for the UK, officials thought that it would be unwise to volunteer that London still preferred to link test suspension with a cut-off.

The Cabinet invited the Foreign Secretary to arrange for consultations to be opened with allied governments to consider the nature of the public response to the Soviet initiative in March 1958 on the suspension of tests.[72] The Foreign Office intended to suggest that the UK and US might make a statement of their intention to suspend all nuclear tests for a period of years from 1 January 1959, provided that the USSR did the same. Other conditions required that the suspension was policed effectively by international inspection, that there was agreement between the three governments concerned to discuss, during the currency of the suspension, the establishment of a cut-off and that agreed exceptions should be made for the continuance of a limited number of tests of small yield atomic weapons for defensive purposes. From the point of view of the British weapon development programme, the preference would be for a limit on small explosions set at 50 kilotons. Continuance of tests up to this level would enable development of small weapons as well as important development work for megaton weapons for some time without needing a full yield trial.[73] The terms of an agreement on these lines would be negotiated with the Soviet government at a meeting of Heads of Government. There would also be a separate agreement with the US to the effect that British acceptance of this arrangement was conditional that, when authorised by legislation, the US would provide the UK with the technical information that could no longer be obtained by testing. If any further concessions were needed to meet the Soviet initiative, the least damaging idea would be to offer to suspend nuclear tests from a convenient date. This could not be any earlier that 1 January 1959, so that the UK could conduct the further test series planned for early autumn – Grapple Z.

The most that London could concede would be an offer to discuss the establishment of a cut-off. However, in view of the serious effects a cut-off would have on UK ambitions, Ministers and officials thought that it would be inadvisable to take too prominent a position on this. On the other hand, the linkage was intended to provide a means of determining the duration of the tests suspension. The right to test small yield weapons was necessary for defensive and tactical purposes; and this right would need to be maintained in the face of Soviet or even domestic

71 TNA CAB 131/16–20, Cabinet Defence Committee D (58) 5th Meeting, Item 2 Disarmament, 20 March 1958.

72 TNA CAB 130/130, GEN.604/8th Meeting, Item 1 Suspension of Nuclear Tests, 2 April 1958.

73 TNA DEFE 7/923, saving telegram no. 2043 from Foreign Office to Washington, 2 May 1958.

opposition.[74] However, there were wider defence programme interests at stake here too. Without any agreement to exempt small yield tests, it would be difficult to keep together the AWRE research teams, a difficulty which was thought to apply in the US as well, albeit less acutely.[75] This is the first occasion that we see any reference in the UK to concerns that a test ban might undermine the scientific basis that underpins a nuclear weapons programme. Furthermore, while the suspension of tests from 1 January 1959 would not prejudice the independent deterrent, the defence programme would be retarded during a suspension of kiloton tests unless the US provided all the relevant information that it had. Unfortunately UK officials thought that they could not be sure that the US would be willing to provide this. The break-up of the AWRE research team might mean that this short-term reliance on the US might well become a longterm one. Setting aside the technical imperatives, there was also a pressing concern that any reluctance to accept a suspension would play into the hands of Soviet propaganda. Western powers simply could not afford to be tardy in reacting to suspension proposals. UK foreign and nuclear defence policy in the first nine months of 1958 was to avoid being impaled on the horns of this dilemma. By the autumn however the Chiefs of Staff at least saw three clear advantages to a test suspension; the preservation of the general lead over Russia; the prevention of the development of a reliable nuclear defence system, and the effective prevention of any new members joining the nuclear club. However, there could be no illusion that once the UK had agreed to a suspension, it would find it very difficult to resume if the wish to do so existed.[76]

Amendments to the US Atomic Energy Act: Impact on UK Testing Plans

Although the talks in Congress on amendment of the US Atomic Energy Act to allow transfer of knowledge to the UK had been going well, it was still impossible to forecast in mid March 1958 how far the implementation of such an agreement would reduce UK needs for testing; a great deal on what might be transmitted to the UK would be left to a Presidential Determination. In fact, officials doubted that the information that might be provided would be sufficient for the UK to halt testing altogether.[77] However, by early April, things took a turn for the worse and Foster Dulles himself decided to go before the Joint Congressional Committee on Atomic Energy in order to press the case for amendment of the Act. The Committee

---

74 TNA CAB 130/130, GEN.604/9th Meeting, Item 5 Suspension of Nuclear Tests, 24 April 1958.

75 TNA CAB 130/130, GEN.604/13th Meeting, Suspension of Nuclear Tests, 11 July 1958.

76 TNA AIR 2/13753, Chiefs of Staff Committee, Extract from minutes of COS (58) 87th Meeting, 14 October 1958.

77 TNA PREM 11/2567, D.S. Laskey, Foreign Office to P.F. de Zulueta, No. 10 Downing Street, 19 March 1958.

was bothered by the Fourth Country problem. The US might come under pressure to provide information to other states and might find it difficult to refuse.[78] The UK Ambassador in Washington recommended that Macmillan write to Dulles to provide further encouragement and this he did and laid it on thick about the level of anxiety that this issue generated. He wrote that, 'I only want to say once more to you how much depends upon it for the future of the thing which I care about more than anything else – Anglo-American cooperation to guide and steer the efforts of the free world'.[79]

The US advised the UK that it would not share some information, but it was not clear what this would include but it was apparent that the US would not share information on the vulnerabilities of US weapons to Soviet counter-measures. London felt that if this extended to antimissile missiles, then the UK might be in trouble. Dulles was told that it would be unreasonable, in view of the UK's geographic position and density of population, for the US to expect the UK to contract out of all attempts to develop a defence against nuclear attack. Dulles apparently took the point.[80] For Penney at least in advising Macmillan in mid August 1958, this meant that the UK would need to ensure that the bilateral talks went as far as possible on the US side in the expectation that this would remove the UK's need for tests in 1959. However, when the USSR announced a halt to its test programme in March 1958, the political pressure on the UK to follow suit increased. Macmillan wrote to Eisenhower on 27 March to tell him that the UK's position on testing would be determined by the help that the US could provide following amendment of the McMahon Act.[81] The Cabinet considered the issue on 1 April 1958 and immediately recognised that this turn of events would put the US and UK at some disadvantage in the eyes of world opinion.[82] Neither was prepared at that stage to forgo the scientific knowledge that some further tests would provide. It would be especially difficult for the UK to give up its next test series given its objectives before it could discover exactly what information they might expect from the US after amendment of the McMahon Act. There was still a risk that this would not come into force until February 1959, but there was also a good chance that things could be expedited before then with Macmillan's visit to Washington in June.

As we saw above testing firing plans were apt to be a bit of a moveable feast: by the end of May the new testing schedule called for: the first attempt at a kiloton trigger, a megaton test derived from the 28 April shot, a second attempt at a

[78] TNA PREM 11/2567, Sir H. Caccia to Prime Minister, 10 April 1958.

[79] TNA PREM 11/2567, Macmillan to Foster Dulles, Prime Minister's Personal Telegram T. 121/58, Foreign Office telegram no. 1980 to Washington, 14 April 1958.

[80] TNA PREM 11/2567, Washington telegram no. 822 to Foreign Office, April 9 1958.

[81] TNA PREM 11/2554, Prime Minister's Personal Telegram Serial No. T. 93/58, Foreign Office telegram no. 1667 to Washington, 27 March 1958.

[82] TNA CAB 128/32, Cabinet C.C. 28 (58), 1 April 1958.

kiloton trigger and the test of a modified megaton weapon. The sequence was important because if the first two were successful, there would no need for the remaining two.[83] It was essential to ensure that the right conclusions were drawn, so therefore the first megaton test was important and had to be conducted before the second proposed megaton shot. The Prime Minister, Foreign Secretary and Defence Minister agreed that Lloyd should make clear to Dulles when he saw him in the margins of a NATO Foreign Ministers' meeting in Copenhagen on 5 to 7 May that there should be no suspension of tests before 31 October.[84] Macmillan was confident that everything could be completed by 31 October and probably before – holding out against public and political pressure would still be an issue. He was acutely aware that if the UK gave up at that point, it would not have a reliable weapon.[85] Furthermore, any suspension at least during the proposed initial six months, should apply to megaton tests only. The UK did not want to be committed to the principle, of suspending small tests as well.[86] Washington had reported that Dulles was working on the assumption that HMG would be prepared to accept the suspension of tests linked to discussion of a cut-off in the production of fissile material provided the US made up the gap in knowledge which suspension would entail.[87] Suspension of larger tests was thought to be sufficient to address public anxieties over safety and permit continued testing of defensive weapons.

London was very concerned that any precipitous announcement ahead of the June meeting between Macmillan and Eisenhower would prejudice the UK autumn test programme by increasing the prospect of public opposition.[88] Lloyd was also to make clear that there should be no final decision reached on proposals for a suspension of tests until after this meeting with Eisenhower. Nor did the UK want to make any public announcement about possible suspensions before or during the technical experts meeting in Geneva on test-ban verification, which was to begin in July. There were fears that this might give the Russians some advantage before they had signed up to the principle of international control.[89] Moreover, the UK only favoured a suspension for six months limited to megaton weapons only. There were also concerns that suspension of small tests for two years would

83 TNA DEFE 13/150, Brundrett to Minister, 28 May 1958.

84 TNA PREM 11/2565, Note of Meeting on 1 May 1958.

85 Macmillan, *Riding the Storm*, pp. 489–90.

86 TNA CAB 130/147, GEN 649/2/1, 4 June 1958. Prime Minister's Visit to Washington and Ottawa June 1958. Disarmament, Nuclear Questions and United States Atomic Energy (McMahon) Act, Brief by the Foreign Office.

87 TNA FO 371/135549, ZE 120/21, Washington telegram no. 1018 to Foreign Office, 29 April 1958.

88 TNA PREM 11/2566, Foreign Office telegram no. 3160 to Washington, 28 May 1958.

89 TNA PREM 11/2566, Foreign Office telegram no. 4583 to Washington, 16 July 1958; Macmillan, *Riding the Storm*, p. 563.

prevent AWRE from retaining its research team on nuclear weapons.[90] Securing amendment of the McMahon Act was crucial and if this worked out as hoped, then the UK would be unlikely to need any more tests – except perhaps the Grapple Z series, which would improve the UK's bargaining position.[91] In October 1958 the MOD still argued that small tests should be exempted from the scope of the ban, but they were however not prepared to press the point.[92] There were certain technical aspects of nuclear weapons, such as making weapons immune from countermeasures, which the UK wanted to discuss with the US before reaching a final decision on these proposals.[93] UK agreement would be contingent upon the US atomic energy legislation being amended to allow for the necessary exchange of information and fissile material for military purposes. As Macmillan was to write later, 'we would also have to be fully satisfied as to the continuing information which we would receive from our American friends'.[94] Separately, Macmillan told the New Zealand Prime Minister, Nash, that if the UK were to forgo the tests planned for later in 1958, it would lose most of the benefits of the Grapple X and Grapple Y tests.[95]

Lloyd saw Dulles again on 19 August.[96] Foster told him that following a meeting the day before in Washington, the President had overruled USAEC and Defence Department objections to a suspension. The US now planned to make an announcement on suspension immediately on conclusion of the Geneva experts' meeting. Initial US thinking had contemplated making an announcement to that effect immediately before the start of the UNGA on 16 September; in other words before Grapple Z was completed. As Macmillan noted for the UK this, 'was something up with which we cannot put'.[97] The proposal thus did not go down too well either with the British and was greeted with 'near consternation'.

---

90 TNA CAB 130/130, GEN 604/13th Meeting Suspension of Nuclear Tests, 11 July 1958.

91 Macmillan, *Riding the Storm*, pp. 489–90.

92 TNA FO 371/132686, IAD 122/25, Hainworth to O'Neill, 10 October 1958. In March the following year the MOD were still hankering after a system that would only limit those tests that the then present national detection system could detect. In practice, this would mean nuclear tests above 25 kilotons. There was a military value to the West in being able to continue with the smaller yield explosions. TNA FO 371/1400516, IAD 27/2, Nuclear Tests Conference Brief for the PM's visit to Washington, March 1959.

93 TNA PREM 11/2566, Foreign Office telegram no. 3161 to Washington, 28 May 1958.

94 Macmillan, *Riding the Storm*, p. 560.

95 TNA PREM 11/2567, CRO telegram no. 44 to Wellington, Personal Message from Prime Minister to Mr Nash, 1 May 1958.

96 TNA PREM 11/2566, New York telegram no. 901 to Foreign Office, 19 August 1958.

97 TNA FO 371/135550, ZE 120/74G, G.G. Brown to H.T. Morgan, 5 July 1958; Washington telegram no. 1819 to Foreign Office, 3 July 1958.

Macmillan had made clear to Eisenhower that it was essential, if the UK was obliged to suspend tests, to secure in return the installation of some control system on Soviet territory. Announcing a suspension beforehand would be to surrender the whole prospect of advance that the Geneva Conference had opened up. It would be much better to wait until the end of the conference before making any public announcement on suspension.[98] London felt that the UK and US had kept in step on testing for two years; it would be a shame to part company now, which was what would have happened had the suspension announcement been made as proposed. Lloyd told Foster that he did not think that it was necessary to make any such statement so soon; time was needed to study the Geneva Scientific Experts' report. He emphasised that the UK could not make up its own mind on suspension until the next series of UK tests had been completed, and in addition, the UK did not yet know the amount of information to be exchanged under the recently concluded bilateral agreement. A UK-US experts' meeting had been scheduled for 27 August; it would be better to await its results. Earlier Penney had advised Macmillan and Lloyd that whilst the UKAEA hoped that the bilateral talks would provide some clarification on the UK's need for further tests, he feared that the curtain would not be raised enough for AWRE to say that there would no longer be any point in doing the particular tests that were planned for 1959. It was thus important to stress to the US the importance of the bilateral talks going as far as possible in the expectation that this would remove the UK need for tests in 1959.[99] In Penney's view, UK support for suspension was given because the UK was relying on getting the information from America that it had intended to acquire for itself.[100]

This proved to be the key meeting, but Lloyd was not to know that. Dulles reassured Lloyd that the Americans could be trusted to deliver on the information. Lloyd subsequently told Macmillan that he thought Foster had been impressed by the argument that the US seemed to be moving too quickly. Nevertheless, it was quite clear that Eisenhower felt that he would be under enormous pressure to respond to the Geneva Experts' report and he wanted to get ahead of public opinion, not follow it. Lloyd suggested that Macmillan might want to send a personal message to Eisenhower urging delay. He advised strongly against giving any impression that the UK doubted the US promises on the extent and nature of nuclear information exchanges. Lloyd felt that Dulles was proceeding on the assumption that the UK would obtain all the details that it needed. This was borne out by a rapid response from Eisenhower himself; he wrote to Macmillan on 19 August in light of the Dulles-Lloyd meeting.[101] In view of some of the doubts

---

[98] TNA FO 371/135550, ZE 120/74/G, Foreign Office telegram no. 4583 to Washington, 16 July 1958.

[99] TNA FO 371/135551, ZE 120/84, Geneva telegram no. 463 (from Penney) to Foreign Office, 18 August 1958.

[100] TNA FO 371/135551, ZE 120/84, Geneva telegram no. 474 (from Penney) to Foreign Office, 20 August 1958.

[101] TNA PREM 11/2566, Eisenhower to Macmillan, 19 August 1958.

expressed by Lloyd, he wanted to reassure Macmillan that it was the US purpose to be 'as completely generous with your government in the matter of passing information, as the law will permit'.

Macmillan sent two messages to the President in response: one on the information-sharing issue, the second on the political aspects of the proposed US statement.[102] In the first part of his message Macmillan thanked the President for his assurance that it was his intention to be completely generous with the UK in the matter of passing information. Macmillan gladly accepted this assurance and emphasised that he had every confidence in Eisenhower's intention to do his utmost for the UK. He emphasised too that the UK had made immense efforts to develop nuclear weapons so that the British could play an effective part in the defence of the free world. He added that the UK was planning some more or less immediate tests looking towards the solution of two design problems, one of which Macmillan described as 'invulnerability' (the R-I problem) and the second as a 'reduction in weight',[103] the answers to which were essential.[104] If the US could provide the UK with the information that it already had and would have in the future, then he could agree to suspension of tests with a clear conscience. Suspension for him in the UK case really meant abandonment. He really wanted to be assured that 'the law will permit' transfer of this necessary data if the UK was prevented from obtaining it for itself through a further series of tests. If Eisenhower could give this assurance, then Macmillan could rest assured that the essential defence interests of the UK would be sufficiently safeguarded.

Eisenhower was quick to respond. He wrote on 21 August that with respect to the question on invulnerability and reduction of weight, he was permitted by the revised Atomic Energy Act to convey to the UK any information needed so long as that information would not endanger US security. The law required that the President make a determination to this effect. Since the US-UK joint purpose was to make certain that the weapons that they manufactured were for use by the free world in the common defence, there would be no difficulty for him to make the necessary certification for this type of exchange.[105] Macmillan was very glad to have this assurance, including on the two points that he had specifically referred. He was now completely satisfied and relieved, in view of the long and chequered

---

[102] TNA PREM 11/2566, Foreign Office telegram no. 5946 Prime Minister's Personal telegram T434/58 to Washington, 20 August 1958.

[103] Jan Melissen, *The Struggle for Nuclear Partnership. Britain, the United States and the Making of an Ambiguous Alliance 1952–1959* (Groningen, 1993), p. 50.

[104] TNA PREM 11/2566, Foreign Office to Washington telegram no. 5954, Prime Minister's Personal Telegram Serial No T434/58, 20 August 1958.

[105] TNA PREM 11/2566, Eisenhower to Macmillan Prime Minister's Personal Telegram Serial No. T.438/5820 August 20, 1958.

history of this issue, that there was now a clear and explicit understanding between the UK and US.[106]

In the second letter Macmillan made clear that he did not doubt the President's judgement that a response was called for, only that it was not necessary to move so quickly to a suspension of testing. It was true that the report would indicate the theoretical possibility of controlling a suspension of tests and that was something that ought to be welcomed. However, it was not yet possible to say that the practical difficulties could be overcome by putting the theoretical possibilities for verification into practice. Macmillan was concerned that the Russians would pocket the suspension announcement as a concession and not give any indication that they would accept international control, which was the key prize in his view. It would then be very difficult for the UK and US to resume testing if they needed to. Instead Macmillan suggested that the UK and US should limit a proposal to those tests that could be detected by national means only, namely those tests over 25 kilotons. He had earlier said that 'small tests' had a part to play in meeting UK needs.[107] The Americans were more advanced than the UK and we must presume that Macmillan was alluding to small high performance 'primaries' as triggers for thermonuclear weapons as well as 'defensive warheads'. This would be consistent with past Western proposals that had emphasised control and would also address public anxieties on fall out and its impact on health. As such, Macmillan urged the President to defer any announcement to the UNGA in the autumn, not the end of the Geneva Experts' Conference in August. He added that the UK tests, which had just been announced, would not be completed until the middle of October. Macmillan himself was also worried that it would be a mistake at that stage to concede suspension without first securing Russian acceptance of an international control system.[108]

Dulles and Lloyd had another discussion in New York on 20 August on the terms of any announcement and it is clear that this was a difficult meeting.[109] Dulles seemed to accept the argument that an announcement was not immediately called for upon conclusion of the Geneva Experts' meeting. Lloyd reiterated the Prime Minister's position that it would be absolute folly to move first without a prior Soviet commitment to international control. It would also make the previous UK-US insistence on such a commitment appear as a cynical ploy. Dulles appeared to accept these arguments and after several calls to Washington came back with a revised draft announcement that seemed to meet UK concerns. It inserted the key requirement that between the final report and the start of a suspension, there had

---

[106] TNA PREM 11/2566, Foreign Office to Washington telegram no. 5985 Prime Minister's Personal Telegram Serial No T445/58, 21 August 21 1958.

[107] TNA PREM 11/2566, Foreign Office telegram no. 1792 Prime Minister's Personal Telegram Serial No. T415/58 to New York, 17 August 1958.

[108] TNA PREM 11/2566, Foreign Office to Washington telegram no. 5986 Prime Minister's Personal Telegram Serial No T446/58, 21 August 1958.

[109] TNA PREM 11/2566, New York telegram no. 925 to Foreign Office, 20 August 1958.

to be agreement amongst all the states that had conducted tests that international control was an agreed point of departure. Significantly, it also put back the start date for the suspension until 31 October in order to accommodate the UK's test programme requirements.

Lloyd detected some irritation in Dulles's demeanour as he was at the receiving end of a lecture on just how much stick the US had endured on delaying announcement on test suspensions until the 1958 Mutual Defence Agreement protected the UK position.[110] Dulles said that the UK had no idea of just how difficult it had been for the Administration to push the bilateral agreement through Congress. He really hoped that Lloyd would not now make things more difficult and wrote to Macmillan to emphasise that the UK really should not ask the US to take further political risks by holding back. Washington did not want to appear in the eyes of the world as a militaristic power and had to reduce the propaganda hits that it had been taking. Dulles went on to explain in a letter to Macmillan that the US had already held back on the suspension matter for several months largely in deference to the US desire not to confront the UK with a possible test suspension before the US could give Britain the benefit of its knowledge. Moreover, the Administration had through great exertion obtained an amendment of the Atomic Energy Act in the UK's favour, and the President had undertaken to act generously under it. For this reason he and Eisenhower wondered in light of this whether the UK should ask the US to take further serious risks in relation to world opinion.[111]

Eisenhower was also clearly bothered by the PR problem and felt that the US had to make a movement on this issue and he told Macmillan that publication of the Geneva Experts' report marked an opportunity for the West to take the initiative. The US could not wait any longer; delay would give the Soviet propaganda machine a field day.[112] The Americans felt under pressure to issue a statement promising an immediate suspension of its nuclear tests. Eisenhower and Dulles thus believed that it would be imperative for the US to respond ideally before the Geneva Experts presented their final report; they did not want the US to be seen dragging its feet on this issue.

Macmillan pleaded that he was motivated not just by the need to protect the interests of the UK weapons programme; he was also very mindful of the international dimension and the need to nail the Russians down to acceptance of international control of any test-ban agreement.[113] Macmillan was however grateful for the changes that had been made to the draft announcement to accommodate UK concerns and as such he told Eisenhower that he was now ready to join the US and

---

110 TNA PREM 11/2566, Secretary of State to Harold Macmillan, 20 August 1958.

111 TNA FO 371/135551, ZE 120/89, New York telegram no. 926 to Foreign Office, 21 August 1958.

112 TNA PREM 11/2566, Eisenhower to Macmillan, Prime Minister's Personal Telegram, Serial T438/58, 20 August 1958.

113 TNA PREM 11/2566, Foreign Office telegram no. 1906 to Washington Prime Minister's Personal Telegram Serial No T444/58, 21 August 1958.

would make a similar and separate announcement on suspension to support the US position.[114] The US originally wanted the negotiations to begin on 1 October, but because of Grapple Z, the UK suggested 31 October instead and the US agreed. HMG announced, having agreed with the US the terms of separate statements, on 22 August that once the Grapple Z series was completed, the UK would from the date that negotiations began on an 'suspension of nuclear tests under effective international control' refrain from testing for a period of one year. However, this was done on the understanding that the object of the negotiations was agreed by all participants and as long as the USSR did not resume testing in the period. The UK was also willing to extend this moratorium for period of one year. That said, this was contingent on the USSR following suit and that satisfactory progress made towards the installation of an effective system of international control over the suspension of nuclear tests and towards the adoption and execution of measures of real disarmament.[115]

Khrushchev replied on 30 August. This was considered at a ministerial and officials meeting on 1 September. Macmillan considered that, despite its unsatisfactory features, the UK's reply should state that it would be ready to negotiate on 31 October and suspend its tests for a year from that date.[116] This point was reinforced again in a note to the Soviet government delivered in Moscow on 20 October 1958.[117] Sir David Ormsby-Gore, Minister of State in the Foreign Office, revealingly informed the Secretary of State in early November that he had been, 'assured by our own people and by the Americans that a halt would leave us at a military advantage'.[118] We might reasonably infer that the UK would not have been ready to pursue a test ban if to do so would have resulted in its inability to develop or finalise its own designs, or left the USSR with an appreciable advantage. In fact, in a press briefing prepared for the 22 August moratorium announcement, the Foreign Office conceded that the UK decision had been made easier by the amendment of the 1954 US Atomic Energy Act to permit the exchange of information about the manufacture of nuclear weapons

[114] TNA PREM 11/2566, Foreign Office telegram no. 5986 to Washington, 21 August 1958.

[115] TNA FO 371/135552, ZE 120/109, Statement issued by No 10 Downing Street on 22 August 1958.

[116] TNA FO 371/132685, IAD 122/4, Summary of Developments during the Secretary of State's Absence, C. O'Neill to Sir F. Hoyer Millar, 9 September 1958; Cmnd. 551, *Report of the conference of Experts*, Annex 3 p. 42. Statement by Her Majesty's Government issued from No.10 Downing Street, 22 August 1958.

[117] TNA FO 371/132698, IAD 123/29, UK policy on nuclear weapons and test suspension, 1958.

[118] TNA FO 371/132698, IAD 123/34, Geneva telegram no. 597 to Foreign Office, 11 November 1958.

and the conclusion of a new bilateral agreement.[119] It is instructive that prior to the announcement the UK was extremely cagey about a moratorium. However, once this was agreed the UK became much more flexible than the US on the conditions for its continuation and this quickly became a bone of contention between the US and UK as they prepared for the start of the trilateral negotiations on a test ban. The Americans required progress on all parts of the disarmament package to enable them to agree to a second or subsequent year's suspension of tests. The UK on the other hand was prepared to settle for a good deal less.[120]

Anglo-American Cooperation in Action: The First Fruits and Future Testing Plans

Eisenhower authorised that fabrication prints, design prints, material specifications on the US Mark 28 thermonuclear weapon should be released to the UK on 6 September 1958.[121] This included a wealth of detail on manufacturing techniques and manuals on such diverse topics as beryllium and solid fusion materials. In November 1958 the DRPC (AES) agreed that a UK version of the Mark 28 should be developed for production at the highest priority as a megaton weapon.[122] This was to become the Red Snow warhead to the British and incorporated into the Yellow Sun Mark II free-fall bomb. It was also the warhead for the Blue Steel powered bomb and would have been the warhead for the aborted ballistic missile Blue Streak. Sir Edwin Plowden (Chairman, UKAEA) told the Cabinet Defence Committee in November 1958 that the information obtained from the US about warhead design would greatly expand the UK's capacity to manufacture nuclear weapons from its existing supplies of fissile material.[123] However, US designs

[119] TNA FO 371/135551, ZE 120/94, Foreign Office telegram no. 6022 to Washington, 22 August 1958.

[120] TNA FO 371/135554, ZE 120/168, New York telegram no. 1113 to Foreign Office, 24 September 1958

[121] Norris, Burrows, Fieldhouse, *Nuclear Weapons*, pp. 48–9 cite a declassified USAEC document.

[122] TNA AVIA 65/1116, Confidential Annex to D.R.P. (A.E.S)/M (58), Development of US Warheads for British Weapons, 11 November 1958; TNA AIR 6/117, Progress Report on Nuclear Weapons Appendix to A.C. (59) 7, 13 January 1959; TNA AIR 2/3682, Extract from DRP (AES)/M (58) 2 held 11 November 1958. Red Snow would the common warhead for Yellow Sun Mark 2 and Blue Streak .This meeting also appears to have decided that there was no need for Grapple M, however if political and other circumstances permitted, it was thought probable that proposals would be submitted by D/AWRE for a test series Grapple O in about March 1960. TNA AB 16/2371, Atomic Weapons Trials Executive Christmas Island Trials Note by the Chairman to Ministry of Supply, 27 November 1958.

[123] TNA CAB 131/16–20, Cabinet Defence Committee D (58) 28th Meeting Item 2 Fissile Material, 26 November 1958.

required a considerable quantity of a 'special material' (tritium), but the UK might be able to secure supplies of this up until 1963 in exchange for plutonium.

An agreement to suspend all tests however at this stage prevented the UK from developing the small tactical nuclear weapons (devices with yields less than one kiloton) that would be important in a limited war, although London thought that it might be possible to secure supplies of these from the US. It seems that the UK thought that if the means for detecting small explosions could not be improved within a year, the UK should reserve the right to resume tests of less than one kiloton.[124] An agreement on suspension might also introduce an element of uncertainty about the efficacy of the weapons that would be made in the UK in the future or acquired from the US. Although it was unlikely that there would be revolutionary developments in nuclear science applications to warfare, there would be scope for improvement of existing techniques and it would not be possible, if tests were suspended, to verify the practical effects of such improvements. However, on balance there were wider advantages arising from a suspension, especially if the USSR had not yet mastered any defensive weapons. Politically it was important to expose the hollowness of Soviet intentions and, if any breakdown in talks were to occur, it would be essential that the Russians should take the blame after the West had gone the extra mile in its efforts to reach agreement. Most certainly the UK should not be seen as the one to impede progress; therefore the West should drop its linkage with disarmament as a condition for progress on testing issues. If the talks did break down, the UK might announce an abandonment of large tests (those over one kiloton) as this would provide the freedom to develop smaller tactical weapons that were needed and could be conducted without danger to health. The NTPC agreed that this should be the UK policy that Ministers might endorse.

Although the UK announced that the two Soviet tests in November 1958 lifted the obligation on the UK from its 22 August undertaking to refrain from testing for a year, from an internal political point of view officials and Ministers realised that it would be very difficult for the UK to resume testing itself.[125] However, the UK line would depend on the US. It was possible that the US might wish to resume itself and in such circumstances it would be most unwise for the UK to adopt a different position. There was an undoubted concern that to do so might make the US less willing to continue to share its nuclear defence information if the UK were not willing to contribute itself. Instead the best line to adopt would be to state that the UK had no current plans to test, which might not be too difficult. In any case, the UK would not be able to mount any large-scale tests until early 1960. Despite this, there were still pressing military requirements to meet through testing. The UK needed to test for four main reasons: megaton tests were required for anti-missile purposes; this was perhaps the most important and could not be addressed by only small yield

---

[124] TNA CAB 131/16–20, Cabinet Defence Committee D (58) 21st Meeting, Suspension of Nuclear Weapons Tests, 15 October 1958.

[125] TNA CAB 130/130, GEN.604/14th Meeting, Suspension of Nuclear Tests, 7 November 1958.

tests. Second, further tests in the fractional kiloton ranges were required for the development of tactical weapons; third, the UK had to test weapons manufactured from adaptation of US designs and finally the UK had from time to time to test its own operational weapon requirements of all sizes. That at least was the assessment in late 1958. As it transpired, none of these requirements were ever met after that date. As long as the bilateral exchange worked well, the need to conduct independent UK nuclear tests were reduced, but not eliminated. Needs could be met through joint US-UK trials. On balance, officials felt that a suspension of nuclear tests would not at that stage work to the advantage of the West, but despite this that is exactly what the West did do – a self imposed moratorium lasted until mid September 1961 for the US and March 1962 for the UK.

At the end of October 1958, as the UK moratorium was about to start, Sir William Cook told DGAW at the Ministry of Supply that Aldermaston's present thoughts were that it would not need to carry out Grapple M earlier than June 1959 and that it would probably not need to include any megaton weapons. Instead it might consist of up to four kiloton bursts – two balloon bursts, the other two being airdrops.[126] These latter two depended on the assumption that the services would wish AWRE to carry out proving rounds of Red Beard. As for Assessment Tests (minor trials that did not involve nuclear yields), more were needed for 1959. In fact the Grapple M programme was dependent on the successful achievement of this series.

Long-term test suspension had implications for AWRE programmes and staff.[127] Penney informed the Aldermaston workforce on the announcement of the UK moratorium that information acquired from the UK trials, including those recently completed at Christmas Island, plus the information from the US provided a reservoir of knowledge on which to draw for some time to come. There would be plenty to do, even if that meant refocusing on civil work. Penney insisted that staff should be reassured and that they need not worry about their continued employment in the UKAEA. In case of the development and production of instrumentation used to monitor the atmospheric tests and on staff morale, Maddock (Senior Superintendent of Electronics Trials Division, AWRE) observed that action in the future depended almost entirely on one basic policy assumption, which was a choice between two mutually exclusive options. Option one would be to assume that a one or two year suspension of tests implied a recommencement of testing at some future date.[128] On this basis the intervening time would need to be used to press forward with techniques and schemes (which had been handicapped by frequent trials in the past) so that when tests were resumed the personnel would be 'up to date'. Alternately if the suspension was a step towards a complete abolition of tests, then it would be wasteful to pursue the techniques beyond a

---

[126] TNA ES 1/957, P.W.B. Brooking to E.S. Jackson, DGAW, Ministry of Supply, 30 October 1958.

[127] TNA ES 1/1138, Suspension of Atomic Weapon Trials, Personal Message from Sir William Penney to AWRE Staff, September 1958.

[128] TNA ES 1/1138, I. Maddock, SSLT to Mr C.A. Adams, CT, 5 September 1958.

certain point, and much more profitable to engage in new fields of research.[129] Maddock was concerned that a compromise decision would be made which would amount to keeping their hand in without spending too much money. As Sir Solly Zuckerman (Chief Scientific Adviser Ministry of Defence and later CSA in the Cabinet Office) recalled, no matter what contrary views they might have held, no scientific civil servant could have campaigned openly against HMG's declared policy – not that there is any available archival evidence to suggest that such thoughts were harboured.[130]

Planning for future testing had to continue notwithstanding the test-ban talks in Geneva. The NTPC reviewed the then current requirements for future UK testing at the end of 1958, early 1959. Sandys sought Macmillan's approval to discuss with the US the possibility of establishing joint facilities for carrying out nuclear tests if the Geneva Conference on the Discontinuance of Nuclear Weapon Tests did not lead to an indefinite suspension.[131] The Australian government would need to be told that there would now be no kiloton tests in October 1959 at Maralinga, and that the earliest date for holding another series would be October 1960, depending on the outcome of the Geneva conference. Finally, a series of assessment trials would be needed in March 1959 at Maralinga in order to test the high explosive systems incorporated in nuclear weapons.[132] Macmillan agreed.[133]

Options and requirements for future UK testing programmes were set out in a minute for the Prime Minister on 23 June 1959.[134] This noted that whatever the outcome of the Geneva talks, the UK should assume that for planning purposes, it would indefinitely cease to carry out tests in the atmosphere. If international agreement did not ban tests underground or in outer space, the UK should seek to cooperate with the Americans in experiments of this kind. Maralinga would still need to be retained for assessment and other non-nuclear experiments.[135] The MOD noted that the consequences of this decision would mean that weapons produced for delivery to the Services could not be proved in the normal way and that any tests that might be required for the development of ABM warheads would

[129] TNA ES 1/1138, I. Maddock to Mr C.A. Adams, CT, 5 September 1958.

[130] Solly Zuckerman, *Monkeys, Men and Missiles An Autobiography 1946–1988* (London, 1988), pp. 307–8.

[131] See Chapter 5 for detailed discussion of the Conference.

[132] TNA DEFE 13/150, Duncan Sandys to Prime Minister, Nuclear Tests, 9 January 1959.

[133] TNA AVIA 65/870, Policy Copy of a minute by the Prime Minister to Minister of Defence, 11 January 1959.

[134] TNA CAB 134/2275, Cabinet Nuclear Tests Policy Committee, N.T. (59) 2, Cabinet Tests Policy Committee Future Programme of Nuclear Tests: Draft Submission to the Prime Minister, Note by the Secretaries, 12 June 1959. TNA DEFE 13/150, A.D. Heskett, V.H.B. Macklen; final version Duncan Sandys to Prime Minister, 23 June 1959.

[135] TNA DEFE 13/150, Minute to Prime Minister, Future Plans for Nuclear Tests, 23 June 1959.

have to be carried out underground or in outer space. As the weapons for service use were virtually copies of US tested designs, presumably Red Snow, Penney considered that the risks arising from an inability to carry out normal proving trials were acceptable. However, if underground tests were not prohibited, it would be important for the UK to conduct such trials, preferably jointly with the US given the costs and for general political reasons.[136]

Macmillan, although agreeing that the UK did not need any further atmospheric tests, asked that the other issues be discussed in a restricted Cabinet Defence Committee meeting.[137] Even if all nuclear tests were banned, there was still some weapon development work that could be done and, unless the UK abandoned nuclear weapons altogether, the UK could not afford to stop such trials. At that time – mid summer 1959 – the UK and US were planning a comprehensive and integrated programme of R&D, and this was a programme which AWRE hoped the Americans would agree to leave some of the work for the UK to undertake. If the UK abandoned its test facilities, this would seriously affect the confidence built up in these bilateral negotiations.[138] But as Macmillan pointed out, UK plans would be reconsidered if the Geneva conference failed to reach a comprehensive agreement for the cessation of all tests.[139] However, for planning purposes, the UK would assume the indefinite suspension of all British nuclear tests in all environments.[140]

Engineering works at Christmas Island were in hand for completion in December 1959 to enable a megaton range test in about March 1960. The Committee reviewed these plans. They had three factors to take into account: the progress of the Geneva negotiations, the attitude of the public in the UK to the health hazards alleged to arise from nuclear explosions and defence programme requirements. Officials recognised that planning would have to be based on the assumption that the UK would not be able to carry out further tests in the atmosphere, on the surface or under water. The consequences of accepting this conclusion were that the current production weapons could not be proved by a series test and that tests necessary for the development of suitable warheads for anti-missile systems would have to be carried out underground or in outer space. However, the draft went on to note that current production weapons were effectively copies of tested US designs and the risk of omitting the final proving test in the then political circumstances was acceptable in Penney's view. The non-nuclear assessment tests involving high explosives (HE), however, were essential if the weapons at present planned for

[136] TNA AVIA 65/1217, Nuclear Tests, Ministry of Defence, submitted to Prime Minister on 23 June 1959

[137] TNA DEFE 13/150, Bligh to Sabatini, MOD, 26 June 1959.

[138] TNA DEFE 13/150, Defence Committee Future plans for Nuclear Tests, Brief for the Minister, 7 July 1959.

[139] TNA DEFE 13/150, Prime Minister's Personal Minute, Serial No. M251/59 to Minister of Defence, 1959.

[140] TNA AVIA 65/1217, Macmillan to Minister of Defence, Future Plans for Nuclear Tests, 9 July 1959.

production were to be approved without a final full-scale test. In addition to the weapons-related work, the Committee also flagged the need for the UK to make a positive contribution to the seismic research (see Chapter 6) that would be required to clarify and identify underground events. This could entail underground HE tests of up to 500 to 1000 tons and these could be undertaken at Maralinga or elsewhere in Australia. Macmillan agreed but with certain qualifications, which the Cabinet subsequently endorsed.[141]

Ministers met on 7 July 1959 to consider future testing options. It was not essential to inform either the US or Australian governments of UK intentions.[142] The Cabinet Defence Committee agreed for planning purposes that the indefinite suspension of all UK tests should be assumed. However, plans would be reconsidered in the event that the Geneva negotiations failed to reach agreement on establishment of a detection system in light of the Geneva experts' findings and recent US experimental results and the 'large hole theory' – see Chapter 5. The risk of undetected explosions in the five to ten kiloton range might be politically acceptable to the Americans and they might feel that the Russians might be deterred if there were adequate allowance for on-site inspections. However, the Minister of Defence worried that the effectiveness of the Western deterrent might be gradually impaired and the balance of power seriously upset to UK disadvantage by the acceptance of any agreement which the Russians could evade, but which would prevent the US and UK from developing anti-missile missiles and fractional yield kiloton devices. In further discussion, there was a general agreement amongst Ministers that if the US were prepared to take such risks, then there was not much the UK could do to object. Macmillan approved NTPC recommendations forwarded by the Minister of Defence that, for planning purposes, it should be assumed that the UK would not carry out further nuclear tests whether in or above the atmosphere, under water or underground. Furthermore, these plans would be reconsidered if the Geneva conference failed to reach a comprehensive agreement for the cessation of all nuclear tests.[143] In view of the present state of the negotiations in Geneva and the UK's position *vis-à-vis* the US, no public statement on the change of policy would be made at that stage, nor would the US or Australian governments be informed until Ministers had been able to consider the outcome in Geneva.

---

[141] TNA AVIA 65/870, Prime Minister to Minister of Defence, Prime Minister's Personal Minute M. 251/9 undated. See Chapter 6 for detailed discussion of UK forensic seismology research..

[142] TNA CAB 131/22, Cabinet Defence Committee D (59) 23, Future Plans for British Nuclear Tests, 7 July 1959.

[143] TNA AIR 2/13753, DGAW to DCNS et al., Future Plans for Nuclear Tests, 10 July 1959.

At the end of 1959 the Ministry of Supply had managed to secure Treasury agreement for four balloons for weapons tests during Grapple M.[144] Officials agreed that conducting a trial in March 1960 should be taken as a planning assumption – Grapple M.[145] This was certainly the directive that emanated from the NTPC.[146] There were still contingencies in UK planning assumptions in autumn 1958. The MOD's thinking, building on AWRE advice in light of the flood of information that had come about from the implementation of the 1958 agreement, was that only kiloton trials related to Red Beard and possibly army surface to surface guided weapons requirements had any justification in 1959.[147] AWRE appear to have had three main reasons for further trials on Red Beard, but only two of these are in the public record. The first of these was the necessity for a full-scale proving trials of a service round as produced for the stockpile and the need for two development improvements – a small loading tube which improved the service handling of weapon and an improved initiator based on US information, which would improve shelf life and ease the maintenance problem. Until that time AWRE and MOD had resisted RAF demands for a full service weapon test because research and development were by far the most pressing requirement; cost factors were also a consideration. However, since AWRE now had more information than it could digest, these counter arguments were losing their force. Penney was prepared to agree to a service test provided the Services paid for it. However, such trials could only take place at Christmas Island since the Australian authorities did not permit airdrops of high yield weapons, and this would have knock on effects on the target response programme on which the Service Departments set great store. AWRE Divisional heads learned on 11 December 1958 that there would be no major trial in 1959 at either Maralinga or Christmas Island. However, subject to overriding political decisions, the next major trial would be Operation Grapple O at Christmas Island in March 1960. (Penney had earlier noted that AWRE would wish to carry out megaton trials at this time provided the political situation permitted this. The DRPC had also earmarked this series as an opportunity to test as a matter of

[144] TNA T225/1079, G.M.P. Myers, Ministry of Supply to I. de L. Radice, Treasury Chambers, 29 December 1959.

[145] TNA T225/1079, Conclusion of a Meeting Held in St Giles Court, Christmas Island: Activities in 1959, 18 November 1959 and see also TNA AVIA 65/1440, Outline of General Policy.

[146] It is instructive to note that AWRE estimates of the costs of trials at Maralinga and Christmas Island included costs for weapons component design, assembly and provision (including fabrication and supply of boosting components. These were £560,000 and £850,000 in total. TNA T225/1079, V.H.E. Cole, UKAEA to L.H. Curzon, Ministry of Supply, 28 October 1958.

[147] TNA DEFE 7/1888, DRP (AES) Meeting to be held on Tuesday, 11 November, V.H.B. Macklen to Sir Frederick Brundrett, 7 November 1958.

priority the UK version of the Mark 28.[148]) Operation Lighthouse was postponed from twelve months from its previous planned date, September 1959.[149] Therefore, one option would be to keep Christmas Island on a care and maintenance basis so that within nine months of the Ministry of Supply stating a requirement, a nuclear test using an airdrop and/or balloon supported weapon could be executed.[150]

It seems that in early November more tests were needed for four reasons: tests in the megaton range of anti-missile missiles, which were the most important; fractional kiloton range tests for tactical weapons (which the US was not permitted to provide to the UK); tests of weapons manufactured from adaptations of US designs and from time to time tests of UK operational weapons of all sizes.[151] This latter point appears to have been the only time in the entire period covered by this book when testing of in-service weapons was contemplated. If the 1958 Mutual Defence Agreement worked, as it should, then the requirement for independent UK tests would diminish, but the need would not be eliminated entirely. Better still, joint testing would be the ideal arrangement. Here we see the crux of the problem. Britain's ability to continue with its weapons programme and being able to follow the disarmament road was by no means certain as late as the summer of 1958. It is clear that officials and Ministers saw that the autumn trials might well be the last opportunity to carry out vital tests.[152]

---

[148] TNA AVIA 65/1116, D.R.P. (A.E.S)/M (58) 2nd Meeting Item 3, 11 November 1958.

[149] TNA ES 1/1138, N. Pearce, Superintendent, Trials Planning Branch, Major Trials Programme, 11 December 1958; see also TNA ES 1/1098, P.W.B. Brooking (Admiral) retd. AWRE to E.S. Jackson, DGAW, Ministry of Supply, 14 October 1958; Lighthouse was cancelled and the Air Ministry had to look for alternatives as one of the purposes of these trials had been to expose portions of V-bomber fuselages to a nuclear explosion to extend knowledge of aircraft lethality. TNA AIR 2/14999, H.J. Dodson for DDOR2 to ACAS (OR) participation in Suffield Vulnerability Trials; Visit by OR 27, 16 June 1960. The original plans had called for a one kiloton ground burst weapon (or five to ten kiloton tower burst weapon) for fall-out investigations, and a 20 to 40 kiloton round burst from balloons at about 1000 feet for a target response blast, thermal and other tests. AWRE plans however could not be finalised by the end of June although the first shot would one kiloton, a forecast for the yields of the other two rounds could not be made until after the conclusion of the Grapple Z tests. TNA AVIA 65/1820, D.R.P (A.E.S)/P (58) 8 DRPC Sub-Committee on Atomic Energy Progress Report on the Interdepartmental Target Response Working Party, 25 June 1958. There were also proposals from Director RAE Farnborough to mount *ad hoc* tests of the fuzing and electronic components of British atomic weapons to see how they could cope with radio flash and nuclear radiation.

[150] TNA ES 1/1138, Staff requirements for the Nuclear Test base at Christmas Island, Care and Maintenance State, 20 September 1958.

[151] TNA CAB 130/130, Suspension of Nuclear Tests, GEN 604/14th Meeting, 7 November 1958.

[152] TNA T225/1079, CRO telegram no. 1125 to Ottawa, 10 July 1958.

The DRPC (AES) was certainly of the view that Lighthouse should take place if political considerations permitted.[153]

At the NTPC's final meeting of 1958, E.S. Jackson sought decisions, in the absence of any international agreement, for trials in the kiloton range (four shots and with objectives similar to the Buffalo series[154]) at Maralinga in October 1959 (Lighthouse) and the megaton range in March 1960 (Grapple O), which had been recommended by the DRPC (AES).[155] Grapple O would only go ahead subject to considerations of whether there was to be any further international suspension of trials.[156] This was the earliest date that AWRE could manage after it had studied and applied the technical data provided by the US with a view to the development of US designs in the UK.[157] These trials were not optional extras; the DRPC (AES) had agreed on 11 November that a megaton test series was a high priority. When it came time to consider whether the earlier requirement for an atmospheric megaton range test in March 1960 was still required, AWRE indicated that a later date, such as September 1960, would be preferable. Dr Nyman Levin, now Director of AWRE, told the Committee that Red Beard and Red Snow should be in service by autumn 1960 and the fruits of collaboration with the US would have had time to ripen.[158] The resource problems presented by 'anglicisation' were at the root of this: AWRE just had so much scientific and engineering experience to go around. Levin later noted in May 1959 that a March 1960 test would undoubtedly embarrass AWRE in meeting current service weapons programmes. It would be better to have a later date, possibly September 1960.[159] Furthermore, he thought that as far as AWRE trials planning was concerned, it would be much more profitable to direct this towards underground rather than atmospheric testing. There would be such an outcry against atmospheric tests that Ministers would not agree even preparation for such tests.

---

153 TNA AVIA 65/1820, Atomic Weapons Trials Executive, Note by the Chairman, Ministry of Supply, 27 November 1958.

154 TNA ES 1/1098, Operation Lighthouse, Initial Summary Statement, 31 March 1958.

155 TNA AIR 2/1890, G.M.P. Myers to M.E. Quinlan, Air Ministry, Future Nuclear Weapons Trials Programme, 26 November 1958.

156 TNA AB 16/237, H.C. Hudspith to P.A. Bundy, Chief Accountant AWRE, Future Trials Programmes, 1 December 1958.

157 TNA CAB 134/2274, Cabinet Nuclear Tests Policy Committee, N.T. (58) 7th Meeting, 5 December 1958.

158 TNA ADM 1/31023, Nuclear Weapons, undated minute. By some point in 1962 the Royal Navy had 17 Red Beards (yield of about 15 kilotons) dispersed as follows: eight on HMS Ark Royal, five on HMS Hermes and four on HMS Centaur. The planned total was 28 and stowage on the carriers for this number existed.

159 TNA AVIA 65/1217, N. Levin, AWRE to E.S. Jackson, DGAW, Ministry of Supply, 1 May 1959.

It was still too soon to say where the Geneva talks would lead and this, along with other factors, made it impracticable to take a long-term view in mid December 1958, on the future of the Maralinga range.[160] Sir Richard Powell was advising the Air Ministry by late spring 1959 that it should assume that there would be no more tests up to 1965.[161] Now the really interesting thing here is that by mid summer 1959 AWRE was confident that it could develop a warhead with a yield in the range of ten to 100 tons without undertaking any further nuclear tests; first deliveries of such a weapon would probably be available in 1962 or 1963. This clearly suggests that the feeling was that a test moratorium was not going to place too many restrictions now on Aldermaston's design capabilities.[162] It would therefore not impinge on the UK's ability to produce small yield weapons for surface to air guided weapons for air defence purposes. As we will see later this confidence in the ability to produce fractional yield weapons appears to have been misplaced.

By the summer of 1959, however, the government decided that it should be assumed that the UK would not carry out further nuclear tests in or above the atmosphere, under water or underground. However, Christmas Island should not be totally abandoned and that with the withdrawals of men and equipment, the base could still, depending on future decisions, be re-activated if needed for full scale testing at 18 months' notice.[163] In view of the then state of affairs in the Geneva test-ban negotiations and the UK relationship with the US, no public statement on the change of policy indicated would be made. Furthermore, neither would the US or Australia be informed until Ministers had been able to consider the outcome of the Geneva negotiations. The Grapple base on Christmas Island was reduced from an operational state to a holding state between October 1958 to April 1959. During this period construction work reduced in scale. In light of Ministerial decisions, the MOD instructed the Task Force Commander to maintain the base, but to reduce it to a minimum holding state.[164] Options were thus kept open and would be so until the mid 1960s.

---

160 TNA AIR 2/13753, Note of a meeting of the Sub-Committee of the Atomic Weapons Trials Executive, 11 December 1958.

161 TNA AIR 2/13753, ACAS (Ops) to D.Air Plans 'Grapple', 4 May 1959.

162 TNA AVIA 65/1050, Low-yield Nuclear Weapons for the Army, Defence Research Policy Committee Sub-Committee on Atomic Energy minutes of meeting held on 28 July 1959. Dr Levin D/AWRE made these claims.

163 TNA ES 1/1138, Extract from a minute to the Deputy Chiefs of Staff, Future Plans for Nuclear Tests, 17 July 1959.

164 TNA T225/1080, Joint Plan for the Reduction of the Grapple Base to a Minimum Holding State, 20 August 1959.

Conclusion: 1958: A Decisive Year

Grapples X and Y put the UK well down the road to meeting its nuclear defence requirements. There were further developments and improvements that still needed to be made in UK design capabilities, irrespective of the promise of US information, which finally became a reality in the late summer and early autumn of 1958. Grapple Z was clearly very necessary in this regard. It is clear that Macmillan and the UK managed to give proof to John Foster Dulles' infamous adage that diplomacy was about keeping three balls up in the air at the same time whilst hanging on to your own. Macmillan had pestered the US about amending its 1954 Atomic Energy Act to permit closer collaboration with the UK's weapons programme; he sought unambiguous assurances that key data would flow as a result; balanced the domestic and international pressures for a test ban, whilst energetically pursuing the UK's design requirements, which were at that time seen as fundamental to the UK's ability to produce a megaton range warhead, economical in its use of fissile material and immune to the R-I effect. This warhead was for its planned ballistic missile (Blue Streak), propelled bomb (Blue Steel) and free-fall bomb (Yellow Sun Mark 1). Had there been a test ban or moratorium in force from the autumn of 1957, it is clear that the UK would have failed in its pursuit of these objectives, certainly at least until such times as atmospheric testing became possible again; and it is equally clear that the flow of US information under the 1958 Agreement might not have been achieved. However, what impact would a testing moratorium have on the UK programme once its core objectives had been met? We can now turn to this question in the next chapter. The UK could now wholeheartedly pursue a test-ban treaty.

Chapter 4

The Testing Moratorium and Its Constraints on UK Weapons: 1958–1963

Introduction

Once the UK had completed its own series of atmospheric tests in the autumn of 1958, which turned out to be the last ever conducted by the UK, and US nuclear data started to flow under the 1958 Agreement, the next issue facing the weapons programme was translating art into article: the arcane art of weaponisation. But was this a simple task in the absence of full-scale testing? Indeed what testing did the UK need in the near-term? And just how would the voluntary moratorium, not to mention the prospect that that a test-ban treaty might actually be agreed in the Conference on the Discontinuance of Nuclear Weapon Tests in Geneva, affect the content and direction of UK weapons programme in the immediate years after 1958?[1]

Assessment Tests (Minor Trials)[2]

In August and September 1958 we see for the first time some concerns that, notwithstanding the imminent suspension of atmospheric tests, the UK should still be able to conduct minor trials. These had been underway in Australia since 1955 and had included initiator trials.[3] The DRPC had noted in November 1956 that such trials were playing, even then, an increasing part in atomic weapons development.[4] They were to assume an even greater role in the UK weapons programme post 1958, as we shall see below. London asked AWRE to consider the political propriety of continuing minor trials at Maralinga after whatever date the UK, US and USSR might agree for the suspension of trials. AWRE also prepared a paper for the Ministry of Supply showing why such trials had to continue.[5] Meanwhile

---

1 See Chapter 5 for a detailed discussion on the Conference on the Discontinuance of Nuclear Weapons Tests.

2 Arnold and Smith, *Britain, Australia and the Bomb*, Chapter 11 *Kittens, Rats* and *Vixens*, pp. 215–34.

3 See TNA AVIA 65/1109, Tim Series No 2, 1954–55. Initiators were used to introduce neutrons to the imploding core at just the right moment to help the chain reaction.

4 TNA AVIA 65/1114, Staffing of The Permanent Proving Ground Memorandum by DGAW, DRP (A.E.S)/P (56) 19, 26 November 1956.

5 TNA ES 1/957, P.W.B. Brooking to Sir William Cook, 29 August 1959.

the UKAEA instructed the AWRE contingent at Maralinga to concentrate on the most important shots first before any temporary suspension might be required.[6] The view was that the proposed suspension of atmospheric tests did not debar HE tests involving radioactive materials but which did not involve nuclear yields.[7] Penney stressed that AWRE had given up using the words 'minor trials' and used 'assessment tests' instead to avoid the possibility that minor trials might be perceived in the public mind as trials of small yield nuclear weapons.[8] AWRE changed the name for political reasons in order to prevent possible misinterpretation.[9]

Given growing UK concerns that the test-ban talks might constrain the assessment tests, Ministers tasked the Foreign Office to pursue this issue with the US and ensure that the scope of any test-ban treaty excluded such experiments. An interdepartmental pre-conference test-ban planning meeting on 21 October 1958 agreed that minor assessment trials and safety trials involving the use of radioactive materials and slight local contamination, but not a nuclear explosion, must be excluded from any suspension of tests.[10] In addition, the Foreign Office produced a suggested definition of a nuclear explosion in order to ensure that the UK's minor trials in Australia could continue; this was acceptable to AWRE, as Con O'Neill's (Assistant Under Secretary of State, Foreign Office) definition would allow both tamper movement and equation of state trials to continue.[11] From a purist's point of view, however, AWRE thought that this might rule out 'single point detonation' trials and possibly certain trials that could give rise to small amounts of fission. Aldermaston felt that that such fission was not the intent of the trial and that if it did produce any it would be an accident, which AWRE was unable to guard against. It might be worth, so AWRE thought, to put the Foreign Office definition to the Americans to see whether they would endorse it.[12] Penney agreed and wrote to tell Sir Patrick Dean, Permanent Under Secretary, Foreign Office and offered his own definition: 'The term nuclear test explosion shall be held to mean explosions which are the consequence of the deliberate release of energy by nuclear fission,

---

[6] TNA ES 1/957, P.W.B. Brooking to Director AWRE, 23 September 1958.

[7] TNA CAB 134/2274, Cabinet Tests Policy Committee, N.T. (58) 6th Meeting, 11 September 1957; see also Lorna Arnold, *A Very Special Relationship British Atomic Weapon Trials in Australia* (London, 1987), pp. 204–7.

[8] TNA ES 1/1138, W.G. Penney to E.S. Jackson, D.G.A.W, Ministry of Supply, 31 October 1958.

[9] TNA ES 1/957, ADD to All Band B Officers and above 'Minor Trials' – Change of Name, 1 October 1958.

[10] TNA DEFE 19/5, Geneva Nuclear Test Negotiations Meeting held in Foreign Office, 21 October 1958.

[11] Tamper movement trials were concerned with efficient functioning of a key non-nuclear component in the implosion system of a nuclear weapon; equation of state calculations concerns how materials respond under dynamic loading which provides the pressure and energy of each individual material as a function of density and temperature.

[12] TNA ES 1/957, P.W.B. Brooking ADD/AWRE to Director, 29 September 1958.

or nuclear fusion, or both.' If this were acceptable, then it seemed to Penney that 'minor trials' would still be permitted. However, there were some further drafting changes and the proposed definition that appeared in the UK delegation brief was as follows: 'the term nuclear test explosion shall be held to mean nuclear explosions for experimental purposes which are the consequence of the release of energy by nuclear fission, or nuclear fusion, or both'.[13] AWRE modified its autumn 1958 series of Assessment Trials so that they could complete the trials by 31 October. Penney sent a signal to the AWRE contingent at Maralinga instructing them to ensure that all firings involving radioactive material had to cease by 2359 hours 31 October.[14] Aldermaston nevertheless hoped to begin another series in the spring of 1959, but this would of course depend on what the Conference on the Discontinuance of Nuclear Weapon Tests agreed in Geneva.[15] Ministers approved a new series for March 1959, but the Foreign Office was still anxious lest such activities embarrass the UK's position in the Geneva. Secrecy was thus paramount.[16] Foreign Office officials worried by the potential negative implications for the test-ban talks in Geneva felt the need to fire a shot across the MOD's bows. Sir Patrick Dean wrote to Sir Richard Powell on 11 November to state that the Foreign Office had no objections to these trials provided they remained secret and caused no embarrassment for the UK in the Geneva negotiations.[17] Foreign Office officials asked the MOD to consult them a few weeks before the trials were due. For officials on the other side of Whitehall, the point was that if a test-ban agreement were imminent, then they wanted to avoid any risk that the Maralinga trials were misrepresented and damage the prospects for the treaty. As far as their relationship to the Geneva test-ban talks were concerned, Harold Watkinson, Minister of Defence, reassured Macmillan. Whatever the talks' outcome, there was no reason to believe that these experiments could be regarded as an evasion of a Treaty as the amount of nuclear energy released was many times less than the capability of any detection system.[18] At that stage, Aldermaston envisaged about 24 experiments per year, only four of which would involve release of measurable amounts of nuclear energy.

AWRE were not impressed with Foreign Office anxieties and resented the suggestion that there should be any limitation imposed by diplomats. AWRE thought that they should make perfectly clear to the Foreign Office that it could not hold the fact that these trials were taking place a secret. It was necessary to make arrangements such as site preparations, transport, staffing well before the tests

---

[13] TNA FO 371/132687, IAD 122/51/G, Annex C Scope of Tests Suspension, 1958.

[14] TNA ES 1/957, AWRE Aldermaston to AWRE Maralinga, 2 October 1958.

[15] TNA ES 1/957, W.G. Penney to Sir Patrick Dean, Foreign Office, 1 October 1958.

[16] TNA AVIA 65/1111, H.C. Hainworth, Foreign Office to Captain F.B. Lloyd, Ministry of Supply, 20 January 1959.

[17] TNA DEFE 16/605, Sir Patrick Dean to Sir Richard Powell, 11 November 1959.

[18] TNA PREM 11/3157, Minister of Defence to Prime Minister, Nuclear Weapons Safety Experiments, 23 December 1959.

and all completed on an unclassified basis. In addition, the Deputy Director Ted Newley felt that AWRE should make a stand on what the purpose of these were and make clear that they would not be subject to any ban and were just another aspect of the programme of weapons development on which it was patently obvious that Aldermaston remained engaged.[19] The Foreign Office did not want to raise the need for such trials again with Australia until the test-ban talks had finished. AWRE got the Prime Minister involved in an attempt to shift the Foreign Office, which was obviously successful.[20]

There were four different types of minor trials conducted in Australia over the next few years. Tims were a general study of hydrodynamics using natural uranium and HE; Rats had the same objective as Tims but employed a gamma ray source to determine space-time characteristics; Kittens were a study of certain types of neutron initiating systems; and Vixens were the study of the consequences of accidents. Vixen A specifically dealt with the spread of radioactivity or toxic materials such as might result from an accident. Vixen B was the study of inherent safety characteristics of warheads that might become involved in accidents.[21] These latter tests started in 1959 and were designed to obtain essential data, unobtainable elsewhere, on the safety aspects when certain radioactive components were either burned or detonated accidentally.[22]

All in all these appear to have had an absolutely critical role on the development of nuclear weapons, which were 'anglicised' versions of US designs provided via the UK-US exchanges that had taken place as a result of the UK-US Agreement for Cooperation on the Uses of Atomic Energy for Mutual Defence Purposes.[23] It was not possible because of differences in manufacturing methods and materials to produce exact copies of US designed warheads.[24] However, by mid 1959 the NTPC noted that it would be prudent to assume, as a basis for planning, that the UK would carry out no further nuclear tests in the atmosphere. The implications of this assumption were that it would be unable to test operational nuclear weapons

---

19 TNA ES 1/960, E.F. Newley, DD/AWRE to Dr N. Levin, Director, 20 November 1959.

20 TNA ES 1/958, R. Pilgrim to ADD Assessment Tests, 8 January 1959.

21 TNA ES 1/961, extract from draft replies to points made by Professor Titterton in his letter of 24th October 1960. See also Richard Moore, *UK Nuclear History Working Paper No. 1 The Real Meaning of the Words: a Pedantic Glossary of British Nuclear Weapons*: http://www.mcis.soton.ac.uk/Site_Files/pdf/nuclear_history/-Working_Paper_No_1.pdf.

22 TNA AVIA 65/1111, N. Pearce, SPTD, AWRE to Capt F.B. Lloyd, DA/Trials, Ministry of Supply, 26 January 1959.

23 Anglicisation was not a term used at that time by contemporaries. This phrase came much later.

24 TNA DEFE 7/2380, V.H.B. Macklen to Sir R. Powell/Sir F. Brundrett, NTPC Meeting – Friday, 5th December, 4 December 1958.

produced to US designs.[25] The Government had announced on 20 July 1959 that it would not resume testing as long as useful discussions continued in Geneva. Such were the importance of these tests, particularly the Vixen tests, that AWRE asked the MOD for permission to acquire a further 1 kg of plutonium from military stocks for tests in 1960.[26] The MOD agreed, or so it appears.

An MOD paper prepared for the nuclear test committee in December 1959 explained the need for safety experiments for nuclear weapon designs intended for the UK stockpile.[27] AWRE felt that the then current political situation made it likely that there would be no more major trials at Maralinga for some time. However, the expectation was that assessment tests were likely to continue and indeed might increase a little in size and scope.[28] Some officials in the MOD worried that their Foreign Office counterparts had asked to be consulted if there were to be any assessment tests whilst the Geneva Conference was underway. The Foreign Office had even told Victor Macklen in the Chief Scientist's office that the Foreign Secretary would probably object to these trials.[29] Macklen feared that if Ministers endorsed this position, then it would be quite impossible for the UK to produce weapons to American designs.[30]

The requirement for a high-yield, reliable, economical size and lightweight nuclear warhead for missile delivery had produced designs in which it was quite impossible to separate the fissile material from the HE system after initial manufacture. Thus, the original method of safeing a nuclear warhead by actual physical separation of nuclear and HE parts until the last possible moment was no longer possible. By 1959–60, British experimental knowledge of the safety of modern designs derived entirely from US sources, but if British experts were required to give safety assurances to Ministers on British manufactured warheads, some independent British experimentation was essential.

---

25 TNA CAB 134/2274, Cabinet Nuclear Tests Policy Committee, N.T. (59) 1st Meeting, 10 June 1958.

26 TNA DEFE 16/432, Survey of possible site for an underground atomic weapons test: allocation of plutonium for assessment tests, 1958–59 Dr N. Levin, 19 May 1959.

27 TNA CAB 134/2275, Cabinet Nuclear Tests Policy Committee, N.T. (59) Item 3 Need for Nuclear Weapons Safety Experiments, Note by the Ministry of Defence, 3 December 1959.

28 TNA ES 1/390, E.R. Drake Seager for Senior Superintendent Radiation Measurements and Trials, Possible Reorganization of Maralinga Range for Future Trials, 13 May 1959.

29 TNA DEFE 7/2380, V.H.B. Macklen to Sir R .Powell/Sir F. Brundrett, NTPC Meeting, 4 December 1958.

30 It seems that the designs of US warheads and their associated safety and arming features, 'were based on a consideration of the hazards arising, particularly during peace time, from neurotic and unstable personnel'. TNA AVIA 65/1784, A N Christmas D/G.W.Tech to D.G./G.W. Safety of Nuclear Weapon System, 29 January 1960.

Although UK weapons used US designs as far as possible, there were still going to be some unavoidable design changes and, most significantly, British fissile material was of a different specification. It is not clear from the available papers what this difference was, although we do know that the British tested devices with high plutonium 240 content during the Totem trials in 1953.[31] Totem sought to establish certain minimum characteristics of fissile material suitable for use in nuclear weapons. It would seem that under the British definition military grade plutonium contained up to 6 per cent of plutonium 240 (the US definition was 93 per cent); HEU was more than 90 per cent U-235.[32] The new service weapons production programme therefore made safety experiments urgent, especially if modifications proved necessary. The trials would include subjecting weapons containing fissile material to the type of detonation of HE that might occur accidentally. It would also appear that in the 1959 Vixen A trials experiments with weapon assemblies included the destruction of a Red Beard Mark 1 by fire.[33] AWRE planned four detonations and six burnings for the Vixen series in 1959.[34] Ministers agreed on 31 December 1959 that the test-ban treaty should define nuclear explosions to permit safety experiments without infringement of the treaty.[35]

---

[31] *The Submission of the Government of the United Kingdom to the Royal Commission into British Nuclear Tests in Australia 1952*, Treasury Solicitor 1985, p. 125; Arnold, *A Very Special Relationship*, pp. 54–5. On 4 October 1957, officials decided to modify all RAF operational weapons by inserting the explosive within the outer cladding plates of the implosion sphere so that the new position of the cartridge port would be underneath. This change was designed to ease the problem of last minute loading of the cartridge into a bomb previously positioned in the aircraft. Centre sections so modified were then designated Mk 3. All RAF stockpiles were expected to have been modified before April 1958. WO 286/39, A Technical Appreciation of the Effect on Blue Peacock of the RAF Decision to Modify Blue Danube Centre Sections From MK. 2 to MK. 3, 1 November 1957. All this effort despite the recognition in 1957 that Blue Danube would be obsolete by 1959 and discontinued in 1961. See Nuclear Demolition Charge Violet Mist Note by the War Office, 11 May 1960.

[32] TNA CAB 134/2241, Implications of a Possible Cut-Off of Supplies of Fissile Material Report by a Working Party of the Committee, Nuclear Requirements for Defence Committee, ND (64) 5, 27 April 1964. Arnold and Smith, *Britain, Australia and the Bomb*, pp. 49–50.

[33] TNA ES 5/336, Experiments with Weapons Assemblies/Destruction of Red Beard Mk1 by Fire Closed Open description. AWRE File retained under Section 3 (4) of the Public Records Act 1968. It is not clear whether this trial involved a complete service weapon.

[34] TNA ES 12/271, Assessment Tests Minor Trials Maralinga, p. 4, 1959.

[35] TNA AB 16/4550, Nuclear Warhead Safety Experiments Brief by the Atomic Energy Authority, 25 March 1960. The reference to the Ministerial decision was in Cabinet Defence Committee D (60) 1, 31 December.

However, the definition that the UK and US were working to had not been shared with the Russians, far less agreed with them.[36] If the Russians accepted the UK/US definition, then safety experiments would not be a problem, but MOD officials were worried that the Russians would claim that a ten ton explosive limit would allow considerable weapons development. However, whilst officials thought this to be true, the control of such tests was not feasible even with the full verification regime then proposed in Geneva. MOD officials concluded that since such tests should seem to be acceptable under a CTBT, there should be no attempt to conduct such tests clandestinely. On the other hand, however, the UK would not be seeking publicity for such events. Consequently, officials sought Ministers' agreement on five recommendations:

1. That there was an urgent and important requirement for nuclear weapons safety experiments involving the release of a very small amount of nuclear energy;
2. While the maximum cooperation with the US should be sought, there was a valid requirement for independent British work;
3. The way in which 'nuclear test explosion' was defined, or explained in a CTBT, should not include very small yield safety experiments, and that the object of this definition among other things, would be to permit such work to be carried out;
4. AWRE should urgently prepare detailed proposals for carrying out the necessary safety experiments for further consideration by Ministers; and that
5. MOD and AWRE should reconsider the definition of the term 'nuclear explosion' in order to see whether the UK could meet the requirement to exclude the Assessment Tests in such a way as to reduce political difficulties.

The NTPC considered this paper on 12 December 1959.[37] In discussion, officials pointed out that there were US design nuclear weapons that were still not fully engineered. The main design used by the UK was the US Mark 28 thermonuclear

---

[36] TNA CAB 134/2275, Cabinet Nuclear Tests Policy Committee, N.T. (59) Item 3, Need for Nuclear Weapons Safety Experiments, Note by the Ministry of Defence, 3 December 1959. Paragraph 7 on page 3 of this document notes that, 'nuclear explosion, including the words when occurring in the expression "nuclear weapons test explosion", means the release of energy by fission or thermonuclear reaction, in a time less than 10^{-6} second and of an amount greater than the energy yield of 10 tons of TNT (4×10^{17} ergs)'.

[37] TNA CAB 134/2274, Cabinet Nuclear Tests Policy Committee, N.T. (59) 2nd meeting, 18 December 1959.

warhead – known in the UK as Red Snow.[38] This offered savings in space and weight over the Granite type of warhead.[39]

AWRE incorporated a number of modifications into the mechanical and electrical engineering aspects of Red Snow and intended to use it for the Yellow Sun Mark II bomb and Blue Steel.[40] These would require progressive safety testing during production in the UK, either because some alterations were necessary, or because substitute materials, including HE and detonators, of British origin were to be incorporated in US designs.[41] Red Snow required different lens shapes and revised specifications for some 70 to 80 new materials.[42] Tests were necessary to ensure against the possibility of accidents during manufacture and subsequent handling and storage. Until AWRE completed these trials, the MOD's Nuclear Weapons Safety Committee would be unable to give categorical assurances that such weapons were entirely safe for service use, storage, handling and transportation. We can see from this that officials regarded these safety experiments as fundamental to the UK acquiring an operational and certifiably safe nuclear stockpile.[43]

The peak year of activity appears to have been in mid summer 1960 when the number of AWRE scientific staff involved rose to 115.[44] Aldermaston planned about 100 Tims rounds and 12 Vixen detonations in that year.[45] In fact, we can have some idea of the importance of the Vixen B trials from the importance at Aldermaston since such trials carried the same level of priority as the Red Snow project in the workshops, stores contracts and the central services.[46] The Foreign Office continued to fret that if details of these trials leaked out they might complicate efforts in Geneva. In April 1960 the Foreign Office advised the Ministry of Aviation that given the Prime Minister's and President's proposals announced in Camp David

---

38 TNA DEFE 7/2380, V.H.B. Macklen to Sir R. Powell/Sir F. Brundrett, NTPC Meeting, 4 December 1958; see Hansen, *U.S. Nuclear Weapons*, pp. 150–54 for details of the US weapon.

39 TNA AIR 20/1080, Design Changes Related to the Introduction of the Red Snow Warhead, 9 February 1959.

40 TNA AVIA 65/1249, Appendix to AC (59) 88, 1959.

41 Hawkings, *Keeping the Peace*, p. 42.

42 TNA AVIA 65/1064, Notes of a meeting on Inspection of Atomic Weapons, 19 June 1959.

43 John Simpson's supposition in the second edition (1986 p. xxiv) of *The Independent Nuclear State* that these trials, 'may have been as significant in the development of British nuclear weapon designs as the full-scale explosions' was clearly correct.

44 TNA CAB 21/6038, CRO telegram no. 1148 to Canberra, 1959.

45 TNA ES 12/272, MOD agreed that 1.4 kg of plutonium could be taken from military stocks in 1959 and 1960 as the minor trials were an essential part of the military programme. DEFE 16/361, F. Brundrett, MOD to E.S. Jackson, Ministry of Supply, 26 May 1959.

46 TNA ES 1/961, I. Maddock, HFE to Admiral P.W.B. Brooking, ADD, Master–Salt Cellar-Vixen II-Vixen B, 7 March 1960.

(see Chapter 5), that if the situation arose, there might then be a need to interrupt or delay these trials if the events in Geneva seemed to require it.[47]

Manufacturing safety was also a crucial consideration. There were some doubts expressed about the Single Point Safety (SPS) of the HE in the Red Snow warhead to be built in 1960–61 and 1961–62.[48] SPS was to replace the mechanical safety mechanisms used in the first generation of weapons such as the gauntlet containing the fissile material used for Blue Danube and Red Beard. These were kept separate from the HE until it was time for the bombers to take off (so too was the metallic ball in the core of Violet Club) to prevent accidental implosion. There was, however, no UK data and instead US trials data for SPS had satisfied Dr Nyman Levin, Director AWRE, as being applicable for Red Snow. However, if there were such doubts then the Royal Ordnance Factory at Burghfield would simply refuse to manufacture the weapon. The version to be built was going to use US HE, even though the Minister of Aviation wanted to use British explosives.[49] Levin himself had to state categorically that there were no safety doubts about this issue in order to remove any lingering doubt.[50]

Non-nuclear Testing in the UK: The Porton Range

Implosion tests also took place on the ranges at the War Office's Chemical Defence Experimental Establishment at Porton Down.[51] Ministers rejected an original proposal for conducting safety experiments at Porton in the 1950s involving small quantities of radioactive material.[52] However, this firing programme included actual weapons components except for the radioactive ones; UK regulations and policy prohibited the use of HE and radioactive materials, which of course was why these trials had to take place in Australia. The Lord President had ruled in October 1956 that he could not agree to the holding in the UK of trials that would

---

47 TNA DEFE 16/361, H.C. Hainworth, Foreign Office to Captain F.B. Lloyd, Ministry of Aviation, 4 April 1960.

48 SPS means that striking there must be no possibility of any nuclear yield because of such a hypothetical accidental detonation of the HE supercharge at a single point.

49 TNA AVIA 65/1249, Need for Nuclear Weapon Safety Experiments Brief on N.T. (59) 3, 5 January 1960.

50 TNA AVIA 65/1249, N. Levin D/AWRE to Sir Steuart Mitchell, Controller of Guided Weapons and Electronics, Ministry of Aviation, Red Snow – Single Point Safety Paper, NT (59) 3, 5 January 1960.

51 TNA AB 16/3680, Material dictated by Mr Donnelly, Aldermaston 16 March 1961. A Mr W. Shearing, Corn, Forage Merchant and Haulage Contractor, Grateley Station complained of the noise and the damage to his property!

52 Arnold, *A Very Special Relationship*, pp. 203–4.

result in the local dispersion of radioactive materials, even under conditions where the maximum permissible level of radioactivity would not be reached.[53]

At Porton, steel or brass simulated radioactive components. The firings were necessary to provide data on the detailed behaviour of the explosive parts of weapon components, and study possible causes of malfunctioning or inefficient functioning. These trials were essential in adapting nuclear weapons to the means of delivery essential to maintain the deterrent. AWRE faced many design problems that required the full charge weight, or where a substantial part of full weight was necessary. The Porton trials involved explosions confined on the average to around 10 lb shots, but occasionally they went as high as 90 lb when meteorological conditions permitted.[54] This gives us a startlingly good insight into the quantities of HE used in third generation UK nuclear weapons. It may be no more than conjecture given the timings of these trials, but these experiments could well have been connected with the development of Super Octopus (a new multi-point implosion system), which the UKAEA was very keen to test in the autumn of 1961 – see Chapter 7.[55] At that stage, this was being developed mainly with one warhead in mind: Skybolt.[56] This was the planned new air launched missile for the RAF Mark 2 Vulcan bombers.

Assessment Tests and the Test Ban: 1960–1961

Officials continued to keep the relationship between these experiments and the Conference on the Discontinuance of Nuclear Weapon Tests well to the fore. In the summer of 1960, the Nuclear Requirements for Defence Committee (NRDC – formerly the NTPC) was considering requirements for maintaining the facilities at Christmas Island and Maralinga.[57] Even though it was unlikely that the Geneva test-ban negotiations were either going to fail or succeed, any measures related to safety experiments meant that special care would be needed in their presentation so long as it was in progress. The Cabinet Defence Committee agreed on 31 December 1960 that the UK should continue to ensure that safety experiments were not prohibited by a test-ban treaty, possibly by the UK tabling a Treaty Article at Geneva defining nuclear explosions in such a way that safety experiments, which might result in a

---

53 TNA AVIA 65/1699, C.H. Latchford DG/G to US/ROF L&G, 22 July 1959 and DEFE 16/599, M. Allen to Mr Cook, Atomic Weapon Tamper Movement Trials, CH (56) 141 10 October 1956.

54 TNA AB 16/3680, E.W. Cross to Mrs M. Deacon, 14 March 1961.

55 See fuller discussion of Super Octopus in Chapter 7.

56 TNA AVIA 65/1836, Super Octopus, A.S./A.W., 20 November 1961, Policy Meeting of the NRDC 21 November 1961 Super Octopus, A.S./A.W., 20 November 1961.

57 TNA CAB 21/6038, Future Installations at Christmas Island and Maralinga, N.D. (60), 1st Meeting, 28 July 1960.

small nuclear reaction, could continue.[58] However, by early 1961 neither the UK nor US had tabled such an article. Further consideration of the problems involved had led the UK to doubt the wisdom of tabling an Article at all.

Officials considered that the risks of exposure in an attempt to introduce Treaty language masking the UK's real purpose were not worthwhile. They did not want to admit the real reason for the exemptions and give the Russians an excuse to make propaganda claims about ulterior Western motives. Instead, officials thought that it would be better to carry out safety experiments without publicity and to maintain, if publicity occurred, that these safety tests were not explosions of test nuclear weapons. Lord Home, the new Foreign Secretary, wanted authority, assuming Macmillan, the Minister of Defence and Sir Roger Makins (Chairman, UKAEA) agreed, to try to persuade the Americans from persisting in their wish to see such a definition in the Treaty. Macmillan, Watkinson and Makins all agreed that to raise the matter with the Americans, although the Minister of Defence doubted that they would want to give up ideas for a definition of the term.[59] The firm preference was that the correct phrase for the treaty's definition article was 'nuclear weapons test explosion'; the term 'nuclear weapon(s) tests' was not be used as it could be taken to include development of nuclear weapons other than by nuclear explosions.[60] The Americans eventually told the UK delegation in Geneva that the US did not intend to press their views on the definitions article, or to take any action on it for the time being. This was of course a satisfactory outcome for London.[61] However, this definitional question ceased to be an issue as far as the Conference on the Discontinuance of Nuclear Weapon Tests was concerned. As it transpired the definition problem, what was a nuclear explosion, was never resolved during the 1958–62 negotiations; nor is there a definition in the Comprehensive Nuclear Test-Ban Treaty eventually agreed in 1996.

Future Warheads, Testing Requirements and UK-US Collaboration: 1959–1963

Duncan Sandys minuted the Prime Minister on 9 January 1959 to set out what officials had recommended for future testing requirements.[62] AWRE did not need

[58] TNA CAB 131/23, Cabinet Defence Committee D (60) 14th Meeting, Item 2 Nuclear Weapons Safety Experiments, 31 December 1960.

[59] TNA PREM 11/3588, P. de Zulueta to A.C.I. Samuel, Foreign Office, 3 February 1961; Roger Makins to Prime Minister, 6 February 1961; and Watkinson to Prime Minister, 10 February 1961.

[60] TNA FO 371/15077, IAD 22/13, Nuclear Test Treaty Notes for final Editorial Revision of Tex, 25 January 1961.

[61] TNA FO 371/157082, IAD 22/92, D.N. Brinson, UK Delegation Geneva to H. Shepherd, AEDD, Foreign Office, 21 March 1961.

[62] TNA DEFE 7/2380, Nuclear Tests, 9 January 1959.

further megaton tests until the spring of 1960, and no kiloton test would be required until autumn 1960. Certainly, at the end of 1958 MOD officials felt that there would be a very strong requirement for a megaton test based on US designs to establish confidence in their efficacy. Jackson, DGAW in the Ministry of Supply, also thought that the UK should have at least two service tests of Red Beard.[63] AWRE had noted at the same time, in relation to the Army's Blue Water surface to surface missile, which required a warhead of about 75 kilotons – with an associated alternative of ten kilotons, that no warhead other that Tony could be offered unless further nuclear tests were permitted.[64] Officials recommended that the UK should discuss with the US the possibility of joint testing facilities. It would also be necessary to discuss the future use of Maralinga with the Australians; and most significantly, that the UK should hold a series of assessment trials at Maralinga in March 1959 in order to test the HE implosion systems for UK service weapons.

AWRE, reflecting on the critical value of its collaborative with the US, noted in 1960 that the pressure in the US for resumed testing was growing in view of the possible breakdown of the Geneva negotiations. It was thus important for the UK to decide what it needed to do. By that stage, it had been UK policy to contribute, however small, to every field of weapon development (except multi-megaton) to ensure that the UK was kept fully informed of US developments. Christmas Island staffing and facilities had been reduced to a level compatible with retaining the capability to resume megaton tests at about 18 months notice – as approved by Macmillan in July 1959. This decision assumed that for planning purposes the UK would conduct no further nuclear tests and would only reconsider if the Geneva conference failed to reach a comprehensive agreement for the cessation of all tests. Ministers and officials gave no publicity to this change. The NRDC noted in July 1960 that proving tests might be necessary for some of the warheads under the then current warhead programme intended to form part of the weapons systems available to UK forces over the next five years.[65] For Aldermaston the results of nuclear tests directly related to development of weapons. Without US collaboration the UK ran the risk of exclusion from actual weapons test information, which was vital to the development of new weapons. Therefore, AWRE argued that the UK should establish a programme sufficient to maintain this collaboration and justify

---

63 TNA DEFE 7/2380, Sub-Committee on Atomic Energy Grapple M and Lighthouse, Note by the Director General Atomic Weapons, 1958.

64 TNA AVIA 65/1041, Warhead for Blue Water, Notes of a meeting, 23 November 1960. Tony was at that time intended as a common warhead capsule for Blue Water, Seaslug Mark 2 and Bloodhound Mark 3. The design weight of Tony (RO 106) was 170 lb, but 200 lb had been allowed in the design to cater for a possible higher yield warhead. P.G. Wykeham, Acting Assistant COAS (Operations) to the AOC Coastal Command, RAF Northwood, 7 August 1959. There were plans at this Stage to cancel the ten kiloton yield warhead.

65 TNA CAB 21/6038, Item 1 Future Installations at Christmas Island and Maralinga, N.D. (60) 1st Meeting, 28 July 1960.

the UK being kept fully in the picture of US developments as they occurred.[66] Levin suggested to AWRE's Weapon Development Policy Committee that a UK contribution to the US effort could consist of provision of a team to work in the US test facility supplemented as soon as possible by the establishment of a UK test facility, which should be underground, with even tests in outer space being contemplated. Unfortunately for the UK the US would not accept this.

Levin said that notwithstanding the need for close collaboration with the US, there were many reasons for the UK to mount its own weapons testing programmes. First of these was the fact that for many cases such as different materials, methods of manufacture and standards, the UK had to make changes in US weapon designs. If it were possible to test precisely the effect of UK modifications, UK designs could be more economic and more efficient. Second, the UK had often been able to influence the US designs and practice by British knowledge and techniques and these had to be proved independently if they were to be accepted by US experts. Thirdly, in addition to the major weapon tests, AWRE had too many tests to carry out so it could not expect to secure any space in the US programme. Finally, once the UK had its own test facility, it could reasonably assume that it could carry out tests more speedily than in the US. It is not clear what happened to these ideas, but there were no further efforts to explore the feasibility of underground tests in Australia and Christmas Island, where in any case the geology was unsuitable. Furthermore, the Cabinet Defence Committee's reassessment of the numbers and types of nuclear weapons required that took place in 1962 given the Treasury's requirements for substantial cuts in the defence and nuclear programmes would have been a key factor.[67] Nonetheless, AWRE testing plans and their critical relationship to the long-term future of Aldermaston and the Anglo-American nuclear relationship resurfaced in 1964, as we shall see later in Chapter 9.

The UKAEA submitted proposals for future tests in a memorandum in the summer of 1962 to the NRDC. This Committee discussed this at its meeting on

---

66 TNA ES 1/1140, N. Levin Nuclear Test Policy, Paper for Weapons Development Policy Committee, 1961.

67 TNA CAB 131/27, Cabinet Defence Committee D (62) 7th Meeting Item 1 and Cabinet Defence Committee D (62) 33, Nuclear Weapons Programme Memorandum by the Chief Secretary to the Treasury and Paymaster General. 4 June 1962. This latter memo called for large and speedy reductions at Calder Hall, Capenhurst, Chapelcross and Aldermaston. The need for economy was acute and urgent. Although accepting the Minister of Defence's proposals for cuts in the kiloton weapons programme (cancellation of the ten kiloton warhead for Blue Water and the missile itself along with a proposal for an artillery shell for the army he felt that the megaton programme cuts did not go far enough. (This decision was proposed during the seventh meeting, the minutes of which are still closed under Section 3 (4) of the Public Records Act 1967), He proposed that this should be reduced to 45 Mark II Vulcans armed with 45 Skybolt missiles. The Defence Committee however decided in July to acquire 90 Skybolt missiles, but also to defer development work on the high yield Red Beard replacement – see TNA AB 16/489, Future of Capenhurst. Future Weapons Programme Defence Committee Decision H.C. Hudspith, 12 July 1962.

17 July 1962. Aldermaston identified three generic categories of tests: those required for improvements in, or adding to, the knowledge of existing weapon types, the development of new weapons and advances in scientific knowledge. The Ministry of Aviation was most interested in the first two categories.[68] For the Ministry the logical approach seemed to demand a review of current warheads and of the test firings undertaken with them. These warheads comprised Green Grass, the large fission warhead in Yellow Sun Mark I and Red Snow in the megaton range and Red Beard in the kiloton category. Green Grass and Red Beard were closely related to devices tested in UK trials in the 1950s. However, Yellow Sun Mark I was scheduled for withdrawal in the near future and both Red Beard and Red Snow, which was currently used in Yellow Sun Mark II and would be used in Blue Steel, were expected to be in service for not more than three to five years. The intention from 1965 onwards was to base UK weapons for the RAF and Royal Navy on RO 106, Cleo and RE 179.

RO 106 was a British version of the US Tsetse warhead, tested successfully by the Americans and intended for use in Blue Water and Seaslug Mark II. It differed from the US device in that it used British explosive EDC 11 and not the US PBX 9404, which had high graze-sensitiveness and which, when once initiated, usually proceeded to detonation. In contrast, RO 106 was single point safe with a yield estimated at eight and a half kilotons, give or take three-quarters of a kiloton. RO 106 planned as the primary for the Skybolt warhead, but officials decided that its yield was not adequate to ensure reliable functioning of the secondary thermonuclear stage. Cleo, on the other hand, represented an innovation since it used the new principle known as the Super Octopus system.[69] In the form tested, it was comparable to RO 106 in yield, but was more compact and had greater developmental potential, presumably in a range of new weapons. It was also more robust and less critical regarding tolerances in manufacture and assembly. It was also thought adaptable as a successor to Red Beard, if it were decided that this second-generation fission weapon would be restricted to a nominal ten kiloton yield.

RE 179 (Jehu, Klaxon) initially took the place of RO 106 as the warhead for Skybolt, with a variant proposed for the higher yield successors to Red Beard for the RAF. This warhead was a British version of the American XW 59, which the US had tested successfully. This may have been one of the US devices tested at Christmas Island in 1962 and to which the UK had access to design and test results information. However, this warhead employed Tsetse, tested as early as 1958, as the primary and since the UK version RO 106 had a somewhat smaller yield, this would be inadequate to ensure reliable functioning as a primary in a two-stage weapon such as Skybolt. The most attractive solution to AWRE involved development of the Super Octopus principle using a larger HE supercharge than

[68] TNA AVIA 65/1771, Memorandum by the Atomic Weapons Research Establishment Brief for meeting to be held on Tuesday, 17 July, 1962 DGAW, 12 July 1962

[69] National Archives closed file, open description TNA ES 11/16 OCTOPUS: possible implosion system to replace baratol and air lenses 1957 Jan 1–1957 Dec 31.

the device tested in the underground shot in Nevada codenamed Pampas in 1962 – see discussion in Chapter 7. Although easily enough done, any changes to the Pampas device had to be kept within limits imposed by the size of the warhead case and the requirements of single-point safety. In the event of higher yield successors to Red Beard, a downgraded version of RE 179 could serve as an alternative; these modifications would involve a reduction in the quantity of fissile materials in the secondary, but would not concern the primary.

After 1965 the UK nuclear armoury would be reduced to Skybolt, the Red Beard successor (WE177 with the Cleo warhead – a device using Super Octopus) plus smaller numbers of Blue Water and Seaslug Mark II, the latter being essentially tactical weapons based on RO 106.[70] The stockpile would therefore be dependent to a high degree on the device adopted for the primary for Skybolt, and which was also going to be used either as the primary in a high yield warhead or if the yield of the latter was to be limited to ten kilotons, as the sole nuclear device in that weapon. It seems that the DRPC (AES) had thought in 1960 that there was a need to take a new approach to warhead development aimed at reducing the number of different options. DRPC (AES) members thought this desirable because otherwise AWRE would be using a large effort to achieve marginal improvements. One possibility was to use the same primary for all warheads and the Committee asked the Ministry of Aviation to prepare a note on the financial savings and operational limitations involved in such an approach.[71] Since it was unlikely that UK HE safety standards would change, there was no fully approved nuclear device available for these purposes. However, the Ministry of Aviation noted that there were good reasons for supposing that a slightly enlarged version of the design tested at Pampas would serve these purposes adequately. However, since so much depended on this single device, the Ministry of Aviation were unable to see how they could give convincing assurances of performance to the Chiefs of Staff without at least one underground test.

For the Ministry of Aviation, the question of additional underground and atmospheric tests to guide the development of new weapons was complex. A

[70] TNA AVIA 65/1771, Brief for meeting of Nuclear Requirements for Defence Committee, 31 May 1962. This noted that there was a proposal to cut the programme for tactical weapons to 334 Red Beards, by omission of all versions over ten kilotons. In a brief for the Tuesday 17 July NRDC meeting the secretaries noted that one of the warheads for UK weapons post 1965 was RE 179. This was part of the MOD review of megaton and kiloton programmes. Although the paper's recommendations were drawn on intelligence and strategic studies, its immediate motive was to save £130 million or so on the defence budget for 1965/66. The proposal was to reduce the number of Skybolt missiles and warheads from 170 to 100, 70 being the absolute minimum. The corresponding figure for warheads was 158, 90 and 60. The strategic assumption was that ability to inflict 50 per cent damage on 15 Russian cities was credible for up to ten years ahead. All Army weapons – the portable ADM, High Yield Blue Water and nuclear shells were cancelled.

[71] TNA AVIA 65/1116, D.R.P. (A.E.S)/M (60) 1 Meeting, Item 4 O.R.1176 Kiloton Warhead and O.R. 1177 An Improved Kiloton Bomb, 12 July 1960.

lighter weight warhead for Skybolt offered important operational advantages – longer range and/or decoys or other countermeasures. Recent joint US-UK studies had emphasised the attractions of very lightweight high kiloton devices, but both would require underground and atmospheric testing programmes. It was clear that unless the UK began development of new weapons, its capabilities would deteriorate rapidly. US advice and assistance would only be provided if the UK could present evidence of a 'need to know' based on an established UK military requirement, and of technical competence and activity in the field concerned. As we shall see, this factor became a compelling problem for the UK in the late mid and late 1960s. Therefore, the Ministry of Aviation went to the NRDC meeting determined to press strongly for an underground test in 1962 of the Cleo (Super Octopus) device upon which the greater part of the future UK nuclear stockpile would depend after 1965. If HMG decided to accept the principle of further tests in 1963 or later, there were also strong technical grounds for embarking on a moderate series of underground and atmospheric tests. Note that even in mid 1962, admittedly with the Geneva talks going nowhere, there was still support in some quarters of the British nuclear policy community for further atmospheric testing.

By October 1963, following the Partial Test-Ban Treaty, officials in the Ministry of Aviation worked on a paper setting out future UK nuclear weapons tests and experiments. The NRDC considered this on 30 October. Approval existed for the development during the 1963–68 period of a low and high yield warhead (WE177 A and B) as well as Polaris, offered to the UK in 1962 following the US cancellation of Skybolt. Both WE177 types would use the British designed Super Octopus implosion system tested at Nevada in 1962. The necessary safety and other experiments on the design took place at Maralinga in March–April 1963.[72] Since atmospheric tests were now effectively out of the question, and since Christmas Island was not suitable for underground tests and that the Australians probably would not agree to such tests anyway, the only choice for the UK, if tests were needed, was in Nevada. However, the Ministry still recommended that the UK should at least retain Christmas Island for the time being as a possible fall back should national trials be required in future. Minor experiments would also be needed, although it was difficult to predict exactly how many and what types, and if the Vixen B trials were conducted in boreholes, then it would be possible to state clearly that they were fully compliant with the Partial Tests Ban Treaty.

[72] TNA AVIA 65/1771, Nuclear Requirements for Defence Committee, Future of United Kingdom Nuclear Weapon Test and Experiments (Memorandum by the Ministry of Aviation), N.D. (63), 11 October 1963.

Conclusions: The Testing Moratorium and Its Constraints on UK Weapons, 1958–1963

Testing constraints in this period do not appear to have unduly harmed the development of the UK weapons programme; the key assessment trials went ahead in Australia unconstrained by the moratorium, or by any language agreed in the Conference on the Discontinuance of Nuclear Weapon Tests. It does appear however, that future requirements may well have obliged the UK to reconsider its voluntary moratorium. Fortunately for the UK the absence of any prospect of imminent success in Geneva and the Soviet resumption of atmospheric testing meant that it did not have to face painful defence and diplomatic decisions. In any case, the Pampas underground nuclear test in 1962 determined the design of the next generation of UK weapons. The reduced warhead programme agreed by the Cabinet Defence Committee in 1962 also removed any immediate need for new underground nuclear tests directly associated with particular warheads for particular delivery systems.

Chapter 5
Diplomatic Pressures and the Testing Moratorium: 1958–1961

Introduction

By the end of September 1958 the UK had as a result of its own testing efforts acquired sufficient new scientific information to be able to accept a testing moratorium for a year subject to review. In doing so, it had managed to resist considerable domestic and international pressure against continued atmospheric testing, although clearly feeling discomforted in doing so. Moreover, UK efforts had now unlocked the door to US data following conclusion of the Agreement for Cooperation on the Uses of Atomic Energy for Mutual Defence Purposes in July 1958 and the first UK-US meetings on exchange of warhead design and production data in the late summer and autumn of 1958 led to a cornucopia of new information. The UK could now turn to negotiations for a comprehensive nuclear test-ban treaty with the Russians, knowing that the major needs of its weapon programme had been addressed. What we need to establish now is just how committed the UK was to a test ban in the period 1958–61 and how it handled the negotiations in the Geneva Conference on the Discontinuance of Nuclear Weapon Tests, and perhaps more importantly how it worked with the US during these negotiations. We only need to highlight a few of the key issues and developments to do this, primarily the verification problem – detection and identification of underground seismic events as nuclear explosions. We need to see if there is any direct or indirect evidence for any or potential adverse impact on the UK's own weapons programme because of the test ban negotiations and whether there was any internal opposition to pursuit of a comprehensive test-ban treaty.

Conference of Experts to Study the Methods of Detecting Violations of a Possible Agreement on the Suspension of Nuclear Tests: Summer 1958

The first series of negotiations in Geneva on a comprehensive ban on nuclear explosion testing among the then nuclear weapon states (UK, USA and USSR) began in 1958. Seismologists from a number of countries, including Canada, were included in the technical delegations of the 1958 'Conference of experts to study the methods of detecting violations of a possible agreement on the suspension of nuclear tests'. The experts agreed that underground nuclear explosions in the range one to five kilotons could be detected and identified if seismograph stations were established in 170 land-based control posts. Although later studies revised these initial recommendations, research on the appropriate type of seismograph facilities began almost immediately.

This technical work cleared the way for more formal treaty negotiations. On 22 August 1958, the day after the end of the conference Eisenhower issued a statement welcoming its outcome and declaring that the USA was, 'prepared to proceed promptly to negotiate an agreement … for the suspension of nuclear weapons tests and the actual establishment of an international control system on the basis of the Experts' report'. HMG, however, would have preferred to take more time to study the report before making any formal announcement, but judged that the balance of advantage lay in issuing a parallel statement immediately. Although this said that the Experts' Report provided an essential technical basis on which to build further progress, it did not commit the UK to negotiate an agreement on the basis based on the report.[1]

UK thinking however held that a test ban would bring a number of important advantages. First, the conclusion of a treaty would show that it was possible to reach agreement with the USSR on an important and sensitive subject. Second, the existence of a treaty would act as a stabilising influence on East-West relations. Third, verification on Soviet territory would help break down physical barriers between East and West. Fourth, the functioning of verification to the satisfaction of both sides would provide a model for other negotiations in the disarmament field. Finally, a test ban would give a guarantee against the creation of fresh fall-out, which would end anxiety about its possible harmful effects.[2] It was also important at the outset of any negotiations for the UK and US to make clear the terms and conditions or prolonging their annual suspension of tests on a year-by-year basis that would run in parallel with test-ban treaty negotiations. These were progress towards the installation of an effective verification system and progress towards the adoption and execution of measures of real disarmament. Washington was at first reluctant to spell these conditions out, London worried that any failure to do so would give the Russians an excuse to blame the West for any subsequent break down.[3] For the British it was important that the conditions should be made clear and stringent to make it easier for the UK to resume testing if it had to.[4]

---

1 TNA FO 371/149306, IAD 22/370, Nuclear Tests Conference, History of the Conference from its opening on October 31st 1958, 6 October 1960.

2 TNA FO 371/149306, IAD 22/370, Nuclear Tests Conference, History of the Conference from its opening on October 31st 1958, 6 October 1960.

3 TNA FO 371/132686, IAD 122/26, Foreign Office telegram no. 7133 to Washington, 10 October 1958.

4 TNA FO 371/132686, IAD 122/26, Disarmament Policy Memorandum, Foreign Office, 15 September 1958.

'A Long, Dull and Quite Fruitless Meeting': The Conference on the Discontinuance of Nuclear Weapon Tests[5]

The conference eventually opened in Geneva on 31 October 1958. The UK delegation went into this Conference with a detailed brief setting out HMG's objectives. This contained 12 very precise requirements:

1. Form of the treaty: multilateral, incorporating detailed control provisions.
2. Scope of treaty: immediate ban on all tests possibly excepting underground tests of less than one kiloton and an international verification system as in Experts' report.
3. Duration: indefinite, subject to effective controls and link with disarmament.
4. Control Organisation: international organisation; conference with largely formal functions; control commission as the supreme directing body, non party and non veto on ordinary control operations; administrator and technical staff.
5. Detection and identification system: eventually worldwide; 170 control posts within four to six years (19–21 in USSR within three years); prompt reporting of data; routine and special aircraft flights. This was one of the more important original objectives: to ensure that there would be some kind of international or non-Soviet presence within the USSR.[6]
6. On-site Inspection: permanent teams (about 16 eventually); automatic despatch by administrator subject to negative vote of Commission; transport provided by control posts.
7. Voting in Commission: no veto on the functioning of control posts, despatch of inspection teams, reporting of violations; simple majority voting (two thirds for finance).
8. Staffing of System: 20 foreign technicians, ten host country personnel at each control post (or minimum ten foreign, 20 host country).
9. Regional headquarters. Three to four first stage, finally about eight; would supervise control posts and launch inspections.
10. Preparatory Work within two months of signature, a preparatory commission with detailed functions.
11. Peaceful Nuclear Explosions: no initiative to be taken.

---

[5] There was an argument over the very title for the conference. The Russians wanted cessation whilst the US preferred suspension; discontinuance was a British compromise proposal. Freeman, Britain's *Nuclear Arms Control Policy*, p. 80. The quotation comes from a reporting telegram from Geneva on another pointless meeting. TNA FO 371/157095, IAD 22/235 Geneva telegram no. 216 to Foreign Office, 19 May 1961.

[6] TNA FO 371/157095, IAD 22/256 A.W. Brooke Turner to Mr Shepherd, 15 June 1961.

12. Fourth countries: accession to be secured through phasing provisions and diplomatic and moral pressure.[7]

After four weeks of procedural argument, the Russians agreed to a discussion of substance and four draft articles for a treaty.[8] These dealt with the obligations to discontinue tests, the setting up of a control organisation, the components for the organisation and membership of the control commission. Progress was not rapid. When the US introduced its paper outlining a draft treaty, it was quite clear from the head of the Soviet delegation's reactions (S.K. Tsarapkin) that the Russians had neither considered nor intended to allow such a detailed system of control as envisaged by the UK and US. Given the control hurdles that were still to be addressed, it was clear to one participant at least that the Conference was not going to produce a final result in the near future.[9]

New Data: The 'Large Hole Theory' and MOD Reactions

Working with the US delegation presented problems for the UK. There were internal clashes within the US and it became clear that all elements in the US draft treaty had not always been agreed interdepartmentally or were even always very carefully studied. However, this gave the UK some chance to influence the US position. The Conference went into recess on 19 December. During the recess, the US had evaluated further data from their underground tests that showed that the conclusions of the earlier Experts' meeting were not valid; these had only been based on a single underground test. It now appeared to the US that the verification system previously contemplated for a reliable detection of tests down to five kilotons would probably only be effective down to 25 kilotons. This did not go down too well with the Russians who insisted that the conclusions from the Experts' meeting still stood. The Russians suspected an attempt to sabotage the progress already achieved.[10]

Penney met two US experts at the end of January 1959 to consider test-ban technical issues. He reported to Macmillan that the latest view from the US technical community was that it would now be possible to conceal from the verification system an underground nuclear explosion of about 100 kilotons. He added that the new line taken by the US that a change should be made from the 1958 comprehensive verification system to an agreement to suspend tests in

7 TNA PREM 11/2861, Geneva telegram no. 358 to Foreign Office, 13 May 1959.

8 TNA FO 371/140445, IAD 22/33, H.C. Hainworth to Sir F. Hoyer Millar, Nuclear Tests Conference, 9 January 1959.

9 TNA DEFE 19/5, R. Press, UK delegation to V.H.B. Macklen, MOD, 15 November 1958.

10 April Carter, *Success and Failure in Arms Control Negotiations* (Oxford, 1989), p. 49.

the atmosphere only.[11] Penney added that the Russians might not have realised how large an explosion could be concealed. If they did, and if it were really their intention to cheat, they could put the West at a serous disadvantage by accepting the verification system recommended by the 1958 Experts' Meeting. Having listened to Dr Harold Brown (Director Livermore Radiation Laboratory) and Dr Albert Latter (RAND) Penney thought that the Geneva conference would fail because the US had now 'got cold feet' because of the implications of the new seismic data.[12] There was also strong opposition to a Comprehensive Nuclear Test-Ban Treaty (CTBT) inside the Eisenhower Administration with USAEC Chairman, Lewis Strauss and the likes of Edward Teller and Ernest Lawrence from the Lawrence nuclear weapons laboratory arguing that the US needed more tests to develop new weapons.[13]

The result from low-yield underground events in the US series of tests in Nevada in October 1958 had only reinforced the position of the CTBT doubters.[14] They seized on the data to demonstrate why a verifiable ban was impossible.[15] But things were not that simple. Others – such as Caltech scientist Robert F. Bacher – took a less strident view, although recognising the theoretical possibilities. Noting that seismology was 'still in its infancy', he suggested that theories about evading test detection through decoupling 'held forth possibilities which could change the underground detection situation completely'. He was sceptical, however, of the 'complicated' decoupling theory that Latter had advanced because it would 'require an approximately 1 million-cubic meters hole for a ten kiloton explosion' at a cost of $2 to $4 million. Nevertheless, Bacher declared that, despite the 'Latter hole's' apparent impracticality, it 'has stood up against severe theoretical scrutiny'. The Americans did not want to reveal the 'large hole theory' to the Russians because it involved a good deal of information concerned with concealment, which

11 TNA PREM 11/2860, W.G. Penney to Prime Minister, The Control of Nuclear Tests, CH (59), 13 March 1959.

12 TNA FO 371/140449, IAD 22/97, H.C. Hainworth to Mr O'Neill, 26 January 1959.

13 G. Allen Greb, 'Survey of past nuclear test-ban negotiations', in Jozef Goldblat and David Cox (eds), *Nuclear Weapon Tests Prohibition of Limitation* (Oxford, 1988), p. 99; see also Carter, *Success and Failure*, p. 42. Powerful forces in Washington had fought a stubborn rearguard action within the Eisenhower Administration against any test-ban treaty before and during the Experts' Meeting. TNA DEFE 19/18, Sir Michael Wright to Lord Home, 29 December 1961. Eisenhower remarked that the greatest regret of his Presidency was his failure to convert the moratorium into a CTBT, Steve Fetter, *Toward a Comprehensive Test Ban* (Cambridge Mass., 1988), p 6. Glenn Seaborg noted that in matters of testing and test-ban negotiations, from the Eisenhower period forward the UK consistently endeavoured, often with success, to exercise a moderating influence on US policy. Glenn T. Seaborg, *Kennedy, Khrushchev and the Test Ban* (London, 1981), p. 114.

14 Greb, 'Survey of past nuclear test-ban negotiations', p. 99. For a detailed discussion of the Hardtack II results see Romney, *Detecting the Bomb*, pp. 105–20.

15 Carter, *Success and Failure*, p. 50.

may or may not be known to them and which would be of military value if the negotiations broke down altogether; nor did they want to hand over this data until the Russians had shown themselves seriously interested in discussing the technical details involved, which at that point they were not. The MOD's reaction was that if the new data now meant that the verification system described in the Experts' report had a threshold as high as 25–50 kilotons instead of five, then an inspection system that allowed 20 kiloton explosions and under to escape detection, then that would be militarily unacceptable. It could only be made palatable if the treaty contained a clause that permitted testing of low-yield devices on a controlled and limited basis.[16] Brundrett remarked that this had to be understood in relation to the increasingly important military application of low-yield weapons; yields of ten to 100 tons were now quite practicable and would be of the greatest importance in the firepower of a small professional army.[17] This is an interesting statement because we know that AWRE was unable to design fractional yield weapons at this time.[18] The MOD therefore concluded that the West should now only agree to suspend those tests detectable by an agreed control system even if this meant that small explosions were exempted.[19] The 'large hole theory' was to haunt the remainder of the negotiations and never really went away as an evasion possibility throughout the test-ban debates right up until the 1990s.

The Conference Resumes: 1959

When the Conference reconvened in 1959 from its Christmas recess on 5 January, the Russians were reluctant to go any further until decisions were reached on the main outstanding issues.[20] These were mainly concerned with the duration of the treaty. The Russians contended that the UK and US aimed at an elaborate verification organisation for the suspension of tests on a year-to-year basis only. It would, in their view, be pointless to set up such an organisation if the UK/US were to resume tests at short notice because the conditions for a year-to-year suspension were not observed. These conditions were progress towards measures of real

[16] TNA FO 371/132697, IAD 122/326, Brundrett to Sir Richard Powell, 15 December 1958.

[17] TNA FO 371/132697, IAD 122/326. In a later letter to F.A. Bishop at No. 10 Downing Street, 18 December Brundrett quotes ten to 100 kilotons, which must be an error.

[18] See later on in this chapter for the discussion of MOD and AWRE views on the provision of possible UK devices for the Seismic Research Programme.

[19] TNA FO 371/140434, IAD 21/13/G, Richard Powell, MOD to Sir Patrick Dean, 12 January 1959.

[20] TNA FO 371 140450, IAD 22/127, Geneva Nuclear Test-ban Talk, 31 January 1959. This paper was prepared for No. 10 and Ministers on the state of play in the negotiations: see TNA FO 371/140450, IAD 22/128, C. O'Neill to Mr Hainworth, 29 January 1959.

disarmament and installation of the verification system. On the former, the UK had made several approaches to the Americans about dropping this line. Lloyd had raised it with Dulles on 25 November 1958 for instance and noted that he thought that the Russians would probably still not accept the kind of verification system that the UK regarded as essential. Furthermore, exposing the Russians would help the government's public position, which as Lloyd pointed out was much more difficult than for the Eisenhower Administration.[21]

Although the UK and US had proposed a moratorium of a year's duration, the UK at least was ready to suspend tests for a good deal longer if necessary; Macmillan had made this point to John McCone of the USAEC in early November 1958.[22] London and the UK delegation in Geneva became more and more concerned about the lack of US flexibility on this matter as the winter progressed, especially as the Soviet Union was making concessions, which placed them in a much better public position. The UK delegation complained that it was difficult to keep in step with the US when its delegation seemed to have little authority with every word referred back to Washington.[23] The Prime Minister had sent President Eisenhower a message on 1 January 1959 urging flexibility on the duration of the test-ban question.[24] Ike replied that the US would now start an intensive review of the whole matter.[25] Eisenhower eventually accepted British advice after a painful interagency review in which the Secretary of Defense and the Chiefs of Staff opposed any change in the US position.[26] The military extracted a high price and insisted on new language on treaty duration. They now wanted to have the right to withdraw from the treaty at the end of the initial period of two years and then at the end of each subsequent year, not only on the grounds of there being a violation, but also because they did not consider that the verification system was operating properly.[27]

This was not the UK view and the delegation thought that it would be a non-starter with the Russians. Worse still, it would give the Russians a good case to put before international opinion that the West was manifestly being unreasonable. This was a recurrent theme in the UK position; concern with presenting the West in the best possible light in face of continued Soviet intransigence and

21 TNA FO 371/132694, Foreign Office telegram no. 8404 to Washington, 25 November 1958.

22 TNA FO 371/132694, C. O'Neill to Sir F. Hoyer Millar, 12 November 1958.

23 TNA FO 371/132695, Geneva telegram no. 756 to Foreign Office, 7 December 1958.

24 TNA PREM 11/2860, Foreign Office telegram no. 14 to Washington, 1 January 1959.

25 TNA FO 371/132695, Prime Minister's Personal Telegram Serial No. T8/59, Control of suspension of nuclear Tests 1958, Eisenhower to Macmillan, 3 January 1959.

26 TNA PREM 11/2860, Washington telegram no. 86 to Foreign Office, 12 January 1959; TNA FO 371/132695 IAD 21/5/G, Washington telegram no. 87 to Foreign Office, 12 January 1959.

27 TNA PREM 11/2860, Geneva telegram no. 61 to Foreign Office, 20 January 1959.

sheer bloody-mindedness. Macmillan, for example, made it clear to Dulles at a meeting in London on 5 February 1959 that if the conference came to the point of breakdown, the UK and US would have to be sure that it was clear to the world where the blame lay.[28] Equally, US attitudes did not help the cause: the UK was caught between a rock and a hard place. David Ormsby-Gore, Minister of State, was seriously concerned that the US might torpedo the Conference. If the UK turned down an agreement under pressure from the US, he felt that accusations of bad faith would be very difficult to refute.[29]

Macmillan's Personal Involvement

The Prime Minister had a close and continuing involvement in the minutiae of the test-ban debate. Macmillan's personal commitment to address the test-ban problem was of course to prove crucial in the long run.[30] Michael Wright observed that his repeated initiatives took place mainly behind the scenes and that this increased rather than diminished their effectiveness.[31] In January 1959 the Foreign Office told No. 10 that the UK negotiating position was made much more difficult by disputes between the different Agencies in Washington concerned in formulating US policy on nuclear tests. Some of the agencies were thought to be almost certainly opposed to stopping tests and were probably influential in Congressional and other circles. The State Department had considerable difficulty in harmonising the views of the Agencies, and still more in making their own more moderate views prevail. This had been evident in exchanges over the treaty duration article.[32]

Macmillan wrote to Eisenhower on 2 February 1959 to reiterate just how much importance he attached to the Geneva Conference and how much he hoped that it would prove possible for the West to reach an agreement.[33] There were risks and the main one was the handicap an agreement would impose on the UK and US ability to improve their nuclear deterrent forces. However, a treaty would have three advantages: it would reduce tension, hinder the spread of nuclear weapons and provide a pilot scheme for verification in other disarmament fields. Macmillan was prepared to run the risk and accept something less than perfect control. A sufficient deterrent was all that was required, but there would be no

[28] TNA FO 371/140435, IAD 21/35/G, A.E. Davidson to Mr Hainworth, Nuclear Tests Conference, 6 February 1959.

[29] TNA FO 371/140436, IAD 21/46, David Ormsby-Gore to Selwyn Lloyd, 10 February 1959.

[30] This is a key message in Oliver, *Kennedy, Macmillan and the Nuclear Test-Ban.*

[31] Wright, *Disarm and Verify*, p. 130 and p. 136.

[32] TNA PREM 11/2860, Geneva Nuclear Tests Conference, P.A. Rhodes, PS Mr Ormsby-Gore to P.F. de Zulueta, No. 10 Downing Street, 31 January 1959.

[33] TNA PREM 11/2860, Foreign Office telegram no. 1126 to Washington, Prime Minister's Personal Telegram Serial No T93/59, 20 February 1959.

question of compromising on essentials. Deterrence was as much a feature in UK thinking on disarmament as in its rationale for its nuclear weapons programme. In Macmillan's mind, the UK approach sought to nail the Russians down on the veto on the conduct of on-site inspections and to make their position in asking for a veto untenable. He intended to press Khrushchev hard during his forthcoming trip to Moscow on this very point. Eisenhower however still held the view that it would be folly for the US and UK to relax their position on effective inspection; he agreed about the importance of the precedence for other agreements, which made it even more important to stick by the old positions. He felt that setting an upper limit for inspections would lead to negotiations over numbers without agreement on the basic principle of on-site inspection.[34] In essence the US was not prepared to concede on such a core issue, but the US Secretary of State Christian Herter told Harold Caccia that if the Prime Minister found any sign of Russian movement on the veto of on-site inspections in his visit to Moscow, then the US would look again at UK ideas.[35]

Despite US views, Macmillan went ahead anyway and raised the quota idea in a tentative and non-committal fashion (no specific figures, only illustrative ones), and then told Eisenhower that he had done so.[36] It seems that that the essence of the idea was that each side should during a period of three years from entry into force, have the right to demand that inspections should be made in the territory covered by the treaty subject to an annual upper limit of X inspections to prevent the right being abused. The figure X would need to be defined and could be done so by establishing a relationship between what was technically and administratively practical from the point of view of the inspecting side and the minimum number of annual inspections which would create a sufficient degree of a risk for a violator to deter him.[37] Eisenhower's message that he would prefer it if the UK did not put this proposal to the Russians had only reached Macmillan after he had had his first meeting with the Soviet leader. Khrushchev's reaction and attitude seemed to justify an exploratory approach. However, scientific reasoning would have to underpin any quota system; it was never just a question of numbers. Macmillan and Lloyd had made clear to Khrushchev that if the despatch of inspection teams was to remain subject to a veto – the Soviet position – no one could be sure whether underground tests were continuing or not. An indication of the interest of the Soviet side could be seen in the fact that Gromyko subsequently pressed Lloyd hard to say what number of inspections he had in mind for the upper

[34] TNA PREM 11/2860, Prime Minister's Personal Telegram Serial No. T. 102/59, Washington telegram no. 515 to Foreign Office, 23 February 1959.

[35] TNA PREM 11/2860, Washington telegram no. 516 to Foreign Office, 23 February 1959.

[36] TNA PREM 11/2860, Prime Minister's Personal Telegram Serial No. T 104/59, Moscow telegram no. 348 to Foreign Office, 24 February 1959.

[37] TNA PREM 11/2860, Foreign Office telegram no. 941 to Washington, 10 February 1959.

limit. Macmillan emphasised that unless some agreed means could be found of ensuring that inspection teams could in fact be sent out to investigate the nature of a suspicious seismic event, there could be no agreement in Geneva.

Continuing UK-US Differences: Tactics and Substance

British difficulties with the Americans continued in March 1959. Eisenhower wrote to Macmillan on 9 March proposing to suspend the Conference well before the end of the month.[38] The President gave no clear reason why he wanted to do this, but he did say that the Americans needed the pause to enable Washington to clarify its thinking on some aspects of the verification system. US experts were working on the problem of identifying underground and high altitude explosions and their studies were near completion. The conclusions would require careful assessment as to their implications for the verification system agreed upon by 1958 Geneva Experts' meeting. Henry Hainworth in the Foreign Office suspected that those who wanted no agreement at Geneva – possibly the Pentagon and USAEC – had had time to get at Eisenhower since Dulles had disappeared from the scene because of his illness.[39] Worse still, Eisenhower's letter contained an assertion that the Soviet reception of Macmillan's informal quota proposals in Moscow appeared to be suggesting that the Russians interpreted his ideas as signs of Western readiness to compromise on the veto of any in-bound inspection. In Hainworth's view, any response would have to reject firmly the President's assertion. Acceptance of a recess would make the UK's public position rather tricky. A recess was only acceptable if all agreed on a clear date for resumption; to do this, Hainworth argued that it might be possible to smoke out the Americans as to whether they really wanted to continue with the test-ban negotiations.

Con O'Neill, Assistant Under Secretary of State, Foreign Office, agreed. He suspected that among those who had instigated the President's recess proposal were powerful elements in Washington that did not want a test ban. He thought that this group hoped that if the Conference was suspended, they could ensure that it never resumed.[40] Both Hainworth and O'Neill feared that the Russians would find it peculiar if there were no follow-up, within the Conference, to the quota ideas floated by Macmillan with Khrushchev. They would assume that that the UK had put these ideas to the Americans only to have them vetoed. Sir David

---

[38] TNA PREM 11/2860, Prime Minister's Personal Telegram Serial No. T. 148/59, Message from Eisenhower to Macmillan, 10 March 1959.

[39] TNA FO 371/140456, IAD 22/239/G, H.C. Hainworth to Mr O'Neill, Nuclear Tests Conference, 11 March 1959.

[40] TNA FO 371/140456, IAD 22/239/G, H.C. Hainworth to Mr O'Neill, Nuclear Tests Conference, 11 March 1959.

Ormsby Gore feared that O'Neill's fears were probably true.[41] The Foreign Office recommended to the Prime Minister that whatever the merits of the situation, he should accept the proposal for a recess, but only for a period of three weeks centred on the Easter holiday period.[42]

The UK reply went to Washington by telegram on 13 March.[43] Macmillan informed Eisenhower that although he would have preferred to carry on working, he was prepared to agree that the Conference might recess for a while. The recess should occur at a time when it would seem natural such as Easter. It would be critical to ensure that any recess was not misinterpreted by the public, and was incapable of being presented by the Russians, as revealing a UK and US desire to break off negotiations. Macmillan said that he thought both himself and Eisenhower both believed that if an agreement could be reached on acceptable conditions it would be of real advantage. And as recommended by the Foreign Office, Macmillan made clear that he did not think that his proposal to Khrushchev had led the Russians in Geneva to think that the UK-US were weakening their position on the veto of inspection. He stressed that he had emphasised to Khrushchev that he could not accept a veto on the despatch of inspection teams and that his ideas were designed to do away with the veto on inspection. The UK attached the utmost importance to effective verification in disarmament matters; and nor did Macmillan think that his suggestion had in any way compromised the prospects for further disarmament agreements. On-site inspections were to remain contentious throughout the entire negotiation.

UK Reactions to New US Seismic Data

Macmillan hosted a meeting at No. 10 on 16 March 1959 to consider the new seismic data's implications for the test-ban negotiations. The Foreign and Defence Secretaries were present along with Sir Richard Powell, Lord Plowden, Sir William Penney, Sir Frederick Brundrett, Sir Patrick Dean, Fred Bishop (Private Secretary in No. 10) and Con O'Neill. Macmillan concluded after discussion that it was probably now necessary to go for an atmospheric test ban only and with perhaps a ban for two years on underground tests, which would have to be unverified during which time further work could be done on seismic research. The defence officials and scientists usually argued against even a two-year on trust suspension, but the

---

41 TNA FO 371/140456, IAD 22/239/G, H.C. Hainworth to Mr O'Neill, Nuclear Tests Conference, 11 March 1959.

42 TNA FO 371/140456, IAD 22/239/G, H.C. Hainworth to Mr O'Neill, Recess of Nuclear Tests Conference, 12 March 1959, C. O'Neill manuscript minute to Minister of State, 12 March 1959.

43 TNA FO 371/140456, IAD 22/239/G, Foreign Office telegram no. 1483 to Washington, 13 March 1959. Wright argued subsequently (*Disarm and Verify*) that without this intervention the talks would have broken down. p. 136.

Minister of Defence supported Macmillan and that appears to have been the end of any opposition.[44] Penney reiterated his earlier view was that the Russians might not have realised how large an explosion could be concealed. If they did, and if it really was their intention to cheat, they could put the UK and US at a serious disadvantage in the further development of nuclear weapons by accepting the control system recommended by the Geneva experts and deliberately cheating.[45] In any event, further thought was required and these issues would not be put to the US the following week.

Camp David Meeting 1959 and Follow-up

Macmillan and Eisenhower met at Camp David on 21 March 1959. Macmillan reported his discussion with Khrushchev explaining his suggestion of a limited number of on-site inspections, but recorded that the Russians were still critical of the proposal because they thought it was a military espionage plan. The Americans were far from happy that Macmillan had trailed the quota idea with the Russians. It was at this point that the US side described the 'large hole theory' and its impact on US test-ban policy. As a result they had concluded that the US could only continue negotiating a comprehensive test-ban treaty in return for dramatic Soviet concessions.[46] They were not prepared to share this theory with the Russians yet. In these circumstances, Macmillan raised the question as to whether the UK and US should still pursue a test ban, abandon it entirely or pursue a new direction. Both sides could not accept any agreement without what Eisenhower called 'safeguards', by which he meant adequate verification that the USSR would not cheat. Macmillan emphasised the importance of a test ban in the interest of discouraging fourth power development (what would be known later as nuclear non-proliferation) and made something of a plea to reach an agreement even though it might not be fully 'safeguarded' as an essential step in the interests of human welfare. Countering the fourth power problem had been one of the original objectives that the UK and US had set for the negotiations in the summer of 1958.

Eisenhower agreed that the US would make available to UK experts the latest data and conclusions on the practicality of establishing effective control over nuclear tests underground or in outer space. Thereafter UK experts would go to Washington as soon as possible to reach an agreed Anglo-American assessment of the scientific position. The hope was to agree a line in time for the resumption

---

44 TNA FO 371/140436, IAD 21/47/G, C. O'Neill to Mr Ormsby-Gore, 16 March 1959.

45 TNA FO 371/140436, IAD 21/48/G, W.G. Penney to Prime Minister CH (59) 13, The Control of Nuclear Tests, 13 March 1959.

46 TNA FO 371/149306, IAD 22/370, Nuclear Tests Conference. History of the Conference from its opening on October 31st 1958, 6 October 1960.

of the Geneva negotiations on 13 April.[47] However, the US position in view of the scientific data was to aim for a ban on atmospheric tests only; such a ban could be monitored effectively without on-site inspection and in any case, the US had extremely important weapon developments in mind that could be brought to fruition in two or three years by underground testing.[48] Their preference was to continue the Geneva talks in such a way as to lead to an early breakdown on the veto of inspections issue and then offer either agreement on, or a unilateral declaration announcing, a permanent cessation of tests in the atmosphere only, but not in outer space.[49] However, even here there were still unresolved verification issues that needed further study.

Macmillan and Lloyd, while not challenging the technical conclusions, emphasised the importance of continuing in Geneva, at least for the time being. They pointed out the improbability of Soviet acceptance of a ban on atmospheric testing only and suggested that, even if the Russians resumed such tests, the UK-US position in relation to public opinion and in the United Nations would not necessarily be strong. In addition, Macmillan and Lloyd suggested that in the new circumstances, it might be better to change the Geneva talks' objective and to seek agreement that it should attempt to draw up a settlement of the differences between the sides for resolution at a summit conference.[50] However, the US indicated that it could not accept this proposal; namely that the negotiators set out the areas of agreement and disagreement to be dealt with at a summit conference. Eisenhower and Macmillan did agree that the US should offer the USSR two choices. Either it could accept the modification of its position on the veto on on-site inspection and on the discussion of high altitude controls as the price for the continued negotiations by the US of a comprehensive test ban, or it could agree to begin by negotiating a treaty prohibiting tests in the atmosphere up to an altitude of 50 km.[51]

Penney subsequently pointed out that given the information available to the UK, US and USSR, it was not necessary to test at full yield in order to prove a device and that a ten kiloton yield gave the weapons designer considerable scope for developing new weapons in the megaton range.[52] The MOD at that time was not so much concerned about the need to continue tests – though they noted that

---

47 TNA PREM 11/2860, Agreed Minute, Extract Anglo-American Talks: March 1959.

48 TNA FO 371/140516, IAD 27/5, Encloses two records of discussion at Camp David on the Nuclear Tests Conference, H.C. Hainworth to Sir Roger Stevens, 24 March 1959.

49 TNA PREM 11/2860, Agreed Minute, Extract Anglo-American Talks: March 1959.

50 TNA PREM 11/2860, Washington telegram no 782 to Foreign Office, 23 March 1959.

51 TNA FO 371/149306, IAD 22/370, Nuclear Tests Conference. History of the Conference from its opening on October 31st 1958, 6 October 1960.

52 TNA FO 371/140436, IAD 21/48/G, W.G. Penney to David Ormsby-Gore, Foreign Office, 20 March 1959.

there was no doubt that UK nuclear weapons could be greatly improved if AWRE were able to continue testing, even if it was restricted to small underground tests. The MOD did not want to be saddled for political reasons with an ineffective verification system. Such an outcome would set a poor precedent for disarmament and, worse, would leave the UK in a situation in which nuclear weapons were seen internationally as morally bad and that as such it was just as wrong to test them as to use them, thus undermining the whole point of the deterrent.[53]

UK and US officials met afterwards to discuss in further detail other alternative courses, but this revealed nothing new with both sides sticking to their respective positions. Officials nevertheless agreed to await completion of a joint scientific assessment of the issues before determining an Anglo-American position for the resumed talks in Geneva.[54] The only agreement reached was that there would be further scientific and diplomatic discussions prior to the 13 April resumption in Geneva. Penney and Sir Edward Bullard would come to Washington on 2 April for scientific discussions on the Berkner and Wolfgang Panofsky reports (both March 1959) – the former on finding ways to improve seismic verification, the latter on verification (satellite and ground based techniques) and concealment strategies.[55] Both reports were given to the UK and the US hoped that Penney would chair the meeting. The Berkner Panel had recommended amongst other things that practical experiment was needed to determine how large a de-coupling factors could in be obtained from the 'large hole theory'.[56]

Penney and Bullard duly met the Killian panel on 1 and 2 April. This had been set up to review seismology and underground explosions verification questions.[57] Continuing research to find new technical ways of overcoming these problems would be essential and might mitigate both Soviet annoyance and public

[53] TNA FO 371/140437, IAD 21/61/G, Richard Powell, MOD to C.D.W. O'Neill, Foreign Office, 31 March 1959.

[54] TNA FO 371/140516, IAD 27/4, Nuclear Tests minutes of a meeting with State Department, C.D.W. O'Neill, 23 March 1959.

[55] The Berkner report was commissioned by Eisenhower's Science Advisor, Thomas Killian to determine *inter alia* what was needed to upgrade the proposed Geneva system for detection and accurately identifying seismic events, natural and manmade. Bates, Gaskell and Rice, *Geophysics in the Affairs of Man,* pp. 130–31; See also Romney, *Detecting the Bomb*, pp. 121–30 for a detailed review of the work of the Berkner Panel. Dr Lloyd Berkner was President, Associated Universities Inc and a member of the President's Science Advisory Committee. Thanos P. Dokos, *Negotiations for a CTBT 1958–1994* (Lanham, 1995), p. 11; the Berkner report can be found at TNA FO 371/140436, IAD 21/50/G, Report of the Panel on Seismic Improvement, 16 March 1959.

[56] It also recommended that the numbers of seismometers should be increased at each control station from ten to 100; that surface seismic waves should be analysed to help distinguish earthquakes and that unmanned instrumented stations should be placed roughly 170 km apart in earthquake zones. Dokos, *Negotiations*, p. 11.

[57] TNA FO 371/140436, IAD 21/53/G, Washington telegram no. 893 to Foreign Office, 3 April 1959.

disappointment at the UK's inability to accept a comprehensive test-ban treaty at that time. Penney thought that the agencies would put a new position to the President that would accept a test ban on atmospheric tests up to 50 km height, but that underground tests should resume. He also agreed with the conclusion on the desirability of testing these hypotheses by experimental explosions, whether nuclear or not.[58]

In light of all of this the Foreign Office came to the view that the best UK policy would be to seek an agreement that exempted underground tests below a certain threshold. These would be stopped for two years only during which the possibilities of controlling them would be reviewed. This would enable the west to resume small underground tests after two years if no effective verification for them had been devised without withdrawing from the agreement. Such a course avoided the difficult situation that would arise if the West insisted on resuming underground tests at once and the Russians countered by resuming atmospheric tests. The public and United Nations would undoubtedly blame the UK a good deal for such an outcome.[59]

New Proposals Put to the Russians: April 1959

The UK and US eventually agreed to put two alternative proposals to the Russians when the conference resumed in Geneva on 13 April 1959. If the Russians agreed to abandon their insistence on the veto on on-site inspections and agreed to effective verification measures, then the UK and US were prepared to go ahead and sign a comprehensive agreement. If not, the UK and US would offer instead a ban on atmospheric tests up to 50 km under a simplified verification system of control with the hope that work could continue on the technical problems of seismic detection and identification.[60] The Russians accepted that there could be veto-free inspections, but the small number of inspections would be determined by purely political factors, not the technical relationship with the number of unidentified events required by the US and UK. However, they did indicate an interest in Macmillan's informal ideas on inspection quotas and the Russians went on to a make a formal proposal in Geneva based upon this new approach. (Macmillan had mentioned 12, ten, six, five and three per year or per two years and Lloyd subsequently referred to X inspections in discussions with Gromyko.[61]) Having considered the Soviet response, Macmillan wrote to Eisenhower to say that

---

58 TNA FO 371/140436, IAD 21/53/G, Washington telegram no. 911 to Foreign Office, 4 April 1959.

59 TNA FO 371/140437, IAD 21/60/G, H.C. Hainworth to Mr O'Neill, 6 April 1959.

60 TNA FO 371/14043, IAD 21/68/G, H.C. Hainworth to Private Secretary, 13 April 1959.

61 TNA FO 371/157113, IAD 24/54/G, David Ormsby Gore to Sir Michael Wright, UK Delegation Geneva, 12 June 1961. This issue came back to haunt the UK with the Russians

if they could persuade the Russians to accept a reasonable solution on the veto, it might be possible to narrow down the problematic issues to a single question of how many on-site inspections there should be in any one year.[62] On 13 April 1959, Eisenhower and Macmillan wrote to Khrushchev to propose that they agree an atmospheric ban only as an interim step since Soviet ideas on control were inadequate for an effective CTBT, which was still the President's objective.

Other Unresolved Technical Verification Issues and Experts' Work in Geneva

The acquisition of data from the USAEC's high altitude tests in the early autumn of 1958 in the South Atlantic made it possible to revise and expand the Experts' Report on the detection of high altitude explosions. By the end of that year the US suggested to the UK creation of a Technical Working Group for this purpose. This was proposed to the Russians in January 1959. In a letter to Khrushchev of 13 April, Eisenhower made Soviet participation in early discussion of such issues one of the three conditions for continued US negotiation of a test ban. The Russians eventually agreed and a technical group was established on 23 June 1959; it held ten meetings and presented an agreed report to the Conference on 15 July. The group recommended the use of four different methods of detection and identification at ground control posts and the use of four additional methods by satellites. All three participants accepted its conclusions. Sir Michael Wright said that the report was, 'a correct technical assessment of possible techniques for the detection and identification of high altitude nuclear explosions in light of the scientific knowledge available at the time of the Working Group's meeting'.[63]

The US position as it appeared to the UK in August 1959 was as follows. Ever since the emergence of the 'large hole theory' both the USAEC and DOD had been opposed to a comprehensive nuclear test ban. In these agencies' view, control of small tests underground was impossible and that they could not afford to risk cheating by the Russians who thereby would be able to develop new weapons, low yield ones especially, an area where the US believed it possessed a lead.[64] Less avowed but also mentioned from time to time had been the US desire to continue testing, albeit underground, for the perfection of the weapons that they had already developed. London had believed that the President's view was that even given

---

insisting that the Prime Minister had used specific figures, when in fact the Macmillan examples were only ever intended to be illustrative.

[62] TNA PREM 11/2861, Prime Minister's Personal Telegram Serial No. T. 224/59, Foreign Office telegram no. 2123 to Washington, 29 April 1959.

[63] TNA FO 371/149306, IAD 22/370. Nuclear Tests Conference, History of the Conference from its opening on October 31st 1958, 6 October 1960.

[64] TNA FO 371/140518, IAD 27/23G, Summary of our and US position as regards policy towards the suspension of tests, Hainworth to O'Neill, 5 August 1959.

the uncertainties a test ban would be well worth having. UK officials hoped therefore that in due course the Administration would opt for a test ban in which the underground problem would be dealt with by a quota of on-site inspections. However, given the changes in thinking, the US position now seemed to be only prepared to agree to no further underground testing for a fixed period except for those required to provide data to secure an effective monitoring system. But even this position did not last long and the US Secretary of State told Lloyd that the US had not in fact decided to suspend such tests. Lloyd strongly argued in favour of a moratorium to accompany any suspension of atmospheric tests. Macmillan suggested that the UK should put such a proposal to the Americans and that a two-year moratorium on underground tests could be used to give time to develop more effective detection and identification capabilities. The US did not bite. Ormsby-Gore commented that the UK should not allow the Americans to switch suddenly to a new course that could lead to a breakdown in the conference.[65]

Internal UK Doubts?

London also had its own share of test-ban sceptics. Although there is little indication of the sort of opposition to a test ban that bedevilled US policy formulation, there were some rumblings of discontent. We can see this from a paper from MOD and the UKAEA that indicated that in certain circumstances it would be militarily disastrous for the West to sign a CTBT preventing further underground testing.[66] This looked at the implications of the 'large hole theory' for the numbers of inspections that might be required to give a reasonable deterrent effect to cheating. This now concluded that about 50 superficial inspections plus ten full-scale inspections per year were the minimum figures that should be proposed. Such figures would give a 25 per cent chance of detecting a single evasion of an event of five kilotons or greater on the assumption of the use of the Experts' seismic system and assuming no large hole or other tricks. The chance of detecting a single evasion under these conditions for an event of two kilotons or greater would be 7 per cent.[67] The military concern was that the Russians would gain a real advantage if they could conduct tests of the order of ten to 20 kilotons without the serious risk of detection, which is what could happen if the decoupling theory was correct. In these circumstances the Russians could develop in time an effective anti-missile

[65] TNA FO 371/140518, IAD 27/23G, Summary of our and US position as regards policy towards the suspension of tests, Hainworth to O'Neill, 5 August 1959; Ormsby-Gore manuscript comment 6 August 1959.

[66] TNA PREM 11/2861, P. de Zulueta to Prime Minister, 25 May 1959. The paper appears to be retained under Section 3 (4) of the Public Records Act 1968.

[67] TNA CAB 130/130, GEN.604/10 Suspension of Nuclear Tests: Onsite Inspection of Underground Events, Note by the Secretaries, Note by the MOD and UKAEA, On Site Inspections of Underground Events, 12 May 1959.

missile, which would seriously affect the UK's deterrent's viability.[68] A quota system would be valueless in such a situation. It was therefore important that the UK should follow the US in assessing the importance to be attached to the 'large hole theory'. There were military disadvantages in accepting a plan that included tests at one kiloton and below because they could not be verified. There were definite military risks in accepting a plan that allowed only ineffective control below the five kiloton level. Any plan that provided for ineffective control up to the ten to 20 kiloton level might well have disastrous military consequences for the UK.

The UKAEA accepted that there was a theoretical possibility of cheating and developing an anti-missile warhead. Duncan Sandys, Minister of Defence, picked up this theme in a minute to Macmillan in which he argued that nuclear tests were needed for the development of anti-missile missiles and fractional-yield kiloton tactical weapons, both of which were vitally important for defensive purposes. AWRE had been confident in mid 1959 that it could produce warheads with yields in the range of ten to 100 tons for army use by 1962–63 without testing.[69] (This is odd because in 1959–60 AWRE said that it could not produce fractional yield weapons for the seismic research programme, but probably explained by the fact that it could not scale down Blue Danube reliably enough for use in calibration experiments.) These tests could be carried out underground or possibly in outer space where they would cause no harm to health and were then undetectable by any practicable method of inspection. If a ban were placed on such tests and the Russians failed to observe it, then the balance of military power might be seriously upset to Western disadvantage. For Sandys this meant that the UK had to reserve its right to carry out tests underground or in outer space until an effective system of inspection was devised.[70] Sandys and MOD experts had perhaps never wholeheartedly accepted the policy for a test-ban treaty, which is not to say that they adopted the same semi-public opposition to government policy that was so evident in the US. They had the same doubts as defence experts in Washington.[71]

In order to deal with the points raised in the Sandys' minute, a meeting was convened, but on the basis of a detailed paper setting out the state of play on the Geneva negotiations. The Foreign Secretary in response thought that there had been no change in UK defence requirements during the recent past that would justify an alteration in the UK position in Geneva. Nor did he think that the UK position on verification had materially weakened. The UK's objective ought to remain a degree of control and inspection under a satisfactory quota, which would

68 TNA CAB 130/130, GEN.604/11, Suspension of Nuclear Tests; On-site Inspection of Underground Events. Note by the MOD and UKAEA, 19 May 1959.

69 TNA AVIA 65/1116, D.R.P. (A.E.S.) /M (59) 1st Meeting, Item 1, Low-yield Nuclear Weapons for the Army, 28 July 1959.

70 TNA PREM 11/2861, Duncan Sandys to Prime Minister, 26 June 1959.

71 TNA PREM 11/2861, Bishop to Prime Minister, 29 June 1959.

constitute a real deterrent against the holding of a clandestine test.[72] Any change in the UK's negotiating position would be politically awkward; Macmillan agreed.[73] Macmillan, Lloyd and Sandys met again on 7 July to discuss further the Defence Secretary's minute and the state of play in Geneva. In Macmillan's view this meeting agreed that if the Americans were prepared to take the risks involved and agreed a comprehensive test-ban treaty, including underground tests, then the UK would have to follow and assist in negotiating such an agreement.[74] If the US were not prepared to take the risks, then the UK would have to agree with the US to try jointly to persuade the Russians to accept an agreement banning atmospheric, high altitude and underwater tests. They would then need to provide for further discussions on the effective verification and inspection of underground tests, which meanwhile would be subject to a moratorium for two years. If, as was likely, the Russians rejected this, the UK should urge the US to join the UK in making a declaration that they would follow this policy irrespective of anything the Russians did. It followed from this that if Eisenhower was undecided about the proper course of action, the UK should consider whether to make any approach to recommend a specific line. This would be a matter of tactics. London would need to bear in mind that UK pressure in Washington might be counter-productive. Sandys noted that his only point had been that the UK would need to be fully aware that a two year moratorium on underground tests might not be sufficient to find out how to deal with the 'large hole' problem. Consequently, the UK would face the awkward choice of either ending the moratorium on underground tests, or of maintaining it despite the absence of effective inspection.[75]

A New UK Option: A Moratorium and Joint Seismic Research

Macmillan decided that the UK should press the US for pursuit of an atmospheric ban plus a two-year moratorium on underground tests whilst joint seismic experiments on detection with the Russians would be conducted to improve verification capabilities. The MOD, however, were worried that a two year moratorium would give the Russians more time to cheat and therefore steal a march on the West. A one-year ban would be better.[76] The MOD did not think that further work would in fact show that effective inspection and control of underground tests

[72] TNA PREM 11/2861, Lloyd to Prime Minister, Nuclear Test Negotiations, 1 July 1959.

[73] TNA PREM 11/2861, Prime Minister's Personal Minute Serial No. M237/59, Macmillan to Minister of Defence, 1 July 1959.

[74] TNA PREM 11/2861, Prime Minister's Personal Minute Serial No. M263/59, Macmillan to Minister of Defence 12 July 1959.

[75] TNA PREM 11/2861, Sandys to Prime Minister, 16 July 1959.

[76] TNA PREM 11/2861, R. Powell, MOD to T.J. Bligh 10 Downing Street, 20 July 1959; P. de Zulueta to Prime Minister, 21 July 1959.

was feasible. No. 10 took account of MOD's drafting comments expect for the one-year moratorium for the Prime Minister's proposed message to Eisenhower, and on a telegram instructing the Foreign Secretary in Geneva to sound out Herter on this new approach.[77]

This radical idea, which appears to have emerged from an earlier meeting with Penney and Plowden, would entail joint underground tests in which the Russians could observe the assembly of nuclear devices and check the fissile components. In order to avoid Congressional problems over release of nuclear information, the UK would be prepared to provide its own nuclear devices of an obsolete pattern. Scientists could take samples of explosive products and detailed analyses of these would be compared to obtain an agreed value for the yields. Seismic recordings would be made at short and long distances from the test site. The Russians would be given every assistance to enable them to participate in such a trial. This issue, however, ended up on the back burner until 1960.

Release of New Seismic Data to the Russians; A New Approach to the Detection Problem

The Americans finally released two reports on the new seismic data to the Russians in June 1959, and although there was a cryptic reference to decoupling techniques, there was no explicit reference to the 'large hole theory'. Such was US ill-ease over the reliability of any detection system for underground tests, Washington was now moving firmly towards a ban on atmospheric testing only. In fact, the Americans wanted to resume underground tests themselves.[78] However, Washington also wanted to send a team to London to discuss the latest scientific findings on detection and identification.[79] UK and US scientists met in London on 10 and 11 August 1959 to discuss scientific research on the detection of underground tests.[80] State Department officials found these talks useful and helped to improve the chances of a more sober approach in Washington to the underground test problem and in particular to the 'large hole theory'.[81] There was substantial UK-US agreement on the analysis of the technical difficulties. The 'large hole theory' made it impossible to provide a control system capable of detecting underground tests. UK and US experts recommended a joint research

[77] TNA PREM 11/2861, Foreign Office telegram no. 558 to Geneva, 22 July 1959.

[78] TNA PREM 11/2861, Sandys to Prime Minister, 17 July 1959.

[79] TNA PREM 11/2861, Washington telegram no. 1697 to Foreign Office, 31 July 1959.

[80] Romney, *Detecting the Bomb*, p. 129.

[81] TNA PREM 11/2861, Geneva telegram no. 574 to Washington, 14 August 1959.

programme to tackle the problem.[82] Research was needed in nine separate fields.[83] Nevertheless, there were two principal difficulties. The research was unlikely to yield definite results in less than six months or even a year and the critical moment when the Russians would have to be told about the 'large hole theory' could not be postponed so long. Second, while the research programme might provide strong evidence of the validity of the theory, evidence of its invalidity would probably not be conclusive.[84] However, the UK felt strongly that it would be necessary to announce at the same time Western readiness to accept an unverified moratorium on underground tests along with an atmospheric ban for a period of two years, during which joint efforts would continue to work out an acceptable system for controlling a ban on underground tests. For both the UK and US it was essential for the West to make its position as acceptable as possible to public opinion at home and overseas. There were several options or variants on testing constraints that could be considered ranging from a complete ban to a partial ban with a threshold on underground tests.[85]

In the event Macmillan was advised by the Foreign Office to persuade the President to continue to search for a comprehensive test ban and to accept an agreed quota of inspections. If this was unacceptable, it was still important that Macmillan should try to persuade Eisenhower not to break the news about the 'large hole theory' and the atmospheric ban only in such as way as would encourage Khrushchev to break off the talks in Geneva. MOD officials were concerned that insufficient emphasis was placed on the risk that a moratorium was always likely to be extended under pressure of public opinion, thereby resulting in an indefinite

---

[82] TNA PREM 11/2861, Note on Nuclear Tests Conference, C. O'Neill, 20 August 1959.

[83] TNA CAB 130/130, GEN.604/16, 13 August 1959, The Technical Aspects of Detecting Underground Nuclear Weapons Test Explosions, Note by the Secretaries, Statement agreed at the end of the Anglo/US discussion held on 10–11 August. Joint research activities should include: Completing underground high explosive tests already planned; engineering studies of the practicality and cost of constructing large holes; consideration of problems of radiation transport and effects of x-rays and neutrons; planning for a series of large hole tests, including complete plans for nuclear tests consideration of small holes and possible large cavities other than underground, rapid implementation of the Berkner panel reports, on-site inspection of selected earthquakes in New Zealand and California, technical study of the design and feasibility of unmanned stations and continuing studying of new techniques for concealment and for detection.

[84] TNA FO 371/140518, C. O'Neill to Mr Hainworth and Sir P. Dean, 19 August 1959.

[85] TNA FO 371/140439, IAD 21/114/G, Michael Wright to Secretary of State, Nuclear Tests, 4 August 1959.

ban on underground tests without an effective verification system.[86] Macmillan's brief was changed to reflect this.[87]

Macmillan-Eisenhower: Chequers August 1959

Macmillan and Eisenhower plus their foreign secretaries met at Chequers on 30 August. Macmillan stated that in spite of all the technical findings he felt that there were still very real political imperatives in trying to find an agreement with the Russians on a comprehensive test ban. Ike agreed, but expressed real doubts as to whether this was achievable and emphasised how important it would be to begin with an atmospheric ban first.[88] He faced considerable political difficulties in persuading the Senate to ratify an agreement unless the balance of testimony indicated that the agreement could prove an adequate deterrent to Russian cheating. Macmillan pressed the advantages from the point of view of public opinion on the possibility of a longer moratorium on underground testing if the two sides could not agree in Geneva, but Ike said that the real danger of this approach would be that it would give the Russians what they wanted without any verification. The meeting left the issue inconclusive as a result. The President made no commitments to any further unilateral moratorium declaration from the US, but Macmillan stated that he might have to make one himself about no further British tests taking place. Given the Soviet statement that it would not be the first to resume, HMG would find it politically difficult to resume testing.[89]

Although there would be difficulties, Lloyd told Herter that he nevertheless thought that as much as possible of the agreed programme for further research should be carried out as soon as possible.[90] He added that if reliable verification of underground events was impossible, he thought that HMG would accept the present impossibility of a comprehensive test ban since there should be no disarmament without control. The joint UK-US meeting in August did subsequently agree that adequate deterrent control was out of reach largely on account of the 'large

---

86 TNA FO 371/140518, IAD 27/24/G, Sir Richard Powell, MOD to C.D.W. O'Neill, 21 August 1959.

87 TNA FO 371/14051, IAD 27/25/G, Nuclear Tests, Recommendation on Tactics, Appendix III, Arguments Bearing on A Moratorium, 1959.

88 TNA FO 371/140518, Nuclear Test Conference, Record of Conversation between the President, Prime Minister, the Secretary of State and Mr Herter at Chequers on 30 August 1959.

89 TNA FO 371/140518, IAD 27/29/G, Records of conversation during visit of President Eisenhower to UK, 27 August to 2 September 1959.

90 TNA PREM 11/5173, Record of Conversation Between the Secretary of States and Mr Herter at Lancaster House on 1 September 1959.

hole theory'.[91] A violator would have to proceed by steps until he was sure.[92] The Americans finally decided to tell the Russians formally about the large hole theory in mid October 1959, but this revelation was to take five weeks during which time they would also be told that no comprehensive test ban was possible, but they would not be informed as to what sort of limited agreement the West could offer. No. 10 was dismayed by this approach.[93] London feared that this would be taken as evidence of the Western powers' intention to break off the negotiations and resume underground testing in January 1960.[94] This would make the Russians appear in a much better light. The UK saw very little prospect of the Russians agreeing to a limited treaty unless it was accompanied by a moratorium on underground tests while further joint research on detection methods was being carried out. In contrast, the US was not at all comfortable with such a moratorium.[95] Foreign Office officials therefore advised Macmillan that it would be counter-productive to raise the moratorium question again with the US. Thus by autumn 1959 it was now very clear to the British, if had not been already, that the Americans were unwilling to accept a comprehensive test ban in the near future because such a ban could not be adequately verified in different geological conditions. This appears to have been the British position too, although Penney was to note that the only counter arguments were cost, time and difficulty of concealing construction of the large hole needed to decouple the explosion, together with the fact that it was still only a theory. Lloyd told Herter in Geneva that if a joint UK-US scientific assessment showed reasonably reliable verification over underground tests was impossible, he thought that HMG would accept the then present impossibility of a comprehensive ban under the principle of no disarmament without control.

Meanwhile, Sir Michael Wright, the head of the UK delegation in Geneva, had had two private conversations with the Russian delegation head, Tsarapkin in which the latter talked about a possible staged treaty. Soviet opposition to a limited treaty was even stronger than London had anticipated. It looked as if the Russians might consider it if the limited treaty provided for a binding commitment for a moratorium on underground tests for two to three years. This would be coupled with a joint seismic research programme, a UK proposal. The treaty would have to be signed as soon as possible without insistence on research beforehand. However, this proposition went further than London had already suggested to the

---

91 TNA FO 371/140441, IAD 21/128/G, Nuclear Tests Conference, Anglo/American Policy Discussion in Washington (14 October), 9 October 1959.

92 TNA PREM 11/3161, Nuclear Tests Conference, Anglo/American Policy Discussion in Washington (14 October) and TNA PREM 11/3161, Geneva telegram no. 748 to Foreign Office, 3 December 1959.

93 TNA PREM 11/3161, P. de Zulueta to Prime Minister, 14 October 1959.

94 TNA PREM 11/3161, Foreign Office telegram no. 4467 to Washington, 14 October 1959.

95 TNA PREM 11/3161, Washington telegram no. 2230 to Foreign Office, 21 October 1959.

US. London had thought in terms of a voluntary arrangement outside the treaty on a moratorium to last for one or two years.[96] Although Tsarapkin's approach was of great interest, London was most anxious to reassure Washington that this did not constitute an independent UK initiative and that Wright had been studiously careful to protect both the UK and US position.[97] London also had to keep in mind that any unilateral decision to announce a UK moratorium might have an effect on the proposed joint research programme and the future of US weapons collaboration with the UK.[98]

Technical Working Groups in Geneva

There was much debate over the terms of reference for a Technical Working Group in Geneva to look at the seismic issue and related inspection matters. Technical Working Group No. 1 met in June and July 1959 and reached agreement on a set of ground-based and satellite sensors to supplement the Geneva Experts' control system. A second Technical Working Group would meet in November 1959 to look at seismic data issues.[99] As one might have expected the Western scientists acted like scientists and used scientific language, whereas the Russians were almost entirely political. It seemed to the UK delegation that some of the US scientists were using the occasion to magnify the problems in order to make it difficult to move to a test ban. Dr Fisk, leader of the US team, however told the Minister of State, David Ormsby-Gore, in strict confidence that an independent committee appointed by the President had recently come to the conclusion that the additional military knowledge to be gained from further testing was very limited. However, the Pentagon disputed this view, but Fisk thought that the President accepted it. Washington was in no doubt that the UK desired a comprehensive agreement. If, because of difficulties of adequately monitoring underground tests, the UK could not secure a comprehensive test ban but had to opt for a partial one, then in order to have any hope of Russian agreement, then the UK was ready to accept an unverified moratorium on underground tests. Eisenhower had told Plowden in mid November 1959 that it would be difficult for the US to agree to any agreement on suspension, even limited in time, unless it were known that there was to be an adequate verification system.[100] This would be for a period during

[96] TNA PREM 11/3161, Lloyd to Prime Minister, Geneva Tests Conference, 3 November 1959.

[97] TNA PREM 11/3161, Washington telegram no. 2336 to Foreign Office, 5 November 1959.

[98] TNA FO 371/140441, IAD 21/128/G, Nuclear Test Conference, Anglo/American Policy Discussions in Washington (October 14), 9 October 1959.

[99] Romney, *Detecting the Bomb*, pp. 134–5.

[100] TNA FO 371/140441, IAD 21/140/G, P. Wilkinson to Mr O'Neill, 16 November 1959; Plowden Note for the Record, 12 November 1959.

which control would be far less effective than would be preferred.[101] Despite the problems London seemed quite optimistic about the prospect for a comprehensive test ban at this time; it seemed to be within reach and such an agreement would go a long way to preventing proliferation.[102] But this optimism was misplaced.

It became clear by 18 December 1959 that the Technical Working Group No. 2, which had started work in Geneva on 19 November, on reviewing mainly seismic verification matters, faced the real possibility of having to report to the Conference that it saw no prospect of any agreed recommendation and would break up without a date set for any resumption.[103] The US side was insistent on emphasising all the limitations and down playing the areas where further work might lead to improvements in detection capabilities. The UK did its utmost to dissuade the Americans from reporting failure and then breaking off the main negotiations. Penney argued with the US technical group that this course would have serious political consequences and that they ought to be sure that it would meet with political approval on their own side at the highest level.[104] Wright told Wadsworth (US delegation leader) that the UK could not accept a break in the negotiations on 19 December 1959. It would be difficult for the UK to defend scientifically the present extreme formulations of some of the US positions, especially on criteria and it could not be maintained honestly that all the matters at issue had yet been exhaustively explored. Still less could the UK justify a break without seeking agreement on the possible improvements, which even the US admitted ought to be possible to reach.

Fortunately, the US moved enough to enable the group to adjourn and agree at Penney's instigation a brief report with an annex containing agreed conclusions on possible improvements in techniques and instrumentation and three further annexes containing the views of each delegation on all matters discussed.[105] The area of agreement was much less than the UK had hoped. Only the possible improvements of techniques and instrumentation for detecting and identifying underground seismic events were submitted as agreed conclusions.[106] Notwithstanding this modest achievement, the main area of disagreement centred on the criteria by which suspect seismic events would be subject to inspection. Dr Robert Press (MOD) however advised the Foreign Office that so long as the

---

[101] TNA PREM 11/3161, Michael Wilford to P.F. de Zulueta, attaching Background Note on Nuclear Tests Conference for Paris Summit, 14 December 1959.

[102] TNA PREM 11/3161, P. de Zulueta to Prime Minister, 14 December 1959.

[103] TNA PREM 11/3161, Geneva telegram no. 804 to Foreign Office, 18 December 1959.

[104] TNA PREM 11/3161, Geneva telegram no. 805 to Foreign Office, 18 December 1959.

[105] TNA PREM 11/3161, Geneva telegram no. 807 to Foreign Office, 19 December 1959 and Personal communication.

[106] TNA PREM 11/3161, Geneva telegram no. 809 to Foreign Office, 19 December 1959; Romney, *Detecting the Bomb*, pp. 152–3.

decoupling theory was not disproved, and the creation of the necessary large hole was held to be feasible, a strong technical argument could be made to the effect that no underground control was possible.[107]

Penney was to tell a subsequent meeting of the Technical Working Group in January 1960 that the UK was, 'guided by a genuine desire for an agreement on the cessation of nuclear weapons tests provided that there was an adequate monitoring and control system'.[108] However, the UK's conclusions were included in one of the report annexes. They were threefold. First, the 'large hole theory' was sound, but careful work would have to be done before it could be said that the conditions required could be met in practice. Second, the new seismic data showed that the 1958 Experts had overrated the value of the criterion of first motion seismic waves and underrated the number of earthquakes equivalent in signal strength to nuclear explosions for a given yield. Third, no criteria could yet be devised to permit the identification of more than a modest proportion of earthquakes.[109] Technical Working Group No. 2 ended in acrimony with no agreement.[110]

Next Steps: January 1960

In light of this dead end what was next in the negotiations? The US thought that the only way ahead was to go for a threshold agreement with the threshold definition expressed in seismic magnitude terms (4.75 mb) rather than kilotons, in other words the level that could be detected with existing capabilities, setting on one side the 'large hole theory' problem.[111] That said, the UK knew that the Russians would reject any such thing unless it was accompanied by a moratorium of underground tests under the threshold.[112] Macmillan and Lloyd still believed that a moratorium would be an essential part of any sensible proposal. Macmillan felt that the UK should make every effort to persuade the Americans to include it in their plans for the next phase of the Geneva conference. He told Lloyd that he should emphasise to Washington that the UK feeling on this had now strengthened.[113] Lloyd then

---

107 TNA FO 371/149312, IAD 24/1, Underground Tests R. Press, MOD to H. Hainworth, Foreign Office, 1 January 1960; and A. Brooke Turner to Hainworth, 22 January 1960.

108 Conference on the Discontinuance of Nuclear Weapon Tests Technical Working Group 2, GEN/DNT/TWG.2/PV.1 pp. 12–13, 15 January 1960.

109 TNA FO 371/149306, IAD 22/370, Nuclear Tests Conference, History of the Conference from its opening on October 31st 1958, 6 October 1960.

110 Romney, *Detecting the Bomb*, p. 159.

111 TNA PREM 11/3161, Douglas Dillon to Selwyn Lloyd, 6 January 1960.

112 TNA FO 371/149273, IAD 22/17, D. Ormsby-Gore to Secretary of State, 12 January 1960.

113 TNA PREM 11/3161, Prime Minister's Personal Telegram T. 24/60, Nuclear Tests Conference, from Ibadan, 14 January 1960.

wrote to Herter to make this point abundantly clear: a moratorium really ought to be added to US negotiating proposals.[114]

This was followed up by instructions to Sir Harold Caccia to leave Herter in no doubt about the strength of UK feeling about the need for a moratorium in any interim or partial test-ban agreement. London believed that the case for this had steadily strengthened over recent months.[115] Joint research on the remaining technical problems could proceed for the duration of such a moratorium in order to make a test-ban comprehensive. Furthermore, both the UK and US were worried about the odium that would attach to the first state to resume testing, even if only underground testing since international opinion saw no distinction between such tests and atmospheric ones. Both were bad and ought to stop forthwith.[116] Lloyd also told Herter that the indications for the UK were that any improvements derived from further testing might only be marginal, although this does not quite square with views we can see in the NTPC and at AWRE – see Chapter 4. If the West was ahead, why let the Russian catch up? Of course, if this were the perception on the Western side, then the Russians would also know of their own inferior position, which probably explained their secret plans to resume testing, which were to become manifest in September 1961. Khrushchev had outlined a new Soviet military doctrine in early January 1960 that placed the strategic missile as the centrepiece of strategy; the Strategic Rocket Forces were established later that year.[117] In such circumstances, it is easy to see where the pressures for extensive testing would have come from.

The UK had four main differences with US policy at this point.[118] First, London believed that any proposal that would leave open the holding of tests below the threshold as legitimate would be unacceptable to Moscow and possibly cause them to break off the negotiations. Second, if the US left a loophole for underground tests below the threshold, the US military would press to resume and would prevail and if this happened, the Russians would resume atmospheric testing, but the odium would fall on the West. Third, if this were to occur, Soviet testing would work to the disadvantage of the West by allowing the Russians to close the gap in design capabilities. Fourth, although a moratorium was not to be effectively verifiable, as long as it was accompanied by seismic research to improve detection capabilities it would not prejudice the general principle of no disarmament without adequate control. Unfortunately for the UK the US disagreed on every point.

---

[114] TNA FO 371/149273, IAD 22/17, Foreign Office telegram no. 184 to Washington, 14 January 1960.

[115] TNA FO 371/149272, IAD 22/6/G, Foreign Office telegram no. 219 to Washington, 16 January 1960.

[116] TNA PREM 11/3161, Foreign Office telegram no. 184 to Washington, 14 January 1960.

[117] Thomas Cochran, William M. Arkin, Robert S. Norris, Jeffrey I. Sands, *Nuclear Weapons Databook Volume IV Soviet Nuclear Weapons* (New York, 1989), p. 9.

[118] TNA FO 371/149275, IAD 22/35, Hainworth to Sir P. Dean, 25 January 1960.

Washington did not think that the Russians could use its idea of a phased approach to a comprehensive test ban as an excuse to walk out of the Geneva conference. Despite London's pleas, Washington was still opposed to an uncontrolled moratorium to accompany its threshold approach.[119] Herter made this clear in two letters to Lloyd on 23 January and 2 February 1960 in response to earlier appeals from the Foreign Secretary.[120] Such an approach would serve to weaken the essential principle of controlled agreements. In short, the US could not agree to a cessation of tests in environments that could not be verified and this included high altitude tests too. The US would be free to resume testing below the threshold when its national security interests so indicated. This was Eisenhower's opinion too, not just Pentagon or USAEC views. Such arguments however did not convince the Foreign Office.[121]

Lloyd returned to the charge and made a further plea to Herter to change his mind on 27 January.[122] A test ban would halt proliferation and establish the important principle of international verification on Soviet territory. He hoped that Macmillan might weigh in too with a personal message to the President, but he decided against doing so, at least for the present. Sir Norman Brook, Cabinet Secretary, had told him that he doubted very much that the Americans could be budged from their position. Herter's two letters to Lloyd had been quite firm.[123] And a third letter was no different; the US was not going to move from its opposition to a moratorium. When Sir Harold Caccia went to see Herter to collect this letter, officials told him that Herter had fears that any open split between the UK and US on nuclear testing might have repercussions for the 1958 Agreement.[124] Herter pointed out that the Joint Congressional Committee on Atomic Energy (JCCAE) tended to be a law unto itself and its powers were great if not always clearly defined. John McCone had made the same point to Caccia. A very clear warning shot across the bows; do not move out of line because you might prejudice your nuclear links. Ormsby-Gore got the hint; he remarked that the US was extremely touchy on this issue and it could affect the UK across the entire nuclear field. Given the dependence on US assistance for the weapons programme, then there was only one option.

[119] TNA PREM 11/3161, Foreign Office telegram no. 26 to Geneva, 24 January 1960.

[120] TNA FO 371/149275, IAD 22/35, Nuclear Tests Conference Encloses message of January 23 from Secretary of State, Mr C. Herter; and TNA FO 371/149276, IAD 22/48, Nuclear Test Conference, Threshold proposal, Herter to Lloyd, 2 February 1960.

[121] TNA FO 371/149312, IAD 24/8, Hainworth to Mr Ormsby-Gore, 25 January 1960.

[122] TNA PREM 11/3161, Foreign Office telegram no. 378 to Washington, 27 January 1960.

[123] TNA PREM 11/3161, Norman Brook to Prime Minister, 29 January 1960.

[124] TNA PREM 11/3161, Washington telegram no. 197 to Foreign Office, 3 February 1960; Foreign Office telegram no. 493 to Washington, 5 February 1960.

The UK Acquiesces

In the event the UK agreed that the US should present its new proposals in open session in Geneva; and to give the US the most wholehearted support possible,[125] even though it did not think that its proposals went far enough.[126] However, the Foreign Office was worried that the US proposal would make it awkward for its delegation in Geneva: what if the Russians asked at once if the UK could support the new US plan?[127] Herter was relieved and when he discussed the tactics of all of this with Ormsby-Gore, he agreed with the UK that they should be as forthcoming as possible on a joint seismic research programme starting immediately.[128] Herter also mentioned the continuing difficulties that the Administration was having with the JCCAE over the Genie rocket – an air-to air missile that might be made available to the UK. He took pains to say that these did not arise from any feeling that the UK was untrustworthy, or from any dissatisfaction in Congress towards UK attitudes to nuclear tests. The implication was that these difficulties already existed and would naturally increase if the UK parted company with the Americans over nuclear test-ban policy in the very near future. Sir William Cook, who had been visiting Washington at about the same time, was verbally attacked by one of the USAEC Commissioners about British attitudes to nuclear tests and was told that if the UK persisted in its present views the USAEC would make difficulties over exchanging information on nuclear weapons.[129]

Ormsby-Gore reported that Herter had assured him that as far as the JCCAE was concerned he had heard no suggestion that UK attitudes were making the Committee reluctant to pass on weapons information. Fact or fiction made no difference; the risk was enough to make the UK think twice. Macmillan decided not to send a further message to Eisenhower; it was evident to him that the President had taken a position on this and now that it had become public through a leak, it would be difficult for him to retreat from it so soon. Instead, it might be better to reserve any further approach for a summit meeting between them, so that Macmillan could try to work on the President between now and then.[130]

[125] TNA FO 371/149276, IAD 22/48, Nuclear Test Conference Threshold proposal, Ormsby-Gore manuscript comment on minute from Hainworth 4 February 1960 and see also TNA PREM 11/3161 Continuation of negotiations in Geneva on suspension of nuclear tests 1959–60.

[126] TNA FO 371/149314, IAD 24/36, H.C. Hainworth to Mr O'Neill, Geneva Tests Conference, 9 February 1960.

[127] TNA FO 371/149276, IAD 22/42, Message from PM to Foreign Office Secretary concerning Herter's message, Hainworth to O'Neill, 1 February 1960,

[128] TNA PREM 11/3161, Washington telegram no. 235 to Foreign Office, 8 February 1960.

[129] TNA PREM 11/3161, D. Ormsby-Gore to Secretary of State, 24 February 1960.

[130] TNA FO 371/149276, IAD 22/42, Personal message from PM to Foreign Secretary, Inward telegram to Commonwealth Relations Office from Pretoria, Track No. 84, 29

Negotiations Resume in Geneva: February 1960

The US presented its new proposals in Geneva on 11 February 1960.[131] This entailed amongst other things a cessation of atmospheric tests as well as those underwater and in outer space up to the greatest height that could be verified and underground nuclear tests down to the lowest limit of size for which, in US judgement, adequate detection capabilities were feasible. Washington had at least beforehand been prepared to make a few changes to the US statement to take into account some UK views.[132] However, and as the UK expected, the Russians said that the US proposal was unacceptable because it would legalise and therefore lead to the resumption of testing. Tsarapkin immediately fastened on the possible resumption of testing by the US and its consequences, and he concentrated his questions and comment on this issue.[133] However, the Russians also gave some indications that they might accept the threshold proposal if it were coupled with a moratorium.[134] The UK supported the US proposal as an advance on the limited treaty suggested by the US in April 1959 in that it banned those tests underground that could be verified and linked the inspection quota idea to those above the threshold.[135]

Sir Michael Wright reported to the Foreign Office on 8 March that the negotiations seemed to be drying up for want of any material. What was really needed to loosen up matters was some new move on the major substantive issue – the moratorium on underground tests. He accepted that this was an extremely delicate matter, but he thought that there were two possible ways of proceeding, either by a UK proposal for a moratorium, or by a more cautious statement that the UK was now prepared to discuss ways of dealing with the situation below the threshold proposed by the Americans. Sir Michael hoped that when Ormsby-Gore came out to Geneva he would bring some new instructions.[136] London at that stage was ready to accept an imperfect treaty that would prevent the US from resuming tests of any sort and with a measure of verification over Russian activities. There would also be a reasonable probability that the existence of a treaty, and the

January 1960.

131 Conference on the Discontinuance of Nuclear Weapon Tests GEN/DNT/PV.170, 11 February 1960 pp. 5–9.

132 TNA FO 371/149278, IAD 22/62, Washington telegram no. 259 to Foreign Office, 10 February 1960.

133 TNA FO 371/149278, IAD 22/65, Nuclear Test Conference Text of 170th Meeting, Threshold proposals, Geneva telegram no. 67 to Foreign Office, 11 February 1960. (UK comment in IAD 24/37 and GEN/DNT/PV.170.)

134 TNA PREM 11/3161, Selwyn Lloyd to Prime Minister, 23 February 1960.

135 TNA FO 371/149278, IAD 22/66, CRO telegrams nos 192, 237 and 134 to Ottawa, Canberra and Wellington, 9 February 1960.

136 TNA FO 371/149282, IAD 22/123, Hainworth to Sir P. Dean, 9 March 1960.

physical presence of control posts established inside the Soviet Union, would deter the Russians from violating it.[137]

Russians Change Tack: Camp David March 1960

The Russians put forward a moratorium proposal at the Conference on 19 March 1960. This would entail a treaty incorporating a threshold of 4.75 mb with a moratorium below it (within) the treaty and joint seismic research.[138] This was music to UK ears and Macmillan and Lloyd clearly saw it as an important development, particularly with regard to the fourth country problem and the forthcoming Paris summit meeting with Khrushchev.[139] This Russian proposal was effectively the same one that the UK had been urging on the US for the last few months. It was essential that there should be a joint UK-US welcoming reaction.[140]

Eisenhower was not as sure as his officials still had problems with moratoria, but Macmillan was prepared to drop everything and go to Washington to talk it through with his old friend. He felt that the great prize of a test ban was now within reach and it would be important to do everything possible to capitalise on the Soviet move.[141] Perhaps even a moratorium of three years might be possible. Washington noted that the UK had the strong feeling that this counterproposal represented an important breakthrough in the negotiations and wanted to see favourable action on it and soon too.[142] John McCone, USAEC Chairman, however, was adamantly opposed to the proposal of a test ban without proper verification and expressed great disappointment with the British attitude. However, Herter certainly felt that the strong feelings of the UK had to be taken into consideration, but that did not mean that the US could let the UK determine Washington's own course of action. Eisenhower needed advice on what his decision should be. Deputy Secretary of State Dillon suggested that the US needed to focus on the desirability of a counter proposal. Eisenhower however did look at the Soviet proposal very sympathetically and thought that it represented progress, but he was doubtful that

[137] TNA FO 371/149317, IAD 24/81/G, H.C. Hainworth to Sir P. Dean, 9 March 1960.

[138] TNA FO 371/149283, IAD 22/131, Geneva telegram no. 151 to Foreign Office, 19 March 1960.

[139] TNA FO 371/149317, IAD 24/83/G, Foreign Office telegram no. 1183 to Washington, 22 March 1960.

[140] TNA PREM 11/3162, Foreign Office telegram no. 1155 to Washington, 20 March 1960.

[141] TNA PREM 11/3162, Record of Conversation between Prime Minister and President Eisenhower, 21 March 1960.

[142] Department of State Memorandum of Conversation Geneva Nuclear Test Negotiations: Meeting of Principals 22 March 1960 at http://www2.gwu.edu/~nsarchiv/NSAEBB/NSAEBB/tb22.pdf.

he could legally agree to a moratorium as an executive agreement because he himself would be President for only ten more months. Macmillan suggested to him that the moratorium could be included in a treaty that banned atmospheric tests, but the timing of this was also difficult given the need for Senate advice and consent before ratification.

Macmillan wrote to French President Charles de Gaulle on March 24 1960 and noted that when he heard of the new Russian suggestion for a test agreement including a moratorium and joint seismic research programme for small tests he was much encouraged.[143] Indeed he went on to emphasise that although he felt that the Soviet proposal was not absolutely acceptable as it stood, it offered a most helpful possibility of a genuine agreement being reached. Macmillan also told de Gaulle that Eisenhower had invited him to Washington to discuss this issue on 28 and 29 March following a telephone conversation; he thought that, 'it may well be a turning point in world history'. Macmillan also informed the Cabinet on 24 March that there had been indications that the US was reluctant to accept the latest Soviet proposal. He had found this disturbing, for he believed that a satisfactory agreement might be in reach. If the US rejected the Soviet plan, the UK and US might end up publicly divided since it was widely known that the UK had urged the US to include a moratorium on atmospheric testing in its proposals for Geneva.[144] Macmillan was conscious that his arrival in Washington might stiffen the attitude of those in the Administration opposed to a test ban. However, he had concluded that he ought to try to resolve, by personal discussion, such differences as remained between the UK and US. He was certainly aware that the Pentagon and the USAEC would bitterly resent and violently oppose any acceptance of a moratorium on tests below the detectable threshold.[145] Meanwhile, both sides needed to find out exactly what the Russians meant by their proposal; the detail was as usual critically important. The Administration would eventually agree upon a response that would include a moratorium on testing below the threshold for a short specified period whilst coordinated research work continued on the outstanding seismic verification issues. Herter gave Harold Caccia a copy of a US paper setting out its policy to the new Soviet position on 26 March, shortly before Macmillan went to Camp David.[146]

There was a perception in some quarters, including the media in Washington, that Macmillan had inspired the Soviet moratorium offer and that his objective in Washington was to persuade the President to agree to it; and that the urgency of the visit was to prevent the President capitulating to the Pentagon and USAEC

[143] TNA CAB 21/4535, Macmillan to de Gaulle, Prime Minister's Personal Telegram Serial No. T244/60, 24 March 1960.

[144] TNA CAB 128/34, C.C. 20 (60) Item 6 Geneva Conference on Nuclear Tests, 24 March 1960.

[145] Harold Macmillan, *Pointing the Way 1959–1961* (London, 1972), p. 190.

[146] TNA PREM 11/3162, Washington telegrams nos. 650 and 651 to Foreign Office, 26 March 1960.

views.[147] Macmillan did indeed go to Washington with a clear objective in mind: to secure US agreement to proceed on the basis of a moratorium without attaching stringent conditions to their acceptance that would make further progress unlikely.[148] London had tried several times at senior level over the preceding 12 months to urge the US to accept a moratorium on small underground tests, but to no avail. During the Camp David meeting on 28 March, Eisenhower remarked that he was much relieved at Macmillan's attitude since he had been led to believe that he had come to the US in order to press for immediate Western acceptance of the Soviet proposal. Since he did not wish there to be a misunderstanding, Macmillan asked Ike directly why he had made this remark. The President replied that someone, he could not remember who, had told him that there had been some statement which indicated that the British wanted to accept all the Russian proposals. He was very glad that this was not the case. It also seems that Eisenhower had had a very hard time in securing inter-agency agreement to the US position paper, a copy of which went to the UK before the visit. The President said that he had even had to threaten to resign. Macmillan then asked him if he agreed that the object was to stop all tests and that he did not wish to continue even small development tests. Ike quite agreed, although he said that some of his people did not share that view.[149]

The President and Prime Minister issued a Joint Declaration on Nuclear Tests on 29 March after their meeting. This noted that it was their sincere hope that an agreed programme of coordinated scientific research, undertaken by the three countries, could resolve the technical problems in setting up an effective control system that would be able to detect underground nuclear tests below a certain size. In addition, there were several other outstanding negotiating problems – quotas for inspections, composition of the control commission, control post staffing and voting matters in the commission as well as peaceful nuclear explosions. All of these needed urgent attention. Both agreed that as soon as a treaty was signed and arrangements made for a coordinated research programme to improve progressively control methods for events below 4.75 mb, they would be ready to institute a voluntary moratorium of agreed duration on nuclear weapons tests below that threshold. This would be accomplished by unilateral declaration of each of the three powers. In order to expedite this matter, the US and UK agreed to invite the USSR to join at once with their two governments in making arrangements for such a coordinated research programme and putting it into operation.[150] Macmillan and Eisenhower accepted, given the technical advice from their experts, that a joint programme would be slow

---

[147] TNA PREM 11/3162, Prime Minister's Personal Telegram Serial No. T. 253/60, Washington telegram no. 636 to Foreign Office, 24 March 1960.

[148] TNA FO 371/149317, IAD 24/88/G, Brief for the Prime Minister's Talks with President Eisenhower, Nuclear Tests Conference Moratorium Proposal, Foreign Office, 24 March 1960.

[149] TNA PREM 11/3162, Note by the Prime Minister, 28 March 1960.

[150] TNA PREM 11/3162, Washington telegram no. 685 to Foreign Office, 29 March 1960.

and cumbersome; instead, it would be far better if the three countries conducted their own projects as part of a co-ordinated programme.[151]

The moratorium's duration was left deliberately unclear; the UK thought that two years was a fair minimum because that was likely to be the time needed to complete the work, the US could only manage one year given the constitutional need to avoid tying the hands of the next President in an election year. Nor was there any discussion of high altitude tests. Macmillan thought it wiser not to press this issue.[152] The High Altitude Working Party in Geneva had produced agreed recommendations in July 1959 on the ground controls and satellite systems needed to monitor a ban on tests at high altitudes. Since then, the US had had second thoughts about the effectiveness of such controls, particularly against clandestine tests in deep space or behind the sun or moon. London worried about the costs of such a system. Instructions followed accordingly to their respective delegations in Geneva.

Penney and American experts worked out a technical paper in Washington that set out the principles for the coordinated effort on the seismic improvement programme, although much of the rest of the paper set out what the US planned to do anyway.[153] The general principles were that the programme would involve a major research, development and test effort whose objectives were to secure the maximum improvement in seismic detection and identification capabilities and to determine capabilities and limitations of any system that might be adopted. Although coordinated efforts by the three parties were desirable, each party would have to conduct such programme activities it believed to be essential. There would still need to be early coordination so that each participant could adjust its national programme to gain better over-all coverage and to avoid undesirable duplication and so that agreement could be reached on the observation of, and cooperation in, the activities of the others. They would also inform each other of the times of tests and their locations.

UK Contributions to the Seismic Research Programme and Follow-up in Geneva

It was now for the UK to work out what it could contribute, and based on a UKAEA proposal, Macmillan invited his Cabinet to support a project that would establish detection stations in Canada. However, Macmillan emphasised that there would be no expenditure until the Soviet government agreed that arrangements for a co-ordinated programme could be made. The proposed UK contributions

[151] TNA CAB 129/101, C. (60) 74 Nuclear Tests Memorandum by the Prime Minister, 27 April 1960.

[152] TNA PREM 11/3162, Washington telegram no. 693 to Foreign Office, 29 March 1960.

[153] TNA PREM 11/3162, Washington telegram no. 692 to Foreign Office, 29 March 1960.

would amount to £3.25 million over the next three years. If so agreed, it would be necessary to seek approval from the Canadian government and to authorise the UKAEA to order a supply of seismometers from the US. In addition, the Australian and New Zealand governments would be asked to see if they wished to provide technical experts to help run the stations. The Cabinet endorsed these proposals, excepting for the commitment to expenditure. That would have to wait until such times as the UK, US and USSR nuclear powers had reached agreement in Geneva on a co-ordinated research programme.[154] However, by early May the US was trying to move away from the idea of a single coordinated research programme, preferring instead coordinated programmes. The key point for them was that if the Russians did not agree to an adequate number of nuclear research tests, then the US would go ahead anyway with what it saw as necessary.[155] UK policy was still to work for a single coordinated programme. A more detailed discussion of proposed UK research seismic research activities is in Chapter 6.

Within a matter of days of the Camp David meeting Macmillan was starting to get agitated about the lack of concrete progress in Geneva. He worried that the Americans were allowing things to become bogged down. This was important in view of the summit meeting with Khrushchev in Paris scheduled for May. Macmillan wanted to have only a few things to settle at this meeting: the moratorium duration, the number of inspections and whether they applied below the threshold as well as above it. He was afraid that some Americans, annoyed by the President's decision at Camp David, would do everything they could to make the conference fail. He knew that this was not true of Wadsworth, the US delegation head, but it was true of both the Pentagon and USAEC.[156] Geneva reported that the US delegation was not deliberately allowing the Conference to slow down; there were still some useful discussions talking place on the moratorium. Composition of the control body was however holding things up. The Americans had in fact agreed to yield parity to the Russians, but they did not want to make the concession just yet. The Russians had no intention of coming to final agreement on inspection issues until they knew that the governing body was not permanently weighted against them. Lloyd intended to speak to Herter to make clear that resolving this issue was the only way to move negotiations to the position at the Summit where there were only two or three things to settle.[157] Given the 'internecine warfare in the American

[154] TNA CAB 128/34, C.C. (60) 28, Item 2 Nuclear Tests, 28 April 1960; TNA DEFE 19/18, Technical History of the Geneva Nuclear Tests Conference 1957–1962 (R.W. Snelling, Geneva September 1961) paragraph 41 p. 19; TNA CAB 128/34, C.C. 28 (60), Item 2 Nuclear Tests, 28 April 1960.

[155] TNA FO 371/149325, IAD 24/208, Geneva telegram no. 470 to Foreign Office, 10 May 1960.

[156] TNA PREM 11/3162, Prime Minister's Personal Minute Serial No. M.93/60 to Foreign Secretary, 1960.

[157] TNA PREM 11/3162, Lloyd to Prime Minister, Nuclear Test Conference, 9 April 1960.

Administration' on test-ban issues, the only way that Lloyd saw for progress was for the Prime Minister and the President to sort out for themselves how they would handle Khrushchev.[158]

The UK and US met in Paris during the summit to discuss strategy and tactics for handling the test-ban issue with the Russians.[159] Herter made clear that the US could only support a moratorium of a maximum of two years; Lloyd wondered whether there was any give if the research programme needed longer. The Americans thought this dangerous because the Russians would no doubt exploit opportunities for prolonging the work. As for the quotas, the US position was now that this should only apply to events above the threshold, although there might be some experimental inspections of events below the threshold as part of the research programme. Washington felt, following new technical studies on the relationship between unidentified events and the inspection quota from RAND, that 20 inspections a year above the threshold would offer an effective deterrent; this would mean that about 30 per cent of unidentified events in Russia would be liable for inspection. However, if Khrushchev could accept nine new control posts, only four inspections per year would be sufficient.

At the end of the day this was all for naught. The U-2 incident ended any hopes that the Summit would provide any concrete outcome on the test ban and on any other East-West issues.[160] Macmillan was disappointed. He told the House of Commons on 20 May that despite the setback HMG would certainly press on to try to secure an agreement. Meanwhile in Washington Gates and McCone advised Herter that the US should now break off the talks on the test ban and that means should be found for laying the blame on the Russians. Herter disregarded this advice and much to the relief of London new instructions issued to the US delegation to go full speed ahead with the negotiations.[161]

Things were rarely that simple in Washington. Wadsworth told Wright in the strictest confidence that the NSC had recently met and recommended or decided that the US should decline to take part in further negotiations, which the Russians were rendering fruitless, unless there was progress by the end of June. The US might then resume underground testing. Wadsworth commented that this appeared to represent a further major effort by McCone and Gates, such as had occurred in July 1959 and again in December 1959 to wind up the conference and to find some means of pinning the blame on the Russians for procrastination or refusal to

[158] TNA PREM 11/3162, Lloyd to Prime Minister Nuclear Tests Conference, 22 April 1960.

[159] TNA PREM 11/3162, Record of Conversation Between The Secretary of State and Mr Herter at the US Embassy in Paris, 14 May 1960.

[160] Dokos, *Negotiations*, p. 14.

[161] TNA PREM 11/3588, Washington telegram no. 1106 to Foreign Office, 1 June 1960.

make concessions.[162] Wadsworth felt that it was the US that could be so accused now that the plenary records of the Geneva conference were available for public scrutiny. The US had declined to make any proposal about composition of the control commission and to talk about high altitude controls and since February 1960 had refused to discuss the Soviet quota proposals and was unwilling to talk about the length of the moratorium.

All this ran counter to the contents of a recent message from Herter to the Foreign Secretary. He felt disturbed in view of the UK's loyal cooperation throughout the conference that it would be unfair that London should be unaware of what was afoot. Ormsby-Gore thought that it was now very desirable that the UK should try to smoke out exactly what was going on in Washington before the opponents of a treaty could carry things too far.[163] The Foreign Secretary might hang a message on Herter's confirmation that US policy was to go full speed ahead and ask for indication on how the US might handle the major outstanding issues. Lloyd did just this and a message was despatched on 15 June.[164] Herter did not respond until 23 June and noted that the US was working on an approach to the safeguards issue for ensuring that Russian access to nuclear devices for seismic calibration purposes would not compromise sensitive information (a Congressional requirement) and he agreed that it was important to avoid the conference stagnating. He also accepted that the Western failure to put forward high altitude positions was a weak point, but the US was still trying to work this out but held little prospect of an early resolution.[165]

UK Deliberations on the Seismic Research Programme, Use of UK Nuclear Devices and the Safeguards Issue: Summer–Autumn 1960

The Camp David meeting had agreed that nuclear explosions should be part of the seismic research programme and that the US should conduct some of these.[166] However, the Russians insisted that there had to be safeguards to prevent states exploiting the seismic research programme of nuclear tests for weapon development purposes. They wanted a full description and blueprints of the structure of any device intended for use to be made available beforehand and to be able to inspect

---

[162] TNA FO 371/149289, IAD 22/220/G, Geneva telegram no. 608 to Foreign Office, 13 June 1960.

[163] TNA FO 371/149289, IAD 22/221, Geneva telegram no. 609 to Foreign Office, 13 June 1960.

[164] TNA FO 371/149289, IAD 22/221, Foreign Office telegram no. 2676 to Washington, 15 June 1960.

[165] TNA FO 371/149291, IAD 22/239, Washington telegram no. 1238 to Foreign Office, 23 June 1960.

[166] TNA FO 371/149334, IAD 24/338, Result of the co-ordinated scientific research programme proposed in the Camp David Declaration, 29 March 1960.

its internal structure. If either the US and UK went ahead with tests without this, then the Soviet Union would feel free to end its unilateral moratorium and resume testing in order to ensure its security.

Penney suggested in a UK-US expert meeting that an effective programme would need one to four calibration tests to see whether nuclear and HE explosions gave the same signal. McCone pointed out that it would be important to ensure that such tests could not be used for weapon development purposes. The possibility was also mentioned, although it is not clear by whom from the US record, of using UK devices, though McCone warned that the US-UK agreement for cooperation would need to be studied before airing this option. This idea had first been aired in 1959. Sir Frederick Brundrett had also suggested that the UK might provide a weapon for this kind of test. This meeting also agreed instructions for the negotiating teams in Geneva.[167]

At the heart of this was the question of safeguards. One way of doing this in order to overcome Russian objections would be to use British devices, especially if it were technically and politically feasible to produce a simple design that could be shown to the Russians and the Americans.[168] However, this would have implications for the 1958 bilateral agreement and therefore the idea required very careful consideration. AWRE were asked to see whether they could supply five kiloton devices of sufficiently primitive design so that their disclosure to the Russians would not give away any valuable information, and if so when might they be available. Secondly, AWRE were also asked if devices of smaller yields, preferably sub-kiloton, could be provided with the same limitations. AWRE looked at the possibilities for providing UK designs for seismic shots; these included copying US gun type designs with the advantage that this would give nothing given away on later designs. An experimental British design could be proposed, but there were real problems with this approach too. Thirdly, the UK service weapon could be used for the proposed ten to 20 kiloton shots. It could also be scaled down to five kilotons for the other shots in the seismic calibration tests. (AWRE might also be able to go as low as one or two kilotons, but the UK did not have the expertise to go any lower.) This was the least attractive option since it would compromise an in-service weapon since the Russians would, under the proposed scheme of joint research, be entitled to check that the devices were not being used to improve on design information.[169]

---

[167] Department of State Memorandum of Conversation Geneva Nuclear Test Negotiations: Meeting of Principals 22 March 1960 at http://www2.gwu.edu/~nsarchiv/NSAEBB/NSAEBB/tb22.pdf. Department of State Memorandum of Conversation Geneva Nuclear Test Negotiations: Meeting of Principals 22 March 1960 at http://www2.gwu.edu/~nsarchiv/NSAEBB/NSAEBB/tb22.pdf.

[168] TNA PREM 11/3584, Geneva telegram no. 472 to Foreign Office, 10 May 1960. See also TNA FO 371/149325, IAD 24/211, Control Organisation for Discontinuances of Nuclear Tests, 1960.

[169] TNA ES 1/1350, Rounds for Seismic Shots, N. Levin, 6 July 1960.

AWRE saw severe disadvantages in providing Blue Danube as this would give the Russians information on an existing service weapon, with its pointer to improved designs. It would also reveal weapon design information that even then was still in advance of anything that the French had yet tested.[170] Even a scaling down of Blue Danube would result in a diversion of effort that could be ill afforded given the anglicisation work on the Red Snow warhead already underway. Sir Patrick Dean was worried about the political effects of this, as it would appear that the UK was not willing to help the US. However, the MOD made clear that it did not want to scale down an existing British weapon to the order of five kilotons. In the event of political necessity to make a British weapon available for joint UK/US/Soviet seismic research purposes, the MOD was willing to declare the earliest nuclear design obsolete even though that design was still in service use.[171] As it happened, Blue Danube by this stage had only about a year left in service. Even at the time, much of this was an academic exercise because the firm view was that there was no chance that the Russians would ever agree to such a scheme.

The MOD pointed out that even if the US provided design details of the 1945 gun type HEU weapon, there could still be problems for the defence programme's progress and if the ten or so devices were needed, fissile material availability problems would become an issue. Sir Roger Makins, in writing to the Treasury to provide information on the potential costs, used an analogy to characterise what AWRE could provide and what it would all cost. Aldermaston's stock stretched from 'a Palaeolithic axe head to three types of Bronze Age arrow head, of which one was slightly more expensive than the other two'. The full range cost £10 million, £4 million and £3 million respectively. Individual weapons varied in price, according to dimension. A medium sized axe head cost just under £1 million; a medium sized arrow head cost £0.5 million increase, and £300,000 in another. He added that it now looked as if the UK might only need to provide one of the cheaper Bronze Age arrowheads. Therefore a UK contribution would not be the cause of great financial concern.[172]

Wright asked London on 10 May whether the possibility of providing UK devices could be examined. The MOD's view was that unless a high level decision were taken to declassify an in service weapon, the UK would be bound to take the same approach as the US to any examination of the device's innards, in other words no. US ideas on safeguards would have permitted observation of all aspects of the firing and its instrumentation except the internals of the devices.[173] This

[170] TNA DEFE 7/2296, R. Press to Secretary, Use of a British Nuclear Device for Seismic Research, 4 July 1960.

[171] TNA DEFE 7/2296, R. Press to Secretary, Use of a British Nuclear Device for Seismic Research, 5 July 1960.

[172] TNA AB 16/4186, Roger Makins, CH (60) 45, UKAEA to R.W.B. Clarke, HM Treasury, 14 July 1960.

[173] TNA FO 371/149325, IAD 24/217, Geneva telegram no. 479 to Foreign Office, 12 May 1960.

was on the assumption that a five to 15 kiloton device was involved, as the UK could not guarantee to provide anything smaller. AWRE saw no technical merit in the UK making a positive offer, but the Foreign Office felt that if there were urgent political reasons for doing so, then this objection would have less weight. Wadsworth told Wright on 4 June 1960 that the State Department was considering two options: the use of UK devices, or to offer to use Soviet devices, which could be exploded in the US under safeguards proposed by the Americans. State Department worried that the Conference might break down on this issue and thus recognised the need to try and find a way round the problem, recognising that the US would be on weak ground if did fail. Another option was to persuade the JCAE to accept that the Russians could examine old US devices. London's concern was what would happen if the US made a formal request for UK devices.

The Foreign Office knew that the technical and security objections to the use of UK devices would be considerable and as long as it was clear that the US was trying to find its way around the problem, the MOD and AWRE would oppose use of a UK nuclear explosive. Sir Roger Makins told Sir Frank Hoyer Millar, Permanent Under Secretary at the Foreign Office, that there would be grave security disadvantages in furnishing a UK device, as it would have to be a service weapon. The MOD was adamant that the defence objections were overwhelming.[174] Foreign Office officials felt that the best option to counter this would be at the political level if it became apparent that the only way to save the conference was by offering a UK device. Such a decision would have to fall to the Prime Minister. At least for the present there was no need to press the case for offering British devices.

Since the Russians had just walked out of the Ten Nation Disarmament Committee, the UK delegation in Geneva worried that the Russians might be looking for an excuse to leave the nuclear test talks too. For this reason Ormsby-Gore felt that the UK should seek to improve the Western posture as soon as possible. This could only be done by the Americans making a reasonable reply on the safeguards issue and by putting forward a proposal on the length of the moratorium that would enable Wadsworth to challenge Tsarapkin to keep his bargain; they could then discuss the inspection quota issue. Ormsby-Gore thought that the UK should try to persuade the Americans to take this approach; a message from Lloyd to Herter, or from the Prime Minister to the President was now required.[175]

However, the US had problems in addressing the safeguards issue; anything that smacked of revealing secrets on atomic weapons was not calculated to go down at all well with Congress. Efforts to persuade the JCCAE ran into obstacles heightened by the fact that it was an election year.[176] Until the safeguards issue

[174] TNA FO 371/149333, IAD 24/325/G, Sir Edward Playfair, MOD to Sir Frederick Hoyer Millar, 28 June 1960.

[175] TNA PREM 11/3588, Geneva telegram no. 676 to Foreign Office, 28 June 1960.

[176] TNA PREM 11/3588, Washington telegram no. 1270 to Foreign Office, 30 June 1960.

was resolved, Herter saw no point in discussing the moratorium and quota issues. Caccia advised London too that there was no point in pressing the issue until the US could put together a new package. Indeed, he suggested that London might now want to consider further the possibility of using British nuclear devices. If the Americans produced a scheme based on reciprocity and rejected by the Russians, the only remaining alternative was to use British devices.[177] There was no prospect, however, of the Russians accepting any requirement that they would have to contribute their own nuclear devices to the seismic research programme.[178] Lloyd sent a personal telegram to Caccia to encourage him to do his best to ensure that the present impetus on the issue was not lost.[179]

Herter was under pressure from critics of US policy and had been made to carry the responsibility of the U-2 incident. Caccia doubted that any personal message from the Foreign Secretary would make any difference.[180] There was thus little chance of progress. London thought in light of all of this that it might be better to put the test-ban negotiations into recess for four to six weeks. If the US agreed, the UK was happy to put the case to the Russians. However, there were risks in that it might only provoke the Russians to walk out and previous experience had shown that a recess that took the pressure off the Americans simply postponed the taking of decisions.[181] Foreign Office officials were bothered by the fact that when the negotiation's records of what had happened since the end of May were published, they would show that the US had dragged its feet to such an extent that many people would argue that the Americans had no intention of entering into an agreement. If Washington could provide nothing of importance for its delegation to say for a week or more, a recess was the lesser of two evils. Sir Michael Wright did not agree. There was still some modest progress that could keep negotiations ticking over; Tsarapkin had even promised some Soviet responses on a range of issues. He had also exposed weaknesses in the Western position on the composition of the Control Commission, which meant that if the negotiations did break up or go into recess, the West would be left at a disadvantage. If the Russians were planning a walk out, then a UK request for a recess might play into their hands; it would also strengthen anti-treaty forces in Washington.[182]

---

177 TNA PREM 11/3588, Washington telegrams nos 1271 and 1270 to Foreign Office, 30 June 1960.

178 TNA FO 371/49292, IAD 22/252, Ormsby Gore to Atomic Energy and Disarmament Department, 1 July 1960.

179 TNA FO 371/49292, IAD 22/246, Foreign Office telegram no. 2941 to Washington, 30 June 1960.

180 TNA PREM 11/3588, Washington telegram no. 1272 to Foreign Office, 30 June 1960.

181 TNA PREM 11/3588, Foreign Office telegram no. 2964 to Washington, 1 July 1960.

182 TNA PREM 11/3588, Geneva telegram no. 689 to Foreign Office, 2 July 1960.

When the US finally produced its safeguards proposal it did not make for happy reading; it insisted that the USSR had to provide the same number of devices for the research programme as the UK and US and explode them.[183] London thought that such a dictatorial position would ensure that such a proposal would fail, especially since the Russians were not at all keen on the idea of nuclear tests as part of the research programme anyway.[184] Wright realised that the UK would find it difficult to ask Washington to review its approach, especially after the pain that the interagency process had just gone through to come up with the current text. Wadsworth and Wright had thought that if it were linked to Russian ideas for safeguards on Peaceful Nuclear Explosions (PNEs), which also included internal inspection and a review of blueprints, then this would be much better presentationally. Wright advised that the UK should urge the Americans strongly that the West could not afford a negotiation breakdown in circumstances where the Russians could argue that the Americans were concerned merely to obtain information about Soviet nuclear devices and also to oblige others to take part in carrying out nuclear explosions. London also saw serious problems with US ideas and hoped that they would not deploy them until Washington had considered UK views.[185]

The Foreign Office raised the test-ban negotiations at the Cabinet Office Working Group Disarmament meeting in early July 1960.[186] Meanwhile, Macmillan received a message from Eisenhower about the safeguards issue.[187] This set out the offers that Washington wanted to make: a revival of the pool idea – all three participants providing devices or from the US only; both would require Congressional approval if device designs were to be viewed by the Russians. If the Russians rejected these or delayed unreasonably, then the US planned to go ahead anyway with the nuclear seismic calibration tests and invite international observation. Eisenhower stressed that this approach would show US readiness to find a common purpose and cooperative solution. Foreign Office officials concluded that from the nature of the letter it seemed almost certain that the option of providing UK devices had been dropped from US thinking – no doubt to the immense relief of AWRE and the MOD.[188] The MOD had certainly been dragging its feet on this one. As Henry Hainworth commented, this 'removes our embarrassment'.

Nevertheless, London did worry a little about the pool idea – as officials felt sure that it would be rejected by the Russians and that the US was prepared to

---

183 TNA PREM 11/3588, Geneva telegram no. 694 to Foreign Office, 2 July 1960.

184 TNA FO 371/49333, IAD 24/323, Washington telegram no. 1286 to Foreign Office, 2 July 1960.

185 TNA PREM 11/3588, Foreign Office telegrams nos 2991 and 2995 to Washington, 4 July 1960.

186 TNA FO 371/149333, IAD 24/326/G, P. Dean to Sir Roger Makins, UKAEA, 1 July 1960.

187 TNA FO 371/149335, IAD 24/349A/G, Eisenhower to Macmillan, 9 July 1960.

188 TNA FO 371/149335, IAD 24/349A/G, H.C. Hainworth to Mr O'Neill, 11 July 1960.

proceed anyway even if the Russians objected to the second option. This might provoke the Russians to walk out, but on balance the Foreign Office thought that UK should support the US suggestion. Macmillan replied to Eisenhower on 13 July to endorse broadly the US approach.[189] British representations and those of Wadsworth had some effect as Washington dropped the ultimatum aspect.[190] The Administration now wanted to go ahead and make the revised proposal and hoped that the UK could support it. Sir Michael Wright was so authorised, but this did not amount to an unconditional support of the US pool of devices idea.[191] When the Russians rejected this, as they no doubt would, the UK would then fully endorse the fallback position of using US devices only. However, nothing was ever simple when it came to the US and the test ban.

As expected the Russians rejected the pool proposal, but in a manner that was considered fairly reasonable; there were no temper tantrums from Tsarapkin.[192] The next issue for the US delegation was to secure Washington's agreement to table the fallback position and this lack of authority was a matter of some concern to the UK as it would leave the West with little to contribute in Geneva.[193] Unfortunately for the UK the US found that for internal reasons it was still not possible to put the fallback position on the table; it might however be possible to see progress on this question in November after the Presidential and Congressional elections.[194] At that time there was just no way that the Administration could convince Congress to accept the safeguards proposal in the fall-back position.[195] Recess in Geneva was the only option.

Continuing Delays in Washington

On 4 August 1960, the Foreign Office was disturbed to learn that it was far from certain that Wadsworth would be authorised to table the fallback position in

---

[189] TNA FO 371/149336, IAD 24/354/G, Prime Minister's Personal Telegram Serial No. T 396/60, Foreign Office telegram no. 3182 to Washington, 13 July 1960.

[190] TNA PREM 11/3588, Washington telegram no. 1331 to Foreign Office, 7 July 1960.

[191] TNA PREM 11/3588, Foreign Office telegram no. 3142 to Washington, 12 July 1960.

[192] TNA FO 371/149339, IAD 24/390/G, Geneva telegram no. 785 to Foreign Office, 3 August 1960.

[193] TNA FO 371/149339, IAD 24/390/G, Foreign Office telegram no. 3550 to Washington, 4 August 1960.

[194] TNA FO 371/149340, IAD 24/419/G, Washington telegram no. 1628 to Foreign Office, 15 August 1960.

[195] TNA FO 371/149340, IAD 24/421/G, Foreign Office telegram no. 959 to Geneva, 17 August 1960.

Geneva and that instructions would not be sent for a while yet.[196] There were signs too that the Russians might actually buy this proposal, although they would not be providing any devices themselves. This made US inaction even more difficult to understand. Perhaps it was time for further high-level messages. The Embassy in Washington thought not; a Prime Ministerial message at that point would not hasten matters in the direction that the UK wanted.[197] US Congressional approval was required for any arrangement concerning the pooling of nuclear devices and since it was about to go into recess, time was running out for any immediate response to the Soviet position.

Macmillan disregarded the Embassy's advice and urged Eisenhower to make the fall back proposal, but the Administration said that there was no prospect whatsoever of Congress considering this issue during its present session.[198] Moreover, soundings had shown that the fallback proposal was extremely unlikely to secure the necessary support during the remaining days of the session either. There was another problem too: reaching agreement on the number and nature of nuclear explosions in the projected American programme. The Russians said that they could not acquiesce in the US conducting explosions to test the 'large hole theory' and the Americans had made it clear that such experiments were essential. Failure to obtain Congressional approval before the end of its current session would thus delay an agreement with the Russians on the seismic research programme until early 1961. Herter suggested that a small US team come to London to discuss next steps. London welcomed this and asked the Embassy to find out what the Americans were likely to propose at this meeting. Since the Administration could not obtain Congressional endorsement for its second safeguards proposal, it suited the UK best if the Americans were prepared to defer this whole question until 1961, especially the nuclear explosions in their research programme.[199] London feared that if the US went ahead unilaterally, then that would spell the end of the Geneva Conference.[200]

---

[196] TNA PREM 11/3588, Foreign Office telegram no. 3550 to Washington, 4 August 1960.

[197] TNA PREM 11/3588, Washington telegram no. 1549 to Foreign Office, 5 August 1960.

[198] TNA PREM 11/3588, Prime Minister's Personal telegram T. 465/60, Foreign Office telegram no. 3694 to Washington, 11 August 1960 and Washington telegram no. 1606 to Foreign Office, 12 August 1960.

[199] TNA PREM 11/3588, Foreign Office telegram no. 3740 to Washington, 13 August 1960.

[200] TNA PREM 11/3588, Foreign Office telegram no. 952 to Washington, 15 August 1960.

UK-US Differences, Moratoria and the Seismic Research Programme: Autumn 1960

The UK also opposed a US request for a recess in Geneva at that time.[201] Therefore, the question for Ministers and officials now was what would Geneva discuss in the autumn, assuming that the US did not conduct one of the proposed seismic calibration tests. Although it was clear to London that a resumed session would not make decisive progress, there was perhaps a chance to move on other aspects of the draft test-ban treaty such as the Control Commission and staffing, inspection quotas, the Parties Article, cost-sharing and possibly even high altitude controls.[202] Politically the UK could not afford the talks to break down completely. A US team led by McCone arrived in mid August 1960 to discuss the test-ban issue as proposed by Herter. McCone showed some desire to press the Russians again to accept the first US safeguards proposal, but UK officials gave him no encouragement and he dropped the idea. Both then agreed that in these circumstances the West's best interests would be served by going for a recess as soon as possible.[203] The UK and US agreed that they would use the time to prepare Western positions in order that there might be ample material ready for use upon resumption. McCone mentioned that advancing the US fallback position on safeguards would not necessarily become easier after the elections in November; it would in any case be impossible to secure Congressional approval before the end of the year.

In a separate meeting McCone told Macmillan that it had been clear from soundings that the fall back position under which only Western devices would be open to inspection no longer had any chance of approval.[204] This was disappointing; it was particularly annoying in Macmillan's view that a technicality should hold up the negotiations and that the US Administration could not act in the matter of releasing devices for seismic research without Congressional approval. But it was now quite clear to the Foreign Office that a test-ban treaty could not be concluded before a new Administration had been in office for at least two or three months.[205] The Russians did not demur when the UK and US subsequently proposed a recess and the conference duly did so from 23 August 1960 until 27 September.[206]

---

201 TNA PREM 11/3588, Foreign Office telegram no. 3753 to Washington, 14 August 1960.

202 TNA PREM 11/3588, Foreign Office telegram no. 952 to Washington, 15 August 1960.

203 TNA PREM 11/3589, Foreign Office telegram no. 959 to Washington, 17 August 1960.

204 TNA PREM 11/3588, Record of a Conversation at Admiralty House, Wednesday 17 August 1960, Present, PM, Ormsby-Gore and Sir R. Makins, McCone, Merchant and Barbour.

205 TNA FO 371/149299, IAD 22/299, Geneva Conference on Nuclear Tests, 1960.

206 *Conference on the Discontinuance of Nuclear Weapon Tests*, GEN/DNT/PV.246, 22 August 1960, p. 21.

There were four main issues dividing the three delegations and these were by no means trivial. The first of these concerned the duration of the moratorium on the conduct of small tests, which, being below the threshold, were not yet susceptible to verification. Second, the quota for annual inspections saw a gap between the Russian preference for three and the larger Western requirement. Third, the number and nature of nuclear explosions in the projected US seismic research programme was a problem for the Russians; and finally, the composition of the control commission and various staffing questions relating to control posts and inspection teams separated the delegations.[207] A further round of UK-US talks took place in Washington in early September to discuss Western options and tactics for the resumption of the negotiations in Geneva. However, these do not appear to have made too much progress.[208] The UK wanted to have serious discussions in Geneva on quota figures and the duration of the moratorium, but the US was unable to move far enough to permit this.[209]

Much of the intersessional preparatory work required fell to the Americans. US ideas on the handling of the moratorium surfaced first and the UK did not like them. US proposed text on treaty duration would see the moratorium begin from treaty signature for the remaining portion of the two-year seismic research programme, which would begin when announced by the US plus a three-month period to review the programme results. London was worried that the Russians could claim that the Americans would be in a position to declare the beginning of the research programme on the morrow of tabling their formula. Moreover, the Russians knew the research programme included nuclear explosions and this might give them a pretext to break off negotiations.[210] Instead, the UK preferred to emphasise that the first deciding point ought to be from the remaining portion of the seismic research programme, which would be announced tripartitely after a co-ordinated programme was agreed. Macmillan asked Lord Home to press this on Herter in the margins of the UNGA meetings in New York in the autumn, noting that the UK would also raise the matter in Washington at official level. Officials felt that a compromise over the duration of the moratorium at about three years ought to be possible. Ministers still attached great importance to the conclusion of the test-ban treaty negotiations and believed that an agreement was both practicable and worthwhile; they wanted to ensure that Washington understood this message

[207] TNA FO 371/149299, IAD 22/299, Nuclear Tests Conference, 24 August 1960.

[208] TNA FO 371/149303, IAD 22/342, Geneva Conference on Nuclear Tests, 1960.

[209] TNA FO 371/149302, IAD 22/334/G, 6 September 1960; and IAD 22/340 Washington telegram no. 1784 to Foreign Office, 8 September 1960; TNA FO 371/149342, IAD 24/445/G contains the records of the meetings held between 7–9 September; see also IAD 24/450.

[210] TNA PREM 11/3588, Foreign Office telegram no. 1802 to UKMis New York, 21 September 1960.

without appearing to chivvy the Americans unduly.[211] In any case, by late autumn Herter felt that as the Russians were marking time in Geneva until the new US Administration was in office, it therefore made sense to adjourn the negotiations until early February.[212] The Russians agreed and the resumption date was set at 7 February 1961.

UK Devices – The Last Word: Early 1961

The Foreign Secretary was most anxious to avoid any US criticism of the UK proposal by countering with a suggestion that the UK should provide all the nuclear devices needed. Home wondered how to deal with the safeguards issue and the US problem in persuading Congress to accept the notion that devices would be subject to outside scrutiny. If there was no alternative, he thought that the UK should offer the devices required for the seismic research programme, which London believed would meet the Russian safeguards requirements.[213] The UK delegation would only table this in Geneva if the US failed to proceed with its own fallback offer. Home thus wrote to the Prime Minister on 12 January 1961 to suggest that the UK should make available some of its own nuclear weapons to help the seismic research needed to support a test-ban treaty. Although the Chancellor of the Exchequer agreed, he added that at that point the UK should not mention costs. Lloyd hoped that the UK would not have to pay the entire £3.5 million bill.[214] Macmillan thought that the UK would indeed have to make its own devices available, and although he did not think the UK should make it a condition, he hoped that the US would cover the costs.[215]

Meanwhile, Makins was still not convinced that this was a good idea as it posed practical and legal problems for the UKAEA in its dealings with US agencies. Any weapon once on US soil became subject to US law and the fact was that UK devices owed a considerable debt to early work in the US. In his view, the UK should not make this offer unless required to break an impasse.[216] Both the UKAEA and MOD were of the opinion that any UK offer of test devices should only be made to save

[211] TNA FO 371/149309, IAD 22/397, Foreign Office telegram no. 2612 to Washington, 11 November 1960.

[212] TNA PREM 11/3589, Washington telegram no. 2349 to Foreign Office, 25 November 1960.

[213] TNA PREM 11/3590, Home to Prime Minister, Nuclear Tests Conference, 12 January 1961.

[214] TNA CAB 21/4536, Selwyn Lloyd to Prime Minister, Nuclear Test Explosion for Seismological Research, 26 January 1961.

[215] TNA CAB 21/4536, Prime Minister's Personal Minute Serial Non. M 43/61, Macmillan to Foreign Secretary, 27 January 1961.

[216] TNA DEFE 7/2296, Roger Makins to Sir Frederick Hoyer Millar, Foreign Office, 27 January 1961; and also 26 January 1961.

the Geneva negotiations and was acceptable to all parties. There would be problems for AWRE if Blue Danube had to be scaled down from 15 kilotons to five, or even increased.[217] The Foreign Office stressed to Makins that Macmillan's agreement was only to the principle and that the views of the previous summer were not forgotten; the whole point of the current proposal was to be able to table it, if it became apparent it was needed to dig the Americans out of their difficulties.[218] The issue of safeguards against any military development from explosive devices used for seismic research was at the heart of this; the Russians had insisted that if this idea were to fly, then they had to be given full access to design details and to the assembly of any device to be detonated. The Americans could not meet this requirement, hence their approach to the UK.

Having obtained Macmillan's agreement Home circulated this idea to the Chancellor, the Minister of Defence and the Chairman of the UKAEA. They would discuss the proposal again once their views had been received.[219] British technical problems were not to be divulged to the Americans at that stage. The UKAEA had prepared a paper that identified the two most practicable alternatives. First, the UK could remake a past British experimental design. This had the disadvantage that the design was of an advanced character and encroached upon the design of current weapons; in addition, not all yields in the range required for seismic calibration purposes were proven by experiment. Second, the UK could offer a service British weapon, though of comparatively early design. This had obvious security disadvantages; furthermore, scaling to produce different yields from those of the weapon's design might produce inaccurate yields and thus cast doubt on the safety of UK weapons. Whichever course they took, it would cost about £3.5 million for the 11 devices the US wanted to explode in the seismic research programme. The UKAEA hoped that the US would pay for most of this. Either course meant a three-month delay in fissile material fabrication for UK service weapons.[220]

The MOD told the Foreign Office on 22 February 1961, shortly before the Foreign Office Minister of State went to Washington to discuss testing issues, that in the last resort, after the US had proved itself quite unable to overcome its present difficulties over the safeguards issue, and if as a result the Geneva talks were threatened with collapse, it would favour offering a UK device as a means of breaking this impasse. However, this offer would be limited to the UK's first generation nuclear weapon, Blue Danube, while safeguarding the security of ancillary equipment.[221] Moreover, assurance would be needed that scaled

[217] TNA DEFE 7/2296, E.W. Playfair to Sir F. Hoyer Millar, 2 February 1961.

[218] TNA DEFE 7/2296, Sir F. Hoyer Millar to Sir R. Makins, 1 February 1961.

[219] TNA PREM 11/3590, P. de Zulueta to J.O. Wright, Foreign Office, 13 January 1961.

[220] TNA PREM 11/3588, Home to Prime Minister, Nuclear Tests Conference, 12 January 1961.

[221] TNA DEFE 7/2296, E.W. Playfair, MOD to Sir Frederick Hoyer Millar, Foreign Office, 22 February 1961.

models of Blue Danube would be acceptable for the proposed seismic research programme (namely, the proposed yields required for the research programme). The MOD emphasised a key further point. For the first UK weapon, AWRE had been indebted at least in part to early work in the US. For this reason London needed an assurance that the US Administration would not object to this becoming known and would defend the UK against possible Congressional attack when the time came for public inspection of the device offered by the UK.

A New Approach: The Kennedy Administration Takes Office

Shortly after taking office the new Kennedy Administration sought to delay resumption of the test-ban negotiations until 21 March: more time was required to review US positions. As for British thinking, there were now essentially three reasons why the UK was still seeking a test-ban treaty. First of all both the development of new types of nuclear weapons and any increase in the number of countries able to develop their own nuclear weapons might seriously prejudice hopes for disarmament; second, the achievement of a treaty would give a fresh impetus to disarmament measures and its verification system would create a valuable precedent for future verification; and third, if atmospheric testing were to be resumed on a large scale, dangers to health might result.[222] The 'nth' country problem was now top of the list of reasons for a test ban.[223] The UK would not resume testing so long as useful discussions continued at Geneva. A UK team went to Washington at the end of February 1961 to discuss tactics and substance ahead of resumption of the talks in Geneva. After three days of meetings a broad measure of agreement emerged on what needed to done, but it was not entirely clear to the UK what revised position the US proposed to put forward at the Conference when it re-opened.[224]

Despite the problems, agreement was achieved and the UK and US tabled a new draft treaty in Geneva on 18 April 1961. Both felt that this took account of progress achieved to date, and much of the text represented material already agreed or adopted in the negotiations to date. The remainder incorporated material put forward by the UK and US in March in an effort to meet the Soviet position, in particular the new treaty provided for underground tests to be discontinued down to an agreed 'threshold', which was fixed in terms of 4.75 mb or greater. For those tests below this event there would be a voluntary moratorium of an agreed fixed

[222] TNA FO 371/157079, IAD 22/40, Cabinet meeting of Commonwealth Prime Ministers, 1961, United Kingdom Delegation Discontinuance of Nuclear Weapon Tests, Brief by the Foreign Office.

[223] TNA FO 371/157079, IAD 22/42, Attachments to H.C. Hainworth to Mr Wilson, 20 February 1961.

[224] TNA FO 371/157081, IAD 22/63, A. Brooke Turner to Mr Shepherd, 14 March 1961.

period during which intensive research would take place to improve detection and identification capabilities of the control system.[225]

Deadlock in Geneva: Pressures to Resume Atmospheric Testing

Within a little over a month of the negotiations resuming Ormsby-Gore had come to the view that the Conference had become completely stuck. Not only had the USSR made no constructive moves since the resumption on 21 March, they had made virtually no move since the collapse of the Summit the previous May in Paris. Tsarapkin, for instance, had once again spoken against US plans to conduct decoupling experiments in the seismic research programme. In his view that issue was political and consequences would be serious if the US went ahead without Soviet agreement.[226] Furthermore, when the conference had reopened after its recess in April, the UK and US had put forward a series of proposals designed to take care of all the reasonable requirements of the USSR.

Western powers now accepted Soviet demands on safeguards that would assure the Russians that devices used for explosions in the US seismic research programme would not give the Western side a military advantage. They also accepted the Soviet demands that the East and West should have parity on the Control Commission, on procedures for carrying out PNEs and for a ban on tests in outer space subject to the best control then available, extension of the length of the moratorium on small underground tests from 27 months to three years and a reduction in the number of control posts on Soviet territory from 21 to 19.[227] Most of the evidence suggested that the Russians had, at least for the time being, lost interest in a test ban.[228] Arthur Dean, the new head of the US delegation, had reported to Washington that the Russians were showing no signs of making any compromise move at the conference table. Indeed, they were going backwards.[229]

So what next for the UK? Some initiative outside the Conference was needed; there was no point trying to shift Tsarapkin. He was immovable given his lack of any flexibility. Ormsby-Gore had lunch with Tsarapkin at the end of April 1961 and told him bluntly that he had previously thought that the USSR had really wanted a CTBT. He now reluctantly concluded the Russians were no longer interested. How

[225] Cmnd. 1937, *Text of the Draft Treaty on the discontinuance of Nuclear Weapon tests tabled by the Delegations of the United Kingdom and the United States at the 292nd meeting of the Conference on the Discontinuance of Nuclear Weapon Tests at Geneva on April 18, 1961* (London, June 1961), p. 3.

[226] TNA FO 371/157086, IAD 22/147, Geneva telegram no. 93 to Foreign Office, 13 April 1961.

[227] TNA FO 371/157089, IAD 22/171, Nuclear Tests Conference, 12 April 1961.

[228] TNA PREM 11/3590, Ormsby-Gore to Secretary of State, 25 April 1961.

[229] TNA FO 371/157090, IAD 22/186, Geneva telegram no. 134 to Foreign Office, 25 April 1961.

otherwise could one explain the reasonably flexible attitude with the Eisenhower Administration compared with the completely negative and obstructive attitude since March?[230] Ormsby-Gore thought that simultaneous messages from Macmillan and Kennedy to Khrushchev should be sent.[231] The object would be to engage the Soviet premier personally and publicly in the success or failure of the Conference and make it a test of his good faith in the whole disarmament field. Only a fundamental reconsideration of its policy by the USSR at the highest level would save the negotiations from early collapse.

Dean agreed with this assessment. Macmillan concurred too and he wrote to Kennedy making the arguments for such an approach.[232] The Prime Minister emphasised that in his own view the Russians would only be induced to reconsider their negative policy when they had been made to understand that their attitude, if maintained, would lead to a resumption of testing, the consequent spread of nuclear weapons and would make renewed disarmament negotiations worthless. Meanwhile, the US military was champing at the bit to resume atmospheric testing and had asked to make their case directly to the President. They had three main reasons for resumption: to improve the space-to-weight ratio of weapons, to develop a neutron bomb, and to develop an anti-missile fusion warhead. US arguments did not convince the MOD however, though it is not clear whether AWRE shared these doubts.[233] The practicality of a pure fusion bomb was currently under discussion between the UKAEA and USAEC, and it was not certain that it would prove feasible, even if tests did resume.[234] The then current scientific view was that the development of such a warhead in isolation would not make anti-missile defence feasible. In any case, the Russians stood to gain more in a military sense from the resumption of testing than the Americans for the simple reason that they were a considerable way behind in terms of development.

---

230 TNA FO 371/157091, IAD 22/204, D. Ormsby-Gore to Head Atomic Energy and Disarmament Department, 27 April 1961; Ormsby-Gore made the same points to the Soviet Ambassador in London on 4 May 1961. He hoped that when he got back to Geneva the following week that the Russians would have something more constructive to say. TNA FO 371/157092, IAD 22/215, Note to Head AEDD, 1961.

231 TNA FO 371/157091, IAD 22/193, Nuclear Tests Conference: Proposed Approach to Khrushchev, A.D. Wilson to Minister of State, 25 April 1961.

232 TNA PREM 11/3590, Prime Minister's Personal Telegram Serial No. T. 243/61, 27 April 1961.

233 TNA PREM 11/3590, Inward Saving Telegram no. 286 saving from Washington to Foreign Office, 4 May 1961; and Harold Watkinson to Prime Minister, 31 May 1961: 'they are much the same arguments which they have always used, and they do not convince us'. TNA FO 371/157117, IAD 26/6/G, E.W. Playfair, MOD to A.D. Wilson, Draft Minute from Minister of Defence to Prime Minister, 18 May 1961.

234 TNA FO 371/157117, IAD 26/6/G, The US Military Case for Resuming Nuclear Tests, Draft Commentary on Washington saving telegram no. 286 to Foreign Office, 4 May 1961.

Ministers and officials were clearly worried about the internal US pressures to resume testing, and were most anxious to avoid any risk that the West would take the brunt of any international opprobrium. It was clear too that in view of Soviet attitudes there was no chance of progress in Geneva and therefore it would be essential to agree a new US proposal to put the talks into recess. If necessary the UK would be prepared to support a resumption of testing by the US, however if the US did so decide, London wanted to be consulted about the timing and modalities of any resumption and about how it should be announced.[235] Macmillan hoped that the US would pursue with the UK the idea of simultaneously offering a partial test-ban treaty without controls providing for a ban on nuclear tests in the atmosphere or under water.[236] In the event, the US did decide to test again, but that is for discussion in Chapter 7.[237]

The Soviet Testing Aide-Memoire: Propaganda Triumphs over Commitment

The situation following the receipt of an Aide Memoire to the Americans in June 1961 setting out Russian views on testing issues was considered at a meeting of the Prime Minister, Minister of Defence, Commonwealth Secretary and Ormsby-Gore on the 14 June. This document was little more than a diatribe against Western policies. The US replied in a vigorous rebuttal of Soviet arguments on 17 June; it had ended by urging the Russians to reconsider their position. On the question of resumption of testing by the US, the note made clear that for nearly three years the US had accepted the risk of not testing with any certainty that the Russians were not doing so; US national security would not allow this risk to be assumed indefinitely.

The Administration was under considerable internal pressure to resume, but the UK and US agreed that the West should not break off the talks in Geneva, even if the US should decide to resume. If, as seemed certain, the US reply produced no change in Soviet attitudes, the UK believed that the Prime Minister and President should make one final appeal to Khrushchev. If this too was rejected, then the UK and US might offer a ban on tests in the atmosphere and under water only and which could be verified without controls in the USSR. Kennedy however was not in favour of this and Macmillan was prepared, following discussions with Home, to acquiesce for the moment.[238] For the present it would be better to keep the negotiations ticking over somehow until the autumn of 1961.[239] After a further

[235] TNA FO 371/157117, IAD 26/156/G, Nuclear Tests Negotiations, Foreign Office paper, 9 June 1961.

[236] TNA FO 371/157117, IAD 26/17/G, Foreign Office telegram no. 4062 to Washington, 14 June 1961.

[237] TNA FO 371/157098, IAD 22/304, Nuclear Tests Conference, 19 June 1961.

[238] TNA FO 371/157098, IAD 22/307, A. Samuel to Mr Mitchell, 20 June 1961.

[239] TNA FO 371/157100, IAD 22/336/G, Philip de Zulueta, Admiralty House to A.C. Samuel, Foreign Office, 5 July 1961.

round of exchanges of notes between the US and USSR in July, there was no improvement in test-ban prospects. Macmillan was particularly pessimistic. Given the generally poor state of East-West relations at that time, he feared that the UK and the US would either have a war with the USSR or a negotiation in the course of the next six months.[240] All these issues were left otiose when the USSR resumed atmospheric testing on 1 September 1961. Detailed planning must have started well before then given the scale of the testing programme – in two geographically separate locations. Khrushchev had told the US Ambassador in Moscow in May that the Soviet military and scientists were pressing for resumption of testing. More tellingly, he said that the Russians had practical plans for further testing.[241] Tests would be useful to the USSR because their present nuclear weapons used large quantities of fissile material and tests could reduce this by half.[242] Marshall Malinovsky wanted to resume testing and Khrushchev had implored the US Ambassador to believe this. It was now a new ball game, but certainly prior to this point no decision had been yet taken about a possible UK announcement on its freedom to resume testing.[243]

UK-American Differences 1958–1961: A Brief Overview

Despite its title, the Tripartite Conference involved two sides – the UK and US on one side and the Russians on the other. This is not to say that there was absolute harmony all of the time and on all of the issues on the UK-American side: quite the contrary. Even at the outset, it was far from clear to the UK that the US really wanted a test ban at all; Dulles told Harold Caccia on 16 October 1958, about a fortnight before the talks actually began, that he had an awful battle to win acceptance for a test-ban suspension even on the then present conditions. The feeling in his country, unlike the UK, was more alarmed at the prospects of reaching an agreement than of the failure to do so.[244]

Britain took a different line on many issues and, as evidenced above, was more ready to contemplate positions that were more moderate and to urge these on the Americans. If it were only a matter of differences over tactics, then there would

---

240 TNA FO 371/157100, IAD 22/336/G, Philip de Zulueta, Admiralty House to A.C. Samuel, Foreign Office, Draft telegram from the Prime Minister to President Kennedy, 5 July 1961.

241 TNA FO 371/157096, IAD 22/276/G, H.C. Hainworth to Minister of State, 25 May 1961.

242 TNA PREM 11/3590, Moscow telegram no. 1006 to Foreign Office, 27 May 1961.

243 TNA FO 371/157098, IAD 22/293, Foreign Office telegram no. 2921 to Washington, 19 June 1961.

244 TNA FO 371/132686, IAD 122/27, Washington telegram no. 2784 to Foreign Office, 16 October 1958.

not be much of a story to tell. From the outset there was a problem, but this did not become apparent to the UK until later in the negotiations. There was much uncertainty in Washington about what policy would best serve US interests. An important reason for this was that the US Administration, Dulles in particular, offered to participate in a conference in the belief that the USSR would never enter into serious negotiations. When events turned out differently, doubts arose about the wisdom of continuing. Many Americans in and outside government feared that entering into an agreement would limit US freedom of action.[245] In fact, Dulles even privately instructed the US delegation to work for a breakdown of the Conference without informing the UK once the USSR had shown no gratitude for the Western dropping of the link with disarmament.[246]

In a paper prepared for the Prime Minister by the Foreign Office in January 1959, officials noted that it was clear that the UK's own negotiating position was made much more difficult by disputes between the different agencies concerned in formulating US policy.[247] Disagreement over duration was one problem where the US wanted a much easier right to withdraw from the treaty if it was proving 'unsatisfactory'. Con O'Neill, who suggested the need for the paper, had expressly asked that this point be included in it. He asked for something about the difficulties of the American position 'with all their warring factions in Washington'.[248]

David Ormsby-Gore remarked that there was no concerted view in Washington in favour of an agreement of any kind and that even at the best of times the Administration worked so slowly as to be almost medieval.[249] For Ormsby-Gore, the major concern was that unless the negotiations led to a reasonably satisfactory control organisation, there was a risk that the UK would find that testing had

245 TNA FO 371/149306, IAD 22/370, History of the Conference from its opening on October 31st 1958, 6 October 1960.

246 TNA FO 371/157105, IAD 22/426/G, Sir Michael Wright to Lord Home (Received September 27) p. 10, Geneva Nuclear Tests Conference: Review of Events to 9 September 1961.

247 TNA FO 371/140450, IAD 22/127, Geneva Nuclear Test-ban Talks, paragraph 8, 31 January 1959.

248 TNA FO 371/140445, IAD 22/33, H.C. Hainworth to Sir F. Hoyer Millar, Nuclear Tests Conference, paragraph 5, 9 January 1959.

249 TNA FO 371/140450, IAD 22/128, D. Ormsby-Gore to Secretary of State, Conference on Discontinuance of Nuclear Weapon Tests Undated received in Registry on 2 February 1959 but probably written before 5 January since the paper states 'when we reconvene on January 5'. Three years later after the Conference had been suspended *sine die*, officials reflected on the lessons learned and whether to share Sir Michael Wright's despatch from Geneva on the history of the negotiations with President Kennedy. One of these was that the Americans had missed a great opportunity to move forward in 1959 and 1960. TNA FO 371/163111, IAD 22/64, C A.D.F. Pemberton-Piggott to Mr A.D. Wilson, 29 March 1962. In fact, the Embassy in Washington pointed out that Kennedy held no brief for the way in which scientific opinion was marshalled against a test-ban treaty. TNA FCO 37/ 163111, IAD 22/646, Washington telegram no. 834 to Foreign Office, 16 March 1962.

ceased without any control.[250] Since Dulles retired on health grounds in April 1959, prime responsibility for deciding policy and for drawing up instructions for the US delegation was in the hands of a Committee of Principals. The differences between the points of view of the five departments represented explained why inaction was so often preferred to action. Even after a few weeks, it became apparent to officials that there was more to it that just a difference over tactics. Selwyn Lloyd minuted the Minister of Defence in mid November 1958 to state that it might be necessary at the end of the day to part company with the Americans on maintaining the link between disarmament and the suspension of tests. Such a decision would be a grave one to take and one that the UK would have to weigh most carefully. Lloyd hoped that it would not come to this and if the Geneva conference looked like breaking down he was sure that the Americans themselves would in the end agree to drop the link.[251] However, the MOD felt that it was still important to avoid being put in a position of having to accept a continuation of the tests suspension without any adequate *quid pro quo* from the Russians.[252] They did and agreed to do so on 20 November 1958.[253]

Wright observed in 1961 that whilst the Eisenhower Administration lasted, the test-ban negotiations were always the subject of a pitched battle, of varying ferocity, between the proponents of a treaty and its opponents.[254] In Wright's opinion, internal US problems were an appreciable part of the responsibility for the failure of the conference. This was true whether the failure was viewed in terms of a treaty lost or, probably more realistically, in terms of an opportunity missed to discover and demonstrate the real nature of Soviet intentions before the Summit of May 1960. Set against this was the recollection of another UK participant, who doubted that the Russians had been serious about a test ban at any stage of the negotiations.[255] The Eisenhower Administration was never able to make up its mind how much it wanted a treaty, indeed whether it wanted one at all – it delayed the decision to table a complete Western treaty for example. This was the real source of all US hesitations and ineptitudes. Failing a clear policy decision at the top, the disruptive tendencies inherent in the US machine were bound to

---

250 TNA FO 371/140444, IAD 22/18, Geneva to Foreign Office telegram no. 11, 8 January 1959.

251 TNA PREM 11/2565, Selwyn Lloyd to Minister of Defence, 18 November 1958.

252 TNA FO 371/132694, Richard Powell, MOD to C.D.W. O'Neill, Foreign Office, 19 November 1958.

253 TNA PREM 11/2565, Geneva telegram no. 645 to Foreign Office, 20 November 1958.

254 TNA FO 371/157105, IAD 22/426/G, Geneva Nuclear Tests Conference: Review of Events to September 9, 1961 Sir Michael Wright to Lord Home, p. 6, 27 September 1961.

255 Interview, 7 November 2006.

have free play and cause policy to swing back and forth, never able to seize the initiative along a predetermined course.[256]

For London, it was clear that the remaining problems were political rather than technical. If both sides were ready to make the necessary compromises, a satisfactory treaty could be achieved.[257] Looking back on events in 1961 Wright commented to Lord Hume that Dulles had forced the UK into an ill-considered conference that he never thought the Russians would take seriously and which he himself regarded primarily as a cold war exercise. UK policy, for good and inescapable reasons, was broadly to settle for any agreement that might be acceptable to the US. All therefore depended on the US, but the initial US statement warmly welcoming the 1958 Experts' Report was soon regretted, repudiated or abandoned in most of its particulars by US officials.[258]

US Views of the UK

Washington had some reservations too about its negotiating partner and these surfaced more or less from the start. As early as mid November 1958 Dr Hans Bethe, of the President's Science Advisory Committee and member of the US delegation, noted some concern over UK attitudes in a report for the then Under-Secretary of State, Christian Herter. Bethe observed that whereas in the open meetings the UK had been at pains to maintain a united front with the US, its position essentially was quite different and eventually, this would come out. In his view, the British believed that cessation of nuclear tests was inevitable under any circumstances and that any controls that could be had from the USSR would be 'gravy'. Consequently, in the last analysis the UK would be willing to settle for substantially less than the US.[259]

A year later Christian Herter, now Secretary of State, could remark that there was a problem of the UK moving ahead of the US position. Although the UK agreed that the problem of verifying underground tests was a difficult one, and that a phased agreement beginning with the prohibition of atmospheric tests would

256 TNA FO 371/157105, IAD 22/426/G, Geneva Nuclear Tests Conference: Review of Events to 9 September, 1961 Sir Michael Wright to Lord Home, p. 14, 27 September 1961.

257 TNA CAB 21/4535, INTEL no 141, 21 November from Foreign Office to Certain of HM's Representatives.

258 TNA DEFE 19/18, Comments on a Technical History of the Geneva Nuclear Tests Conference, Sir Michael Wright to Lord Home, 29 December 1961.

259 Department of State Memorandum of Conversation, Geneva Nuclear Tests Negotiations November 18, 1958 Present: The Under Secretary, Philip S. Farley and Ronald Spiers, S/AE, Dr James Killian, Special Assistant to the President, Dr Hans Bethe, President's Science Advisory Committee at http://www2.gwu.edu/~nsarchiv/NSAEBB/-NSAEBB94/tb08.pdf.

be easier to police, the UK argued that the only way to get the USSR to go along with this as an outcome would be to couple it with a three year voluntary ban on underground tests whilst more research was put in hand on the control problem. This difference with the UK was now reaching a crucial stage in at least one US official's mind – Phil Farley.[260] Herter himself was aware that the British attached a great deal of importance to achieving inspection in the USSR since this would represent a breakthrough on disarmament. In addition to this, Macmillan keenly felt the strength of British public opinion on this issue and was above all aware of the UK vulnerability to nuclear weapons. Farley said that in short the UK felt that the technical uncertainties of the situation – the new seismic data from Hardtack II – did not outweigh the potential advantages arising from a test ban with less than perfect verification provisions. Herter recognised that it would be difficult for the US to pursue its preferred course of action without UK support. In fact in the following month, Herter told his Cabinet colleagues that the British had continually indicated that they would be prepared to accept a more limited system than the US desired.[261] It was clear that the UK had been exerting strong pressure for a positive response.

Perhaps there should be no surprise in this. One academic observer has noted that there is a general impression that Macmillan was more eager to make concessions on the testing issue, for the sake of agreement, than was the US Administration.[262] But this would be unfair; it is also clear from the record, that whilst Macmillan was determined to secure agreement, he was not going to throw out the verification baby with the treaty bath water. Verification of some description was always part of the British approach. In November 1960, for example, the Foreign Office sent out a briefing telegram to selected posts. This emphasised that the UK certainly desired a treaty, but also insisted that it must provide adequate guarantees that no violation of its provisions would be likely to escape detection. The UK would not accept spurious verification measures.

Conclusions: Diplomatic Pressures and the Testing Moratorium, 1958–1961

We must infer that the UK was quite prepared to accept a self-imposed testing moratorium; it must have taken the political and scientific decision that it did not require further testing even though there are signs that neither the UKAEA nor the MOD was entirely convinced of the wisdom of such a policy. There is no evidence for either AWRE or MOD opposition to Macmillan's energetic pursuit of a test-

---

260 Department of State Memorandum of Conversation, 11 November 1959, Meeting of the Secretary's Disarmament Advisers.

261 United States Atomic Energy Commission Washington 25 D.C. record of Cabinet meeting, 11 December 1959 Consideration of Test Moratorium Negotiations, 14 December 1959 in http://www2gwu.edu/~nsarchiv/NSAEBB/-NASAEBB94/tb20.pdf.

262 Clark, *Nuclear Diplomacy*, p. 203.

ban treaty. It is certainly possible that scientists may have preferred privately to continue testing, but there was nothing equivalent to the concerted opposition that the treaty faced in the US. As we saw in Chapter 4 there were certainly contingency plans for further UK testing. It is reasonable to suggest that the British civil service ethos at that time would have made such a course unthinkable. In any case, access to US designs post 1958 meant that the future of the UK deterrent was secure with the adoption of the Red Snow warhead for both the Yellow Sun Mark II free fall bombs for the RAF as well as the self-propelled guided weapon – Blue Steel. We also see very clearly that the UK strove mightily for a test-ban treaty, so much so that it would have been prepared, if push had come to shove, to provide British nuclear devices for the seismic research calibration programme even though this would have been a major inconvenience for its own weapons programme. Although it is clear too that Macmillan wanted a verifiable test ban for its own sake and his concerns over an arms race out of control, atmospheric pollution and nuclear weapons proliferation were genuine, we cannot overlook the fact that there was a good deal of concern over appearances and propaganda. The West had to look good in relation to Soviet behaviour, but this does not detract from the determined UK effort in the 1958–62 Conference on the Discontinuance of Nuclear Weapon Tests to secure a comprehensive test-ban treaty.

Chapter 6

Test-Ban Treaty Verification and the Role of Seismology: UK Efforts 1958–1965

Introduction

As we have seen in the preceding chapter, the core technical problem in test-ban treaty verification was detection and identification of low-yield underground nuclear tests in differing geological formations. Cracking this problem would open the door to a treaty since it would help ensure that its signatories were observing the provisions of the ban. Moreover, effective remote verification could also address the political problem of Soviet objections to the large numbers of on-site inspections that both the UK and US believed were essential given the then limitations of remote sensing. Therefore, what could the UK do and where could it turn to for scientific and technical advice? Could the technical expertise developed to record the UK's atmospheric nuclear tests play a part here? What research and experimental work would be required and how could the UK work with the Commonwealth on this problem? What would it cost and how long would it take? This chapter will review UK seismic research and its impact on the detection and identification problem. If we want to understand the extent to which the UK was prepared to go to achieve a test ban, then we must look at the detailed technical work commissioned to help achieve that objective.

UK Seismic Verification Research: First Steps

Until the late 1950s, the British contribution to seismology had been limited to discussion of observational data collected by others.[1] It had none of its own and the absence, however, of observational and experimental work meant that there was little stimulus for theoretical investigations. In fact, the UK's only real experience in recording large explosions was that obtained in 1947 when army

---

1 A Revd R. Taylor and a Revd S. Gilfillan, Comrie, Perthshire had noted that the village had suffered a number of earthquakes and these two worthies began to record them. Other locals joined them and they became know as the 'Comrie Pioneers' and resulted in the establishment in 1839 of a Committee for the Investigation of Scottish and Irish earthquakes. One of its members, a Mr Milne, devised the term 'seismometer' and its associated words. Christopher Winn, *I Never Knew that About Scotland* (Chatham, 2007), pp. 189–90.

engineers destroyed wartime German coastal defences in Heligoland and the recordings of the Maralinga tests in 1956 obtained by the Dominion Observatory.[2] There were few trained seismologists resident in the UK at that time, most of those who studied geophysics at British universities taking up employment overseas.[3] In the late 1950s UK officials were grappling with what would become a familiar set of questions in future years. What was enough verification? What was necessary politically? What was achievable and/or practicable? If inspections were to be a deterrent, what did this imply in practice? Reducing the uncertainties in detection and identification for the detector whilst increasing them proportionally for the violator was the heart of the test-ban treaty verification conundrum, then as it is now. The inherent uncertainties over knowing when an inspection could take place might be enough to deter a would-be cheat, but no one could ever be certain. On-site inspection would always be necessary: remote detection and identification could not guarantee to explain all events.

As we saw in the previous chapter, one of the very first evasion scenarios that emerged was decoupling. A nuclear test might be conducted in a cavity. This would reduce the seismic signature and make it more difficult if not impossible to detect and identify the event as a nuclear explosion. Seismic signals might in fact be several hundred times smaller than those that might be expected as result of the observations from the Rainier shot.[4] However, scientific opinion was divided. In the US, Edward Teller insisted that such a scenario was indeed possible, whereas John Carson Mark did not.[5] As the UK had virtually no experience in observational seismology, the British delegation had been able to make little contribution to the seismic verification discussions at the 1958 Experts' Meeting. As one of its members explained, the UK team went into the meetings 'seismically naked but

[2] The 1956 Maralinga seismological experiments provided very successful results. Ten seismograms of importance were obtained from the atomic detonations. These readings were used to determine for the first time the thickness of the Earth's crust in the Australian continent. It was concluded that the Western Australian crust had a thickness of around 35 km in depth. The depth of crust and seismic properties were similar to other continental shield areas such as Canada, South Africa and Siberia. http://www.allshookup.org/quakes/atomic.htm; P.L. Willmore, 'Seismic Experiments on the North German Explosions, 1946 to 1947', in Philosophical Transactions of the Royal Society of London. Series A. Mathematical and Physical Sciences, vol. 242, Issue 843, pp. 123–51, August 1949; Personal Correspondence.

[3] TNA FO 371/149319, IAD 24/112, Commonwealth Collaboration in Seismology, UKAEA, 18 March 1960.

[4] Atomic Weapons Research Establishment, *The Detection and Recognition of Underground Explosions*, p. 91. The Pascal A, 26 July 1957 shot in the Rainier test series was the only underground nuclear test by 1958 and data from it had been used to help design the 1958 Geneva experts' report.

[5] http://www2.gwu.edu/~nsarchiv/NSAEBB/NSAEBB94/tb05.pdf, Department of State Memorandum of Conversation, US Policy on Nuclear Tests, 13 August 1957, p. 4.

with a lot of ability and wit'.[6] It was heavily outnumbered and overwhelmed by US and Soviet seismic data and experience. So the first job on returning home was to establish a research programme to correct this.

Once the decoupling scenario was presented to the Russians in Geneva in the Conference on the Discontinuance of Nuclear Weapon Tests, Penney instructed AWRE Foulness to see whether this theory could be verified. The Foulness Ground Shock team, which mostly dealt with basic ground shock and air blast effects of UK nuclear explosions at that time, quickly set up a programme of four coupled and decoupled chemical explosions, recording the ground velocity from each explosion. Foulness had a good fundamental understanding of scaling laws, and the tests involved charges of two ounces of high explosives fired three feet below the surface, with subsequent shots in the cavity created by the first explosion.[7] Decoupling factors of ten to 20 were easily attained and the UK delegation presented the results to the Conference.[8] This gave the experts at Foulness a higher profile both within AWRE and with Penney. Consequently, some of them were transferred to Aldermaston and later Blacknest to continue the work.[9] Harold Watkinson, Minister of Defence, told the Foreign Secretary and Prime Minister in December 1959 that the UK experiments had supported the strong probability that a 20 kiloton device could be exploded in an underground cavity, with only a slight risk of detection. Furthermore, since nothing bigger than 20 kilotons was likely to be needed for weapon development purposes, the consequences of this for a treaty were therefore very significant.[10] Whilst excavations of the scale needed to create a sufficiently large cavity could not be concealed in the Western world, inside the USSR was another matter. There was a risk that the Russians could engage in clandestine nuclear weapons testing without fear of certain detection by the control systems currently envisaged and available.

AWRE Involvement

In 1959 the Atomic Weapons Research Establishment at Aldermaston was the clear choice as the centre for the UK studies into seismological methods of detecting

---

6 Personal correspondence. Sir Edward Bullard, a geophysicist at Cambridge University, however did represent the UK for some of the time.

7 TNA ES 1/1349, Some Preliminary Notes on the Seismological Problem, J.K. Wright, 2 June 1959.

8 P.D. Marshall, *UK Contribution to Regional Seismic Discrimination* in AWE Report 632/01, *Test-ban treaty verification research: Seismic signals recorded at regional distances* by D. Bowers, December, 2001, p. 2.

9 Oral History Seminar, AWE Blacknest 10 January 2006: Peter Marshall, Dr Alan Douglas, Dr Eric Carpenter and Frank Whiteway.

10 TNA PREM 11/3584, Harold Watkinson to Foreign Secretary, 7 December 1959.

underground nuclear explosions.[11] The operational centre for all UK seismic research was located at Blacknest, a country house a few miles from the main AWRE site. Penney thought that it was important that this centre should be outside the high security regime at Aldermaston, so that academic and foreign experts could readily visit it. Instrument development studies and record analysis would take place at this site, and it would serve as a data collection and analysis centre. Aldermaston was already engaged on studies of other methods of detection, and a body of instrumentation experts and their resources were available because of the nuclear test moratorium.[12] Aldermaston was also examining, as part of its detection work, changes in atmospheric pressures produced by nuclear explosions at heights below 100,000 feet and auroral effects in the upper atmosphere from an explosion at heights of many hundreds of miles.[13] Some 40 to 50 scientists were eventually deployed on seismological studies, although initially the AWRE team did not have any nationally recognised seismologists.[14]

AWRE had to learn fast. The problems faced were very different from those encountered in traditional seismology and new equipment and approaches had to be developed. Instead of worrying about recording events that lasted only nanoseconds, there was now a requirement to deal with very low frequency signals. Yet John Wright one of the leading scientists involved at that point had already concluded that seismometers placed in boreholes would be the solution to the sound to noise ratio problem.[15] Diagnostic techniques used during the UK's own testing programme such as microbarograph, electromagnetic, seismic, radiochemical, ionospheric and optical detection methods would help tackle the detection of atmospheric and high altitude tests. AWRE also had experience in data recording, data processing and telemetry.[16] What was now taking place at Aldermaston was a unique case of the redirection of resources towards verification requirements within the UK defence scientific community. Almost all those involved had no previous experience in seismology. AWRE consulted the

11 Atomic Weapons Research Establishment, *The Detection and Recognition of Underground Explosions A Special Report of the United Kingdom Atomic Energy Authority*, Published by the United Kingdom Atomic Energy Authority, 11 Charles II Street, London S.W.1, December 1965, p. 16.

12 Oral History Seminar, AWE Blacknest, 10 January 2006; Bates, Gaskell and Rice, *Geophysics in the Affairs of Man*, p. 194.

13 United Kingdom Atomic Energy Authority Sixth Annual Report 1959–60, 13 July, 1960 (London, 1960), p. 37; see also United Kingdom Atomic Energy Authority Tenth Annual Report 1963–1964, 15 July, 1964 (London, 1964), p. 47 for work on detecting nuclear explosions in the upper atmosphere and space by means of ground based stations.

14 Bates, Gaskell and Rice, *Geophysics in the Affairs of Man*, p. 194.

15 Oral History Seminar, AWE Blacknest 10 January 2006: Peter Marshall, Dr Alan Douglas, Dr Eric Carpenter and Frank Whiteway.

16 TNA ES 4/405, AWRE Report No. O. 30/60, Control and Detection of Nuclear Explosions. An Estimate of Cost and Technical Complexity, I. Maddock, July 1960.

Institute of Oceanography for advice. Initially there were only two seismologists on the AWRE team – Hal Thirlaway, although recruited from the outside to lead the project, he had advised the Secret Intelligence Service on test detection, and Pat Willmore. Others had to learn on the job. The seismic problem was examined purely as one of exact measurement, with special emphasis on the nature of the signal form and what happened to it along its transmission path.[17] Further research and empirical testing were clearly required given the scale of the difference in the yield of the US Rainier nuclear test in October 1957.

Orpheus

To satisfy the requirement for empirical testing, the UK and US went on to conduct further experimental test explosions using conventional high explosives to check the hypothesis that the seismic signal from an underground nuclear explosion could be greatly reduced by conducting the explosion in a large cavity – projects Orpheus and Cowboy respectively.[18] These projects were fully complementary to avoid duplication.[19] A small UK team also participated in the US tests. The first Orpheus experiment, Phase A, took place in a spherical chamber in a limestone quarry at Caldonlow, Derbyshire in January 1959. The second, Phase B, eventually taking place at Glenridding, Cumberland in the Greenside lead mine in April 1959. The latter entailed decoupled shots of a few thousands of pounds of explosives and took place in a sphere of around 30 feet in diameter.[20] Laboratory work simulated

---

[17] Bates, Gaskell and Rice, *Geophysics in the Affairs of Man*, p. 194.

[18] Preliminary evaluation from the 'Cowboy' series showed that decoupling through underground testing in salt was practical and that a considerable reduction in signal could be obtained even by a system of decoupling which was not completely successful. See Department of Memorandum of Conversation Geneva Nuclear Test-ban Negotiations: meeting of Principles 22 March 1960, p. 4 in http://www2.gwu.edu/~nsarchiv/NSAEBB/NSAEBB94/tb22.pdf. The experiments were to test the decoupling theory as set out later publicly by A.L. Latter, R.E. Lelevier, E.A. Martinelli and W.G. McMillan in *Journal of Geophysical Research* vol. 66, 3 March, 1961; see Romney, *Detecting the Bomb*, p. 146.

[19] TNA DEFE 7/2289, D.E.H. Peirson, UKAEA to F.F. Turnbull, Atomic Energy Office, 30 October 1959; TNA AB 16/2971, Atomic Energy Authority UK Experimental Work on Underground Explosions, Note by Sir William Penney, 30 October 1959. A joint programme was agreed by the Cabinet Working Group on Disarmament GEN 604/16, 13 August 1959. Other proposed UKAEA activities included improvement of instrumentation and inspection, the study of on-site inspection of selected earthquakes; the study of the problems of radiation transport and the effects of x-rays and neutrons produced by a nuclear explosion in a large cavity; and further studies on determining new techniques for decoupling and for concealment and for detection and inspection.

[20] TNA ES 5/285, AWRE Report No. T 11/60 Operation Orpheus Phase C, The Seismic Signals Recorded from a Decoupled 3000 lb and Coupled 1000 lb Charge, E.E. Carpenter, J.E. Uppard, March 1961, p. 4.

a nuclear explosion at depths of 1000 feet; this entailed the firing of an ounce of chemical explosive in a concrete cube with dimensions of only six feet.[21] Further small-scale tests took place at Callington, Cornwall in the autumn 1959 following a survey of possible locations in the UK in the summer 1960 and a request to the Duchy of Cornwall for permission to proceed with the experiments in June.[22] In all 70 shots were fired; 40 decoupled and 39 coupled.[23] Three large shots were fired in Orpheus Phase C, a calibration shot of 100 lb, one decoupled shot of 3010 lb and one coupled shot of 1160 lb.[24] Orpheus was intended to obtain preliminary data about seismic decoupling for the 1959 Geneva Technical Working Group No. 2. This was convened to discuss the problems involved in detecting underground explosions. Independent work was important because it enabled the UK to talk on the subject area with as much insight and original data as the other nuclear powers.[25] Indeed one of the reasons given to the Prime Minister for conducting these tests was the need to acquire independent UK information in order to be able to exercise a much greater influence on US attitudes towards the Geneva talks.[26] The results provided ample evidence that decoupling was a real phenomenon that was observable in practice. The only questions of principle that remained were whether a decoupled nuclear explosion would behave as the theory predicted and was decoupling a practical engineering proposition.[27] When the UK presented the results to the Conference's Technical Working Group No. 2 between 25 November

---

21 United Kingdom Atomic Energy Authority Sixth Annual Report 1959–60, 13th July (London, 1960), p. 37.

22 TNA ES 1/353, Non-Nuclear Underground Explosions Possible Sites in the UK, E.R. Drake Seager, SSRT, AWRE, June 1959; TNA ES 5/313 AWRE Report No. T12/62 Operation Orpheus Phase A, January 1963 Mining and Engineering Report No. T12/62, H.D.M. Jager, E.H. Yeo, J.P. Withers, January 1963.

23 TNA ES 5/306, AWRE Report No. T 5/62 Project Orpheus, February 1962 Phases A and C, Operational Report, E.R. Drake Seager, February 1962.

24 TNA ES 5/300, AWRE Report T 11/61 Project Orpheus Phase A 1962, The Seismic Signals at Surface Zero from Coupled and Decoupled Charges, R.J.N. Timothy and S. Melville, January 1962. A involved operations with small charges in 6 ft diameter cavities in granite and a metamorphosed shale at depths of 100 to 300 feet in Excelsior tunnel, Callington, Cornwall; Phase B also entailed participation in the US Project Cowboy with charges up to 3000 lb in 30 ft diameter cavities at depths of 800 ft in rock salt at Carey salt mine, Louisiana whilst Phase C used charges of 1100 and 3000 lb at depths of 1700 ft and a site at Greenside Lead Mine Glenridding, Westmoreland; TNA ES 5/306, AWRE Report No. T 5/62 Project Orpheus February 1962 Phases A and C, Operational Report, E.R. Drake Seager, February 1962.

25 TNA DEFE 7/2296, Underground Nuclear and High Explosive Tests in UK, Comparison of Status of COWBOY and Orpheus Programmes, October 1959.

26 TNA PREM 11/2839, Plowden to Prime Minister, Underground Explosions in Cornwall 20 July 1959; UKAEA Press Release, 11 September 1959.

27 Atomic Weapons Research Establishment, *The Detection and Recognition of Underground*, p. 93.

and 18 December 1959, UK experts supported the US proposition that the Hardtack underground tests had proved the 1958 Geneva Experts' Group conclusions had been over optimistic in setting the detection requirement for an effective control system.[28] Their results were based on a single test. We should also recall that at this point that there was a great dearth of instruments for precise seismic measurement. AWRE therefore launched a development programme in the autumn of 1959 to produce instruments with more sensitive capabilities.[29] Existing instrumentation was antiquated and modern recording and data handling techniques were not applied in this area. Much needed to be done, both in sifting through existing data and in creating an integrated modern recording system that would enable the disputed facts in detection and identification to be defined more accurately.[30]

Seagull and MacSeagull

Once the three phases of Orpheus concluded, AWRE turned its attention to the next phase of the UK's seismic research programme: Operation Seagull Phase I. Planning for this started at the end of 1960. Seagull was a field experiment designed to learn more about large explosion phenomenology, in particular the variation of seismic signal with yield as a function of frequency and distance. It would also address methods of improving the detection of seismic signals by putting instruments down boreholes in arrays.[31] In order to provide a source of seismic signals whose size, location and time of burst could be controlled for array experiments, the Royal Navy collaborated in firing a regular pattern of nine 300 lb depth charges 14 miles south of Portland Bill in the spring of 1961. This operation also gave an opportunity to study the way in which the immediate location and the basement rock-type of a station influenced the character of a signal. Accordingly, 13 simple seismometer stations were positioned at locations across the south of England (Oakhampton Range Dartmoor) and Wales. Seagull was singularly successful in demonstrating that the array principle actually worked. The records from this exercise were extremely satisfactory and were beyond the expectations of those scientists involved.[32] AWRE's aim was now to establish and operate a seismic station with a performance as good as, or better than, the standard set by

---

[28] TNA DEFE 19/18, Technical History of the Geneva Nuclear Tests Conference 1957–1962 (R.W. Snelling, Geneva September 1961) paragraphs 34 and 35, pp. 16–17.

[29] TNA AB 16/3666, Note by Sir William Penney, draft paper, Programme of Underground Non-Nuclear Explosions 'Operation Orpheus', October 1959.

[30] TNA ES 4/399, AWRE Report O-22/60 Detection of Underground Nuclear Explosions, I. Maddock, p. 19, April 1960, p. 19.

[31] TNA AB 16/3446, E.R. Drake Seager (AWRE) to D. Ostle, Geological Survey of Great Britain, 2 December 1960.

[32] TNA FO 371/157112, IAD 24/24, Mr E.R. Drake Seager to Dr Levin, Aldermaston, 9 March 1961; TNA ES 5/293, AWRE Report No. T.3/61 Operation Seagull Phase 1,

the 1958 Geneva Experts' Meeting. Blacknest scientists also wanted to look at the feasibility and practical problems involved in operating unmanned seismic stations for short range. This was known as zone one, which covered distances out to 500 km away from the signal source. It had been apparent from the last round of test-ban negotiations in Geneva that such things might be more acceptable politically to the USSR.[33] Eskdalemuir in Scotland was conceived initially with this in mind. AWRE also wanted to acquire data to help distinguish explosions from earthquakes.

Seagull, however, was not the only element in UK seismic research. In 1961, there were five other main elements to the UK seismic programme. First, seismic recording stations were operated near Aldermaston and Eskdalemuir, at which latter station a Geneva-type instrument system and other new systems, including a long-period seismometer, were to be subjected to long-term tests. Experiments with a borehole seismograph were also to be undertaken. Second, seismograms from the Canadian seismological network and from Aldermaston and Eskdalemuir were to be processed at Blacknest. Third, AWRE planned to develop a new long-period seismometer and the existing short-period and long-period devices would be modified to make them conform to the 1958 Experts' report recommendations. Fourth, development of digital and frequency-modulated tape recording systems would be completed.[34] Finally, research into the problems with the existing seismometers together with the associated recording systems was started at about this time and a study of the methods for improving the signal to noise ratio began.[35] AWRE's research programme sought to accumulate the critical data on background signals and to develop criteria by which they could be recognised. This was needed to ensure that detection systems were not overburdened with data so that false alarms could be identified and discarded.[36]

One of the major problems was provision of suitable seismometers that could reduce the signal to noise ratio. UKAEA experts considered three projects to address this problem. The first of these would look at the possibility of increasing the number of seismometers at a seismic station and deploying them in some form of array; the second would place seismometers down deep boreholes, the third was to conduct some borehole experiments to discover the optimum depth at which

Operational Report, W.P. Cook, November 1961. There is some confusion as whether there were 12 or 13 stations as two different reports give different numbers.

33 TNA AB 16/3446, Atomic Energy Executive Weapons Group Seismic Research Programme, Note by Sir Claude Pelly, 19 December 1960.

34 TNA DEFE 19/18, Technical History of the Geneva Nuclear Tests Conference 1957–1962 (R.W. Snelling, Geneva September 1961) paragraphs 56, pp. 29–30.

35 AWRE Pamphlet No. 2, United Kingdom Atomic Energy Authority, Atomic Weapons Research Establishment, Data Processing Facilities and Data Available at the UKAEA Data Analysis Centre for Seismology, AWRE, Aldermaston, Berks, July 1967, p. 1.

36 United Kingdom Atomic Energy Authority Seventh Annual Report 1960–61, 7 July 1960 (London, 1960), p. 46.

to place seismometers.[37] The AWRE Weapons Group undertook this work. This resulted in work done out at depths of 300 and 500 feet in boreholes made available by BP and the University of Durham; and thirdly AWRE would seek sites where the noise levels were much better than those used in the various systems studied so far. As part of this work, the UKAEA established a small array at Porton Down on Salisbury Plain to act as the prototype of the Eskdalemuir array.[38]

Seagull Phase II, or MacSeagull as it was known, took place in March when AWRE piggybacked on the Naval Construction Research Establishment's experimental programme with ground mines in Loch Striven, Argyllshire.[39] Sixteen charges of 1600 lb – much larger than Seagull – were detonated on the loch bed. Some 18 seismic stations, some on different geological strata, monitored the signals.[40] Two of these were in Wales, four in Scotland itself and the other stations were those used in Phase I and for Orpheus Phase C. AWRE saw this programme as an opportunity to put the UK in a strong position in the Geneva negotiations.[41] In fact, the UKAEA's Press Office was able to remark that this programme had generated a good deal of positive publicity for the Authority. Leonard Cina, from the Authority's London Headquarters, remarked that it 'was certainly a revelation to the press to realise that the same people who were hell-bent on producing weapons should be equally furiously engaged on working for nuclear disarmament'.[42]

Commonwealth Seismic Research Efforts: Spring 1960

Following agreement in Washington between Eisenhower and Macmillan at their Camp David meeting on 28 and 29 March 1960 that there should be a joint

37 TNA AB 16/3841, A.W.R.E. Site Requirements for Seismic Testing Stations, W.E. Miller, UKAEA AWRE to F.J. Neary, UKAEA Health and Safety Branch, London, 31 May 1960.

38 Atomic Weapons Research Establishment, *The Detection and Recognition of Underground Explosions*, p. 18.

39 TNA AB 16/3446, E.R. Drake-Seager, AWRE to D. Ostle, Geological Survey of Great Britain, 6 January 1961.

40 TNA ES 5/295, AWRE Report No. T 5/61 Operation MacSeagull Operational Report, W.P. Cook, November 1961. The map in this report actually shows eight Scottish stations at Wauchon, Echt, Comrie, Duns, Earlyburn, Eskdalemuir, High Eldrig and Campbeltown.

41 TNA AB 16/3446, Expanded Seismic Programme (AWRE) Operation Seagull Phase 1, I. Maddock, Head of Field Experimental Division, AWRE to L. Cina, Senior Press Office, UKAEA, London, 19 January 1961; Press Brief March 1961, AWRE Seismic Experiments in Scotland.

42 TNA AB 16/3446, Expanded Seismic Programme (AWRE) Operation Seagull Phase 1, Press Relations, Aspects of AWRE Seismic Programmes, Report by L. Cina, May 1961.

research programme on seismic research with the Russians, Penney set out some ideas on a proposed UK contribution.[43] He had already initiated collaboration with New Zealand in 1959 that entailed the UKAEA equipping seismic stations there and his next proposal was for a large area experiment spaced over 1000 km to monitor the signal produced by HE and nuclear explosions at the Nevada Test Site.[44] Seismic stations for this purpose were needed in Canada. Experts from Australia and New Zealand could be invited to make a modest contribution by seconding staff to establish and operate these stations.[45] Macmillan submitted a paper to the Cabinet at the end of April 1960 setting out UKAEA proposals for further UK work on seismic research, which would implement the offer he had made at Camp David.[46] This obviously built on Penney's ideas, as it repeated the Canadian plan and sought Cabinet agreement for £2.5 million to be allocated for collaborative seismic research work. Canada was now central to British plans for a renewed effort to solve the verification problem.

The UKAEA explained the details of what it hoped to do in Canada to the Canadian High Commissioner on 29 April 1960. One main seismic station was envisaged at a seismologically quiet location, preferably on granite and more than 500 miles from the sea or any big lake. There would also be some mobile posts.[47] Unfortunately, Canadian Ministers were emotionally opposed to anything to do with nuclear explosions and as such were not persuadable. They failed to understand that the UK was seeking an agreement on a significant verification effort designed to ensure the end of nuclear testing, not its promotion. The delays in Ottawa put more and more pressure on the UK. In Geneva the Russians were most interested in seismic measurements in Canada and possibly also New Zealand and Australia and the UK delegation asked for authority to be more specific in its response.[48] Until an answer was forthcoming from Canada, Macmillan was in the invidious position of having nothing to contribute to the joint work on seismic detection.[49] Macmillan wrote on successive days, 10 and 11 May, to Diefenbaker

---

[43] TNA CAB 21/4535, W.G. Penney, A Proposed UK Contribution to the Research on the Seismology of Test Control, 1 April 1960.

[44] TNA CAB 16/3301. Commonwealth Collaboration on Seismology, Extract from A.E.X. (60) 10th meeting, 11 March 1960.

[45] TNA CAB 21/4535, Commonwealth Contribution on Seismology, Roger Makins to Lord Privy Seal, 18 March 1960; TNA PREM 11/3584, Norman Brook to Prime Minister, 28 April 1960.

[46] TNA CAB 129/101 C (60) 74, Nuclear Tests, Memorandum by the Prime Minister, 27 April 1960.

[47] TNA PREM 11/3584, Commonwealth Collaboration on Research Into Nuclear Test Control.

[48] TNA FO 371/149325, IAD 24/216/G, Geneva telegram no. 482 to Foreign Office, 12 May 1960.

[49] TNA FO 371/149325, IAD 24/216/G, Foreign Office telegram no. 633 to Geneva, 12 May 1960.

to urge a positive response to the British request. Time was running out since experts from the UK, the USSR and US were starting work on 11 May to begin discussing a co-ordinated research programme. Macmillan was worried that without Canadian agreement, the UK would be unable to participate in any positive way.[50] When Ottawa did manage to respond, it would only agree to become involved once there was agreement in Geneva on a co-ordinated programme, but not before. This put the cart before the horse thereby making it impossible for the UK to make its proposal public.[51] Macmillan lost patience. He asked for advice from the Cabinet Secretary, Sir Norman Brook, on this 'silly story of Canadian intransigence and humbug … I might denounce Diefenbaker publicly as the man who is holding up the abolition of nuclear tests'.[52] Macmillan of course did not do this: a further measured message asked for a change of heart. The Canadian equivalent of Cabinet Secretary Sir Norman Brook, Norman Robertson, was also fully briefed on all the details and arguments.[53] UK and Canadian officials met subsequently in Ottawa on 31 May to 1 June to discuss deploying a seismic station in Canada. British experts outlined the detailed field requirements and it was immediately apparent that the area around Yellowknife, whilst some 200 km less than the preferred minimum range, coincided with Canadian requirements for a station in their own 28 station network.[54]

Canada finally yielded to the force of the British argument, and notified London on 3 June that its government could inform the US and USSR of Canadian willingness to help with the collaborative seismic research project. However, there would have to be no public announcement until such times as the Canadian government had informed its Parliament.[55] Macmillan was very happy to give such an undertaking. In the meantime, the Russians had started to back slide on the research programme and had not pressed the question of seismic stations in Canada. Sir Michael Wright was nevertheless authorised to tell his Russian and US counterparts in June 1960 in strict confidence of the Canadian decision.[56]

---

50 TNA CAB 21/4535, Prime Minister to John Diefenbaker, 10 May 1960 and 11 May 1960.

51 TNA CAB 21/4535, Canadian High Commissioner to Prime Minister, 18 May 1960.

52 TNA CAB 21/4535, T. Phelps, No.10 to D.R.J. Stephen, Cabinet Office, 23 May 1960.

53 TNA CAB 21/4535, Norman Brook, Cabinet Office to T. Phelps, No 10, 24 May 1960; CRO telegram no. 566 to Ottawa, 26 May 1960.

54 TNA AB 43/196, Seismic Research (including working group on Disarmament) Anglo/Canadian Collaboration in Seismic Research Summary of Discussion in Ottawa, 31 May to 1 June 1962, 1 June 1962.

55 TNA CAB 21/4535, Ottawa telegram no. 536 to Commonwealth Relations Office, 3 June 1960.

56 TNA FO 371/149327, IAD 24/258/G, Foreign Office telegram no. 744 to Geneva, 7 June 1960.

Seismic Research Programme Advisory Group (SRPAG) May 1960: UK Role

The Seismic Research Programme Advisory Group (SRPAG) within the Conference on the Discontinuance of Nuclear Weapon Tests held its first meeting on 11 May, with the last of 13 meetings taking place on 30 May.[57] Dr Henry Hulme (AWRE), UK delegation leader, noted on 11 May 1960 that the, 'purpose of our present meeting is to make sure that adequate work is done in all promising lines of attack on the problem of the detection and identification of possible nuclear explosion'.[58] The USSR and US described their programmes, and the UK team explained its modest proposals.[59] Hulme outlined these in tentative form at the second SRPAG meeting on 12 May and had ten elements. These had resulted from discussions with the US in Washington and in AWRE's Technical Advisory Committee:

1. Cooperation in any programme of nuclear and chemical explosions, and the readiness to build one 'Geneva Type' station and up to six subsidiary seismic stations.
2. Establishment of an experimental 'Geneva Station' in the UK for the testing of instruments, arrays and deep buried instruments.
3. Model scale studies of the mechanisms of generation, propagation and detection of seismic waves.
4. Developmental work on seismological instrumentation.
5. Work on the development of unmanned stations, including the problems of digitalisation of records, transmission of records, power supplies and anti-tampering.
6. Study of the reduction of seismic noise by burying seismometers in deep holes.
7. Study of the means to be employed in the handling and analysis of data from seismic stations.
8. After consultation with other countries of the Commonwealth, studies of seismic noise in at least one other area of the world, together with the associated problem of detecting first motion against noise background in the case of small earthquakes.
9. Theoretical studies of the effectiveness of different criteria for distinguishing earthquakes from possible explosions, evaluation of different kinds of networks and general theoretical work on seismological problems.

---

[57] Romney, *Detecting the Bomb*, p. 165.

[58] Conference on the Discontinuance of Nuclear Weapons Tests Seismic Research Programme Advisory Group GEN/DNT/SRPAG/PV.1, p. 12, 11 May 1960.

[59] Other UK delegates were Dr Sutcliffe, J.K. Wright and Dr H. Wilson from the UKAEA and Dr Maurice Hill from Cambridge University, who was an advisor to HMG on clandestine detection. TNA FO 371/149326, IAD 24/237, Briefing meeting for British Delegates to Geneva, 5 May 1960.

10. Studies of the general conditions controlling the quietness or otherwise of sites for seismic stations.[60]

The UK delegation emphasised that this programme was not fixed; it could be changed in light of discussions in the advisory group.[61] Work would be undertaken on the development of unmanned stations until the associated problems of digitalisation of records, power supplies and anti-tampering precautions could be addressed. The UK did not disclose that authority to proceed with this programme was conditional on agreement on a co-ordinated programme.[62]

At its sixth meeting the group was moving towards an agreed report and the UK delegation presented a paper that tabulated the three national programmes side by side and suggested the most appropriate form of cooperation for each project.[63] Dr Hulme thought that agreement was desirable on a broad range of instruments to be used in the research.[64] The seventh meeting saw the establishment of small informal sub-groups to discuss specific aspects of the research programme in detail.[65] However, the Russians made clear in commenting in detail on the US programme, that they would never agree to the inclusion of research into decoupling as part of the programme.[66] By the time the group adjourned it could have produced an agreed outline of a co-ordinated programme covering all but the following key issues: the number of nuclear explosions in the US programme, de-coupling experiments, coordination methods, safeguards against use of calibration nuclear explosions for weapons development work and the programmes' duration. The wider political environment created by the failed Paris Summit had an adverse impact on the SRPAG since it could not agree to forward any conclusions or recommendations to the main Conference on the Discontinuance of Nuclear Weapon Tests. The Soviet delegation, having received fresh and far more stringent instructions, argued that the up-graded Geneva system agreed in December 1959 was quite adequate as

60 TNA CAB 130/130, GEN.604/20, Notes for the Policy Committee on Seismic Research, Note by the Secretaries, Some Notes by AWRE for the Policy Committee on Seismic Research, 4 July 1960.

61 TNA PREM 11/3584, Geneva telegram no. 490 to Foreign Office, 12 May 1960; TNA FO 371/149325, IAD 24/225, Geneva telegram no. 495 to Foreign Office, 13 May 1960. IAD 24/228, inward saving telegram no. 104 from United Kingdom Delegation Geneva to Foreign Office, 13 May 1960.

62 TNA FO 371/149306, IAD 22/370, Nuclear Tests Conference, History of the Conference from its opening on October 31st 1958, 6 October 1960.

63 Romney, *Detecting the Bomb*, p. 169.

64 TNA FO 371/149326, IAD 24/240, Geneva telegram no. 512 to Foreign Office, 18 May 1960.

65 TNA FO 371/149326, IAD 24/243, Geneva telegram no. 516 to Foreign Office, 19 May 1960.

66 TNA FO 371/149327, IAD 24/250, Geneva telegram no. 528 to Foreign Office, 23 May 1960.

a basis for signing a comprehensive test-ban treaty.[67] Once again it seemed to the Western participants that where necessary the Soviet government expected scientific integrity to be subordinated to political expediency.[68]

UK Research Summer 1960: What Next?

After the hardening of the Soviet position, the MOD were anxious that the Cabinet Working Group on Disarmament should discuss amending the Cabinet's earlier conditional agreement on the UK seismic research programme, involving funding only being released once a joint research programme was agreed between the US, USSR and UK. As the latter was now unlikely, the MOD felt that it was in the UK's own interests to continue its own national investigations if it was to make an effective contribution towards solution of the unresolved research problems.[69] They proposed that the conditionality hanging over the programme should be withdrawn and the UK experts proceed with their planned programme. It therefore made several positive recommendations for further UK seismic research in early July 1960, including committing to expenditure of some £600,000 over two years on seismic research to improve methods of detection and identification.[70]

The UK's emerging technical view was that it should concentrate its research efforts on detection at ranges out to 1000 km.[71] Just over £600,000 was divided as follows. £125,000, £30,000 and £30,000 would go to basic studies of the physics of explosions and seismic statistics at Foulness, universities and AWRE respectively. £350,000 would go on the study of instrumentation, data recording, unmanned stations and field trials, whilst £60,000 would be devoted to investigations into the reduction of seismic noise by the use of deep boreholes. The remaining £25,000 would go to instrumentation for Commonwealth observatories. Approval was sought to build one 'Geneva' station in the UK at the cost of £350,000, £550,000 for full cooperation with the US programme and a final tranche of £500,000 for seismic research in the Commonwealth. The plan for setting up a seismic station in Canada was later dropped, as the Canadians wanted to run it themselves.[72] There

---

67 Bates, Gaskell and Rice, *Geophysics in the Affairs of Man*, p. 173.

68 TNA DEFE 19/18, Technical History of the Geneva Nuclear Tests Conference 1957–1962, (R.W. Snelling, Geneva September 1961), paragraph 46, p. 22.

69 TNA FCO 317/149332, IAD 24/315, Sir Solly Zuckerman to Sir Patrick Dean, Foreign Office, 29 June 1960.

70 TNA CAB 21/4535, Cabinet Working Group on Disarmament, 8 July 1960.

71 TNA CAB 130/130, GEN.604/18th Meeting, Item 2, The Main Features of the United Kingdom Programme, 8 July 1960.

72 TNA CAB 21/4535, R.R.G. Watts, Commonwealth Relations Office to J.S. Orme, Cabinet Office, 19 August 1960; see also TNA PREM 11/3584, Home to Prime Minister, 2 August 1960; TNA AB 16/4186, Claude Pelly, AWRE to Sir Roger Makins, UKAEA, 11 July 1960.

was still the same need, as before, to obtain basic seismic knowledge. Moreover, there remained a major requirement for the UK to develop its own knowledge in this field, especially if it wanted to make an effective contribution towards the solution of the problems separating the Americans and Russians in the Geneva negotiations.[73]

However, the failure of the May 1960 Paris Summit seems to have had a direct negative impact on Macmillan's attitude towards these proposals, as he stated: 'there was now little to be done except to conceal as best I could my disappointment amounting to almost despair – so much attempted, so little achieved'.[74] This seems to have had an adverse impact on his approach to the UK seismic research programme. Having been at the forefront on pressing the Canadians in May, by late summer 1960 Macmillan was now arguing that the money allocated for the work was excessive as the UK had little direct interest in seismology and the purpose of the programme was political.[75] Both the MOD and UKAEA, supported by the Foreign Office, however were still strongly in favour of continuing the work because it would improve the UK's ability to contribute to future debates on the role and capabilities of seismology and test-ban treaty verification. The MOD and the Foreign Office both saw a clear national interest in this, as it would aid evaluating the results of US experiments if it was agreed subsequently to expand the test ban to cover smaller explosions (those producing a seismic signal of less than 4.5 mb), or to defend a continual refusal to do so.[76] Makins discussed the programme again with Macmillan on 18 August 1960 and the Prime Minister agreed that if the UKAEA put up a limited programme for 12 months' work he would look at the project again, but only after Treasury and Foreign Office had considered it.[77]

The Foreign Office duly proposed the revised programme, with costings of £300,000 for the first 12-month phase with a review at the end of the period.[78] Macmillan remained unconvinced and sought Treasury views. The Chancellor felt that so long as there was a prospect of a treaty, and of a tripartite research programme in which the UK should play a part, it was right to continue to plan for

---

73 TNA CAB 130/130, GEN.604/21, United Kingdom Seismic Research Programme, Note by the Secretaries, 4 July 1960.

74 Macmillan, *Pointing the Way*, p. 213.

75 TNA DEFE 19/12, Private Secretary Prime Minister to Private Secretary Foreign Secretary, 4 August 1960.

76 TNA AB 16/4186, R. Press, MOD to Sir Roger Makins, UKAEA, 12 August 1960.

77 TNA AB 16/4186, Philip de Zulueta, 10 Downing Street to A.K. Rawlinson, UKAEA, 18 August 1960; Roger Makins to Sir Hugh Stephenson, Foreign Office, 18 August 1960.

78 TNA AB 16/4186, Oliver Wright, Foreign Office to P.F. de Zulueta, 25 August 1960.

some limited expenditure.[79] In addition, there would also be a need for £50,000 towards the cost of the Canadian seismic network, making for a total revised estimate of £350,000. This programme was the minimum necessary to enable the UK to play a modest role in carrying out the government's policy of ensuring that disarmament treaties were properly verified.[80] The information sought from the programme would ultimately be required in setting up an organisation to control cessation of nuclear tests. Work had to proceed independently of tripartite agreement in Geneva.[81] In fact, the minimum programme proposed specifically focused on the problem of detection.[82] This would include basic research and instrumentation in the UK; a 'Geneva'-type station in the UK; preliminary consultation and collaboration with Canada and the US, work in Canada and US; and up to six mobile stations and instrumentation in a suitable earthquake area, possibly in Pakistan. However, certain major sectors of the programme could not be implemented until the political situation in Geneva had been resolved.[83]

Officials clearly felt that the UK needed to be able to contribute to the debate. Previously, they had been in the invidious position of having to accept seismic data supplied by the Americans or the Russians and being unable to do anything but swallow it even when they had doubts about their accuracy because of the lack of UK experimental evidence. Macmillan agreed on 27 August 1960 to the revised proposal and that the UKAEA should spend £350,000 over the next twelve months.[84] A review would take place in twelve months in the light of the political atmosphere at the time. The earlier figure of £3.5 million had been predicated on the existence of a tripartite seismic research programme feeding its results into the nuclear test-ban conference in Geneva.[85] The work by AWRE's Weapons Group

---

79 TNA PREM 11/3584, Selwyn Lloyd to Prime Minister, United Kingdom Seismic Research Programme, 8 May 1961.

80 TNA DEFE 19/12, Atomic Energy and Disarmament Department, Foreign Office to Private Secretary, Prime Minister, 25 August 1960.

81 TNA CAB 130/130, GEN.604/23, United Kingdom Seismic Research Programme, Note by the Secretaries, 8 August 1960.

82 TNA PREM 11/3584, Oliver Wright, Foreign Office to Philip de Zulueta, 25 August 1960.

83 TNA AB 16/4186, Minimal Programme of Seismic Research, I. Maddock, H.F.E., 1 July 1960.

84 TNA PREM 11/3584, Philip de Zulueta to Prime Minister August 26, 1960. Macmillan's manuscript comments, 27 August and Philip de Zulueta to Oliver Wright, 27 August 1960; TNA FO 371/149341, IAD 24/429/G, H.B. Shepherd, Foreign Office to R.R.G.Watts, Commonwealth Relations Office, 15 September 1960.

85 TNA DEFE 19/12, Private Secretary Prime Minister to J.O. Wright, Foreign Office, 27 August 1960.

was regarded as having started on 8 July – the date of the Cabinet Working Group meeting that had first addressed this topic.[86]

Future Efforts: Spring 1961

By early 1961 the UKAEA was very keen to continue this work.[87] The Foreign Office did not need much convincing and the Authority's arguments found their way via Sir Roger Makins, its chairman, into correspondence from Lord Home to the Prime Minister in April 1961.[88] This asked for agreement on further expenditure on seismic research, the first twelve months having convinced the UKAEA that it would now be possible to improve the effectiveness of a detection system.[89] A failure to continue the programme would waste the earlier time, effort and money. Macmillan was still unsure. Whilst he supported the search for a test-ban treaty, he always seemed to baulk at relatively small expenditures for supporting verification work. Fortunately, for the UKAEA, he was in a minority. Officials and other Ministers, with the exception of the Chancellor, were firmly in favour. Hailsham, Science Minister, noted that despite the uncertainty about the future of the Conference on the Discontinuance of Nuclear Weapon Tests, the UK should continue its work at the same scale.[90]

In response to Macmillan, UKAEA officials made a new request for continuation funding in May 1961, which the Foreign Secretary, the Minister of Science, the Lord President and even the Chancellor now supported.[91] Macmillan was still unconvinced, given the generally poor prospects for the test-ban treaty at that moment. He wanted to defer the decision until mid June.[92] Home, however, disagreed and urged the Prime Minister to agree the continuation of the research programme. It would, amongst other things, help the UK monitor Soviet underground tests in the event that there was no test ban. No doubt the US would continue its own research and a UK effort would allow UKAEA scientists to keep in touch and cooperate with their counterparts across the Atlantic. Moreover,

---

86 TNA AB 16/4186, R.N. Simeone, UKAEA to P.A. Bundy, Chief Accountant AWRE, 19 September 1960.

87 TNA AB 16/4186, Claude Pelly, AWRE to Sir Roger Makins, Seismic Research Programme for the Detection and Control of Nuclear Tests.

88 TNA AB 16/4186, Sir Roger Makins to Sir Hugh Stephenson, Foreign Office, 8 March 1961.

89 TNA PREM 11/3584, Home to Prime Minister, 25 April 1961.

90 TNA CAB 21/4536, Hailsham to Prime Minister, United Kingdom Seismic Research Programme, 3 May 1961.

91 TNA DEFE 19/12, Foreign Secretary to Prime Minister, UK Seismic Research Programme, 25 April 1961; see in same file Lord Hailsham to Prime Minister, 3 May 1961 and S. Lloyd to Prime Minister, 8 May 1961.

92 TNA DEFE 19/12, P. de Zulueta to Wright, May 1960.

improvements in seismological techniques would lead to a significant reduction in the number of earthquakes that would remain unidentified and an improvement in the probability of locating a nuclear explosion. More significantly, this work would help to resolve some of the inspection problems that were delaying progress in Geneva.[93] Unfortunately, Macmillan now took the view that the project was a waste of money and wondered what the Chancellor's view was. Lloyd agreed with his boss, despite himself having previously been Foreign Secretary, closely involved in the test-ban debate and previously in favour of some work.[94] He did however suggest that Ministers discuss the matter; the MOD had since weighed in strongly to support continuation of the programme. Macmillan and Lloyd were in the minority.

Continuing Uncertainty

It would appear that because there was no Ministerial consensus, the decision was remitted to a special Cabinet sub-committee, GEN.604/22, which produced a revised and reduced seismic research programme of £250,000 for the period until 31 March 1962, subject to review if the Geneva Conference reached a definite conclusion.[95] Thus in this initial period, AWRE generally had sufficient funds to sustain the research programme – its cost was small beer (about £300,000 for 1961–62) compared to the weapons programme.[96] The AWRE team had designed recording and analysis techniques that greatly improved the ability to observe weak signals in the presence of noise – a vitally important quality in a detection system. AWRE established that a detection improvement was possible in signal to noise ratio of at least 100 times better than the instruments used by the US in the Hardtack underground test in October 1958. Failure to continue with UK research would require the redeployment of just over 50 personnel, plus the laboratories and equipment that they used for work associated with weapons experiments or basic research. Moreover, if the Geneva talks failed, and this indicated a reversion to the Cold War atmosphere of the mid 1950s, then it would be essential to retain these experts in their existing teams.[97] By summer of 1962, Macmillan eventually accepted the argument in favour of a

[93] TNA DEFE 19/12, Foreign Secretary to Prime Minister, Seismic Research Programme, 2 June 1961.

[94] TNA DEFE 19/12, PS/No.10 to D.F. Hubback, Treasury, 7 June 1961 and Hubback to P. de Zulueta, 22 June 1961.

[95] TNA AB 16/4186, H.C. Hudspith, UKAEA to P.A. Bundy, AWRE, 12 July 1961.

[96] Oral History Seminar, AWE Blacknest, 10 January 2006.

[97] TNA AB 16/4186 , Consequences of Delaying or Terminating the Seismic Research Programme, Roger Makins to Claude Pelly, AWRE Seismic Research Programme, 12 May 1961 and N. Levin, Director AWRE to Sir Roger Makins, Seismic Research Programme, 15 May 1961.

continued programme.[98] Evidence suggests that Macmillan finally agreed to the UKAEA's recommendations in October 1962.

UK Work with Seismic Arrays at Home and Abroad

At this point, the UK technical view appears to have been that a network of 25 seismic stations spread around the world and probably not even in the USSR, would detect and identify the great majority of militarily significant underground nuclear tests inside the USSR. The original 1958 Geneva Experts' system had posited a system of 170 stations, but work in the UK on propagation of seismic signals had suggested a more practical solution involving fewer stations. This depended upon developing techniques for detection at very long ranges – 3000 to 10,000 km from the source. Hal Thirlaway, then the chief seismologist at AWRE, had observed from studies of earthquakes and underground tests at ranges greater than 3000 km that it seemed that at least 90 per cent of earthquakes could be identified from simple inspection of the seismograms of the first motion from one station.[99] This was a key change in thinking and approach to the discrimination problem and was one of the UK's most significant contributions to the development of forensic seismology, which considering the state of UK knowledge in 1958 was quite a remarkable achievement. There was also the 'depth of focus discrimination' in which data reviewed by AWRE to the summer of 1961 suggested that 85 per cent of the natural seismic events would have been eliminated as having focal depths greater than eight kilometres. AWRE saw this as an interesting possible way for distinguishing explosions from earthquakes and this approach was brought to US attention. It had potential uses in resolving the inspection quota problem by reducing the number of suspicious events that might require on-site inspections.[100] More work was needed however.

The UK had first experimented with a small seismological array installed at Pole Mountain, near Laramie Wyoming in the US. This array was about 1000 km from the Nevada Test Site, and had been established with USAEC cooperation. It was used initially to participate in Project Gnome, which involved the explosion of a five kiloton nuclear device 1200 feet below the ground in a salt formation.[101] Pole Mountain also demonstrated that seismic waves from nuclear explosions at distances around 3000–10,000 km could be relatively efficiently detected as it had

[98] TNA DEFE 19/12, S. Zuckerman to Dr N. Levin, AWRE, 31 July 1962.

[99] Alan Douglas, 'Recognising UGTs', *Discovery. The Science Journal of AWE*, Issue 1 March 2000, p. 16.

[100] TNA FO 371/157114, IAD 24/71, Establishment of control, monitoring and inspection procedures nuclear tests under NTBT 1961, Sir Solly Zuckerman, MOD to Dr Jerome B. Weisner.

[101] United Kingdom Atomic Energy Authority, Eighth Annual Report 1961–62 (London, 1962), p. 49. Pole Mountain closed during 1963/64.

picked up one Soviet test at Semipalatinsk on 2 February 1962 and one French low-yield underground test in the Sahara on 2 May 1962 with remarkably clear signals.[102] This provided immense encouragement to pursue the teleseismic line and drop first zone verification research.[103] The UKAEA thought that a network of comparatively few stations could supplant the prohibitively expensive Geneva system.[104] However, although these tests were detected, it was data from the covert US seismic network, supplemented by UK data that gave information on the yields. If it had not been known that the events were in known testing areas, the scientists would have been highly suspicions but not certain that these events were nuclear explosions.[105] The key was the array. Frank Whiteway's team at Blacknest had been instrumental in developing and designing the array approach to detection. In contrast to a standard seismograph station, which housed one or more seismometers at a single location, an array consisted of a number of seismometers spread out over a large area. Computer processing of the recorded data allowed the array to be steered like an antenna not only to enhance detection of seismic signals, but also to estimate independently the locations from which they came. In light of the Pole Mountain experience and other work, the UK focused its research attention on the teleseismic (the third zone) range (3000 to 10,000 km), and re-designed its arrays to have 20 seismometers spread over a region of 25 km in diameter. An array of seismometers sited in an area with low ambient seismic noise level was thought to be capable of detecting, but not necessarily identifying, seismic events of 4 mb or greater at such distances.[106]

Despite this progress both the UK and US were clear, however, that there would still be unidentified events where there would need to be on-site inspection for definitive clarification. Twenty high quality array stations at ranges between 4000 and 10,000 km around the world would enable detection and location of seismic events of 4 mb. At this level, the number of shallow earthquakes in the USSR was estimated to be 170, of which about 100 could be eliminated by application of depth of focus and first motion criteria. The remaining events could probably be further reduced to about 50, most of which would be less than 4.25 mb, which represented about four kilotons equivalent in tuff. Such a capability assumed that there were five stations in the USSR sited in the five principal seismic areas. Without such stations and without any new criteria for identification, about half

[102] Peter Marshall, Alan Douglas, former Heads Forensic Seismology Group, Atomic Weapons Establishment, Blacknest, in Jenifer Mackby and Paul Cornish, *US-UK Nuclear Cooperation After 50 Years* (London, 2008), p. 292.

[103] Bates, Gaskell and Rice, *Geophysics in the Affairs of Man*, p. 196.

[104] TNA AB 43/196, E.F. Newley, Deputy Director AWRE to Mr F. Chadwick, Finance, UKAEA, London, 20 August 1963.

[105] TNA PREM 11/4557, Nuclear Tests: Basic Arguments B. Detection and Identification of Underground Nuclear Tests W.G. Penney, 16 May 1963.

[106] United Kingdom Atomic Energy Authority, Ninth Annual Report 1962–63 (London, 1962), p. 42.

of the seismic events of 4 mb would remain unidentified, of which most would be less than 4.5 mb – about seven kilotons equivalent in Nevada tuff.[107]

Eskdalemuir and the Commonwealth Arrays: 1961–1965

As a direct result of Orpheus, Seagull and MacSeagull and the increased funding for seismic research that flowed from support for the Geneva negotiations, the UKAEA decided to establish a permanent experimental facility in the UK. This would allow development of array techniques, instrumentation proofing and seismological data collection. Much of the work in the UK had been concerned with improving the detection stations' instrumentation, including the use of linear arrays to reduce the signal to noise ratio and as we saw above this became the focus of the work. The main object of the Eskdalemuir array, near the existing Meteorological Station in Dumfriesshire, Scotland, was to explore this development.[108] Site selection was shaped by the idea that an even higher density of stations than that proposed by the 1958 Experts' Meeting might be practicable if the stations were unmanned and could be linked by radio to a central point. Earlier capability studies had shown that station spacing of 500 km improved the overall capacity to detect and identify earthquakes. The UKAEA recognised that study of unmanned stations at a range of 500 km from a recording station was therefore a useful project in support of the nuclear test-ban talks and a search was undertaken for a suitable low noise location 500 km from AWRE. A survey in 1960 had shown that Eskdalemuir was admirably suited to the establishment of a high performance seismological station.[109] This site was favourably placed for recording events from central Russia.[110] Site preparation began in 1961. It consisted of 22 Willmore Mark 1 seismometers sited at intervals of about 1000 yards and arrayed in cruciform pattern. Operation Seagull II took place in order to calibrate the new array in the azimuthal and radial planes.[111] This entailed the dropping of 39 depth charges at precise positions off the east and west coasts of Scotland. This exercise began on 17 July 1962. Its aim was to record the effects of the detonation of known explosive charges at specific distances from

[107] TNA CAB 178/2, Detection and Identification Capability based on Seismic Observations at Distances Between 3000 and 10,000 kilometres, R. Press, H. Thirlaway, 17 August 1962.

[108] TNA AB 43/196, F. Chadwick, UKAEA to M.I. Michaels, Office of the Minister for Science, 22 August 1962; and Oral History Seminar, AWE Blacknest ,10 January 2006.

[109] TNA AB 16/4052, Facilities for Seismic Research, I. Maddock, 30 November 1960.

[110] TNA CAB 178/2, 1965 R. Press to Dr H. Thirlaway, AWRE, Seismic Research – Detection and Identification of Underground Explosions 1958–1966, Notes accompanying table of Soviet and French Underground explosions, 26 November 1965.

[111] TNA AB 48/234, Operation Seagull Phase 2, Minutes of Meeting, 30 March 1962.

the Eskdalemuir array.[112] Timings had to be recorded extremely accurately. The navy dropped 320 lb depth charges in the Irish Sea and North Sea at 48 km intervals on arcs of about 200 km from Eskdalemuir and along radial lines extending from the Solway Firth and Firth of Forth.[113] Once the calibration was completed, the array became fully operational in August 1962.

Commonwealth Arrays

Agreement on the construction of an array at Yellowknife in Canada required the UK to supply and install all equipment and Canada would provide the site, do the necessary construction and, through the Department of Mines and Technical Surveys (now Natural Resources Canada), supply the personnel required to operate the array.[114] Installation of 19 seismometers was completed in November 1962. Plans were also afoot for a seismic station at first Maralinga, then at Alice Springs, in central Australia – later still moved to another location, Warramunga, near Tennant Creek, Northern Territory following initial surveys. An Australian station was necessary to help confirm the teleseismic hypothesis because it was in the Southern Hemisphere and the centre of a large landmass.[115] Treasury approval however was not forthcoming for this seismic installation work since the decision to go ahead with a further test of Super Octopus made it difficult to agree to release extra funds.[116] However, some rearrangement and savings in the budgets elsewhere in the Ministries of Aviation and Defence meant that the UKAEA was able to go ahead with the Australian plan. The Australian National University's Department of Engineering Physics participated in the design and installation of this array at Warramunga. An array was also installed at Gauribidanur, India in cooperation with the Bhabha Atomic Energy Research Centre. The Indian Atomic Energy Commission had asked the UKAEA to help it establish an Eskdalemuir-type array

112 TNA ES 5/327, UKAEA AWRE Report N. T 18/63 Operation Seagull, Phase 2, June 1963.

113 TNA AB 16/3841, A.W.R.E. Site Requirements for Seismic Testing Stations, Calibration of Eskdalemuir. The changes and delays appeared to have caused a fair degree of angst amongst the departments concerned. This programme entailed extensive consultations with, and notifications of, relevant bodies such as the Isle of Man, the Irish governments, the Scottish Office and local authorities. TNA AB 16/3841, A.W.R.E. Site Requirements for Seismic Testing Stations, F.J. Neary, UKAEA Secretariat to E.W. Cross, AWRE, 10 August 1962; TNA ES 5/320, AWRE Report No. T3/63 Operation Seagull, Phase 2 Operational Report, Maj. T.K. Batty, March 1963.

114 http://www.can-ndc.nrcan.gc.ca/yka/history_e.php

115 TNA AB 43/196, Memorandum on Seismic Research Station, Alice Springs, Central Australia, 16 July 1962.

116 TNA AB 43/196, A.D. Peck to M.I. Michaels, Office of the Minister of Science, 11 September 1962. The test codename was later changed from Pampas II to Tendrac.

with the supply of equipment.[117] Despite some initial concerns over funding, the UKAEA was able to meet the Indian request and received data from the station once it was up and running.[118] Gauribidanur started operating in October 1965.[119] The location of these steerable arrays was chosen to enable the maximum number of seismic events to be detected.[120]

These stations were running at the time of the US Longshot event in the Aleutian Islands on 29 October 1965, which had been held in order to see whether UK and US seismologists could really classify correctly the frequent earthquakes in the Kamchatka peninsula of the Soviet Union. They did. British analysts were tremendously impressed by the simplicity and symmetry of the energy radiated by Longshot as compared to the asymmetrical wave patterns generated by earthquakes occurring in the same region.[121] Closer to home Foulness investigated the credibility of using seismic instruments placed in deep boreholes as a means of detecting seismic events. The advantages were similar to those seen for large arrays, namely they might be able to improve the signal to noise ratio and thereby assist in identification of the event. Foulness made use of boreholes drilled by BP, mainly in Lincolnshire and Nottinghamshire at depths varying from 2000 to 4500 feet.[122]

AWRE Research and Consultations with the US: 1962–1965

By the spring of 1962 the results of the seismic research had a direct impact on UK thinking on what would be required for an effective system to verify the test ban. AWRE now believed that it had made such important improvements to instrumentation that far more information could be extracted from seismic

---

[117] TNA AB 43/196, Seismic Research (including Working Group on Disarmament) Homi Bhabba, Chairman, AEC to Sir Roger Makins, 18 December 1962.

[118] TNA AB 43/196, Roger Makins to Dr Homi J. Bhabha, India Atomic Energy Commission, 9 April 1963.

[119] Douglas, Recognising UGTs, p. 24.

[120] TNA ES 4/1089, AWRE Report No. 079/68. A Comparison of the Short Period Seismic Noise at the four UKAEA Type Arrays and an Estimate of their Detection Capabilities, R.F. Burch, January 1969, p. 1. In particular, the need was to satisfy two requirements: world coverage to give information about the azimuthal variation in the seismic signal and to have low ambient seismic signal ('noise').

[121] Bates, Gaskell and Rice, *Geophysics in the Affairs of Man*, p. 204.

[122] TNA AB 48/234, Operation Seagull Phase 2 Boreholes (and seismographs) July 1962; TNA ES 5/320 AWRE Report No. T3/63 Operation Seagull, Phase 2 Operational Report, Maj. T.K. Batty, March 1963 and Oral History Seminar AWE Blacknest, 10 January 2006. A major part of the AWRE's outside collaboration was with the Departments of Geophysics and Applied Mathematics at Cambridge University. In summer 1962 AWRE made use of a 2500 feet borehole drilled by the University of Durham and this was used to record signals from Seagull Phase 2.

observations in the third seismic zone than had been possible hitherto.[123] Zuckerman discussed the prospects with those involved in detection work in the UK on 7 March 1962.[124] Given his strong personal support for a test ban, he was always quick to light upon any progress made at the technical level to solve the detection and identification problem. As more data arrived, the third zone versus first zone hypothesis gained strength and as one contemporary noted, Zuckerman 'swallowed it hook line and sinker'.[125] AWRE's research objective was to see whether it was essential to insist on detection stations on Russian territory. Much more information had become available since the 1958 Experts' Meeting Report and even since the further expert meetings in Geneva in 1960. Whereas the general UK view had been that it was possible to get away with underground explosions up to a yield of 20 kilotons, scientists now knew from observations of the Russian series of tests and the Nevada test carried out by the US, that this was not so. A five kiloton underground explosion in the US had been detected in Japan, Finland, and numerous other places. Furthermore, it was also established that the Hardtack test from which the US took its position over the decoupling theory had in fact taken place in anomalous terrain and one could now extrapolate from this confidently that similar levels of decoupling could be achieved in other types of soils.

Zuckerman told the Minister of Defence that it would be immensely difficult to organise an underground test of something in the range of five to ten kilotons with the certainty that seismic stations well outside a state's own territory would not detect the event. Furthermore, as noted above, the decoupling theory had also taken some knocks. These developments did not mean that a scientist could ever declare with assurance that every underground test, however, small, could be detected. Nevertheless, scientists would be able to detect any underground shots of a magnitude that might have technical or military significance. This meant that an international system of seismic control posts would provide a reasonable degree of assurance for the detection of all underground tests with yields that mattered militarily. Most importantly, it also concluded that this applied to underground shots above an insignificant level, and that the types of shots that were unlikely to be detected would not affect in a serious way the balance of power.[126]

AWRE's seismic group looking at the problem concluded unanimously that it would be possible to construct an effective system of control based on three factors. First, seismic data would be collected from a worldwide chain of national stations, including inside the USSR. Second, these stations would be improved by standardisation of their equipment and the simultaneous development of data

123 TNA CAB 178/2, Research – Detection and Identification of Underground Explosions 1958–1966 US/UK meeting to Discuss the Possibilities of a Modified Control System for Monitoring an International Nuclear Test Ban, 18 March 1962.

124 TNA DEFE 7/930, Zuckerman to Minister, Nuclear Test Ban, 7 March 1962.

125 Personal correspondence.

126 TNA DEFE 7/930, Foreign Office telegram no. 1973 to Washington, 7 March 1962.

handling techniques to deal with the product; and thirdly, through improvement of the 1958 Geneva seismic equipment, as the UK was now doing at its stations in the US and Eskdalemuir. In short, the AWRE seismic group believed that they had made such important improvements to instrumentation that far more information could be extracted from seismic observations in the third teleseismic zone than had been possible hitherto.[127] This made a big impact on Ministers. Watkinson, on the very same day that Zuckerman sent his report, went on to write to and discuss the subject with the Prime Minister and Foreign Secretary.[128] He was now convinced that from the UK perspective it was possible to accept a Treaty of the type described by Zuckerman, if the Prime Minister and Foreign Secretary judged it necessary to do this in order to make progress with disarmament in general. As a result, he immediately told Zuckerman that he would like him, 'To strive as earnestly and as quickly as possible to convince your American opposite numbers of the validity and importance of the points contained in your minute to me of today's date'.[129]

London instructed the Washington Embassy to sound out the Americans on this point. The response was not promising. William Foster, Director of the Arms Control and Disarmament Agency (ACDA), made it clear that the Kennedy Administration could never persuade Congress to accept the abolition of national control posts in a test-ban agreement. Macmillan sent a personal message to Kennedy urging him to show greater flexibility on the need for control posts on Soviet territory.[130] Watkinson however told the Prime Minister that it would be very difficult to convince the US of the new position on detection, but suggested sending Zuckerman and Penney to Washington to make a strong and unequivocal presentation of the UK case on scientific grounds to reinforce Macmillan's personal message to Kennedy.[131] Macmillan seized on this suggestion and Zuckerman and Penney travelled to Washington to meet US experts, led by Dr Long, ACDA and Jerome Wiesner on 18 and 19 March 1962.[132]

Zuckerman reported that at the end of the meeting, at which the UK had set out its views, there was general recognition of a *prima facie* case for establishing

---

[127] TNA DEFE 7/932, Disarmament Suspension of Nuclear Tests US/UK Meeting to Discuss the Possibilities of Modified Control System for Monitoring an International Nuclear Test Ban, 18 March 1962.

[128] TNA DEFE 7/930, Harold Watkinson to Prime Minister, Nuclear Tests Control, 7 March 1962.

[129] TNA DEFE 7/930, Harold Watkinson to Chief Scientific Adviser, Nuclear Test Controls, 7 March 1962.

[130] TNA DEFE 7/930, Washington telegram no. 762 to Foreign Office, 8 March 1962 and Foreign Office telegram no. 2022 to Washington, March 1962.

[131] TNA DEFE 7/930, Harold Watkinson to Prime Minister, Nuclear Tests Detection, 13 March 1962.

[132] TNA DEFE 7/930, Philip de Zulueta to A. Hockaday, MOD, Suspension of Nuclear Tests, 13 March 1962.

quickly whether the UK observations and views meant that earthquakes could be differentiated from low-yield nuclear explosions at great distances. However, as this was not immediately possible, the scientists could not claim on technical grounds alone, other considerations remaining the same, that a major departure from currently tabled treaty proposals was justifiable.[133] Zuckerman noted that the Americans were unable to produce any additional data to supplement or presumably to contradict the British information. Nevertheless, they recognised the potential value of the UK discoveries, and planned to re-orient their research programme and to provide additional data against which UK ideas could be tested further with a minimum of delay. However, one US participant was very clear that the new UK ideas on the role of large arrays had problems and were not the solution to the identification issue.[134]

UK and US experts produced an agreed summary statement on their discussions, which covered detection, location, identification and inspection of seismic events.[135] However, this left certain things unsaid, notably that the US, despite their seismic research effort, had practically no data that could be used to check and amplify UK ideas. UK data was still, however, insufficient to permit firmer statements. It was clear to Zuckerman and Penney that the Administration thought that it could not move Congressional opinion on the need for large numbers of on-site inspections until it was able to produce objective scientific evidence to justify changes of the US position on the need for inspections and placement of seismic stations in Russia. Therefore an impasse existed which could only be removed by political action at the highest level.

UK pressure on the US seemed to touch a raw nerve. Zuckerman was left in no doubt by private remarks outside the main meeting that the President had already made major concessions to the Russians despite heavy opposition within the interagency process as well as from Congress. He was warned that if Kennedy were pressed to abandon his current position, unless there was a real technical breakthrough, at a time when the USSR was rejecting all his concessions, then the results might adversely affect the chances of Kennedy making any other advances on disarmament. Moreover, this was not the only problem. The account of the meeting that filtered through to the US delegation in Geneva did not reflect what the UK team had said. Godber, the Minister of State at the Foreign Office, when in Geneva at the end of March, reported to Home that it was clear from Arthur Dean that the version he had been given of the meeting was that Zuckerman and Penney had nothing to contribute except some hopes.[136] Worse still, Dean believed

133 TNA DEFE 7/930, B.D.S. Washington to Ministry of Defence, London Personal for Minister of Defence, 18 March 1962.

134 Romney, *Detecting the Bomb*, p. 188.

135 TNA DEFE 7/930, S. Zuckerman and W.G. Penney to Prime Minister, 21 March 1962.

136 TNA DEFE 7/931, Geneva telegram no. 133 to Foreign Office, Personal from Minister of State to Secretary of State, 30 March 1962.

that the US team had convinced Zuckerman and Penney of the soundness of the US position.

This misrepresentation provoked a swift response from the Chief Scientific Adviser himself. In a minute to Macmillan he said that it was incorrect that the UK team was convinced of the US position.[137] He added that those US representatives who wished to brush aside the UK observations were identified by their affiliation with those official US groups that had always been opposed, both publicly and privately, to a test-ban treaty. As for the technical side of the argument, Zuckerman told Macmillan that within three years long-range seismic stations equipped with arrays would demonstrate the possibility of identifying the great majority of earthquakes down to 4.75 mb. In Zuckerman's view, this implied that it would be possible to identify the great majority of seismic events of equivalent yields in the range of one to 50 kilotons, depending on the geological environment. The progress in the ability to discriminate earthquakes from explosions at long-range produced a major change in negotiating objectives. UK scientists now thought that the figure of unidentified seismic events in the USSR could be reduced to 25 and that could be done without any un-manned seismic stations in Russia.[138] One purpose of the Washington meeting had been to recommend that the US re-direct its programme and use fewer but larger arrays with 'delay and sum' processing. It seems that this convinced the Americans as the USAEC announced this change in July 1962. The importance of the UK seismic research programme's findings can be seen from the fact that by 1963 the UK and US were able to reduce their demands for an on-site inspection quota of 20 to seven a year.[139]

Military Significance of Low Yield Tests and the Impact of the PTBT

British scientific advice at the end of 1962 had noted that there was little chance that improvements in seismic instrumentation and in the analysis of records could ever give certainty of identifying every earthquake as an earthquake. For this reason, the necessity of on-site inspection would not vanish. On the other hand, the military importance of low-yield underground nuclear tests could be insignificant since both the US and USSR had recently completed extensive tests. However, it could not be said that underground tests were of little or no significance. After another year or two, it was possible that weapons scientists might design a low-yield device to which they might attach a great importance.[140] Penney argued later

[137] TNA DEFE 7/931, S. Zuckerman to Prime Minister, 2 April 1962.

[138] TNA FO 371/171230, 1963 IAD 1115/12/G, H.B. Shepherd to Mr Pemberton-Pigott, 23 January 1963.

[139] TNA CAB 21/6041, Nuclear Test-ban Negotiations – Points at Issue, 16 July 1963.

[140] TNA FO 371/171117, IAD 1021/10/G, R. Makins, UKAEA, (CH (62) 91), to Sir Harold Caccia, Foreign Office, 11 December 1962.

for the benefit of Macmillan that a single test of 20 kilotons or so would not have major significance. It was conceivable that a series of three or more might, but a series was easier to detect than a single shot. Given the then state of weapons knowledge in early spring 1963 in the US and USSR, Penney did not think that anything less than a series of three or more tests would have any significance.[141] This assessment was to colour British thinking on the acceptable military risks in pursuing a test ban for the next few years. By 1964, and with the new data, there was a wider acceptance in the US of the UK claims about the significant improvements that could be achieved by array techniques at ranges of several thousand miles. What was once regarded as unsubstantiated British optimism now seemed to be a possibility. Thirlaway now thought that about 90 to 95 per cent of earthquakes, down to 4 mb, could be identified from records at the British teleseismic stations.[142]

Partial Test-Ban Treaty (PTBT) 1963 UK Monitoring Posts: Crown Royal

Monitoring of the PTBT was left to national systems; no international monitoring effort was contemplated. To obtain US congressional support for the PTBT, the US President agreed to extend the US national system, which had largely been concentrating on the northern hemisphere. In order to address the gaps in the southern hemisphere coverage, the Americans asked the UK to assist. Even though there would be no on-site inspections under the PTBT, the UK still thought in terms of effective monitoring not just of the treaty itself but also of Soviet tests underground. The MOD proposed that the UK should set up a monitoring system composed of five stations at an initial cost of £1 million and with operating costs not exceeding £1.25 million annually, including R&D. The cost would be borne, at least initially, on the defence budget. This would be part of a larger US project in which the Americans would have 17 stations. In addition, the UK might also use its good offices with commonwealth governments to support US approaches for monitoring stations abroad, but at that point, there were no indications of where these would be.[143] The Defence and Overseas Policy (Official) Committee

[141] TNA FO 371/171230, IAD 1115/41/G, Summary of Scientific Position Regarding Nuclear Test Control W.G. Penney, 8 March 1963. This view no doubt was shaped by UK experience during the Grapple, Buffalo and Antler trials. Much would depend on the nature of the device or principles being tested and where the state in question was in its weapons developmental programme. The key here of course is that this was the technical advice given to Macmillan and which helped underpin his policy and acceptance of risk in an imperfect test-ban treaty.

[142] TNA CAB 178/2, Current capabilities in detection and identification of underground explosions, Dr R. Press to Chief Scientific Adviser, 2 November 1964.

[143] TNA CAB 21/6037, P. Rogers to Sir Burke Trend, Monitoring the Test-ban treaty (D.O. (O) (64) 56); TNA AB 48/546, C. Hood to Mr Drake, Monitoring the Test-ban Treaty,

considered MOD proposals in July 1964 and recommended that the Prime Minister should accept them.[144] The MOD proposed that the UK should contribute five out of the 22 monitoring stations that the US considered necessary. Officials thought that UK participation on this scale was sufficient to provide the UK with an independent source of information about nuclear tests carried out by the USSR. They would also enable useful data exchanges with the US. The exchange of information that the MOD envisaged with the US was seen to be of considerable potential value to the UK; moreover, it would strengthen still further the links between the UK and US on scientific intelligence. Furthermore, these could well constitute a useful precedent if the scope of disarmament, and therefore the need for some kind of intelligence support, was subsequently enlarged.

Ministers agreed that officials could negotiate with the Americans for a complementary monitoring arrangement for the PTBT.[145] Following a US proposal that the UK and US should collaborate on monitoring issues and hold technical meetings in Washington to work out the details, the UK and US made an agreement in 1965 to cooperate in a programme to monitor atmospheric nuclear testing.[146] To ensure compatibility with the existing and proposed US network, US detection equipment would be used in the British stations.[147] Some UK monitoring stations were situated in Commonwealth countries and the existence of these sites was avowed – Swaziland, Australia (Perth) and Mauritius. Other locations remained secret and still are.[148] However, another paper in the National Archives from April 1965 listed the locations as Ordfordness, Mauritius, South Africa and Eastern

10 August 1965.

[144] TNA CAB 148/6, D.O.(O)(64) 56 23 June 1964 Cabinet Defence and Oversea Policy (Official) Committee, Monitoring the Test-ban Memorandum by the Ministry of Defence; TNA CAB 21/6037, Burke Trend to Prime Minister, Monitoring the Test-ban Treaty, 28 July 1964; TNA CAB 148/6.4 D.C. (O) (64) Item 1 Monitoring of the Test-ban Treaty, 16th Meeting, 1 July, 1964.

[145] TNA FO 371/176436, IAD 1015/18/G, H.B. Shepherd to Sir H. Beeley, 23 September 1964.

[146] TNA FO 371/176436, IAD 1015/10 G, George Newman, US Embassy to H.B. Shepherd, AEDD, 7 August 1964, and IAD 1015/11/G, H.B. Shepherd to Newman, 26 August 1964; IAD 1015/13/G, Summary Record of a meeting held on August 15.

[147] TNA DEFE 69/501, E.R. Drake Seager, S/Field Projects, AWRE to Lt Col V.H. Martin, Defence Staff (Signals), MOD, 23 April 1965; TNA AB 48/546, Applied Physics Department Note No. S/29, Detection of Nuclear Explosions in Space and Underground, Proposed U.K. Detection System, December 1964, noted that a wide range of detection technologies would be deployed: radar back-scatter at Orfordness, broad-band EMP, seismic noise in addition to existing acoustic and seismic techniques at Peshawar, EMP, VLF Phase anomaly, atmospherics fluorescence, cosmic noise, magneto telluric at Mauritius, the same as Mauritius plus acoustic and ground filter in South Africa, seismic, VLF phase anomaly in East India and EMP, atmospheric fluorescence and cosmic noise at Fiji.

[148] TNA PREM 15/1364, Background Note Test-ban Monitoring, 5 March 1973.

India (Calcutta).[149] Political approval was sought from the governments concerned to operate from these territories, South Africa refused and Swaziland was then looked at as an alternative.[150]

Conclusions: Test Ban Treaty Verification and the Role of Seismology

The UK programme had been very good at anticipating future problems and ways in which they might be addressed; for example, it established that there was no need for national seismic stations in the USSR. All this in part was a result of the nature of the working environment at Blacknest and the support of senior UKAEA staff (Makins, Penney, Maddox, Hulme and Newley for instance). The scientists were left to pursue what they thought would work and drop things that did not. Being aware of what the US was doing, and more importantly, what the US was not doing, was also a factor. Despite the disparity in resources, financial and human, the UK effort led to significant contribution to the science of seismology on the development of arrays and the processing of broadband data for discrimination of events being two good examples. Moreover, all this had been built from scratch. Furthermore, the Blacknest scientists focus and expertise in the science of detection and identification of underground explosions rather than earthquakes gave them a different and more insightful perspective on the problems of test-ban verification.

The PTBT did not put an end to UK seismic research; on the contrary, Whitehall and AWRE saw a clear need for it to continue as a means for checking that the states were fulfilling their obligation to refrain from testing and for monitoring underground explosions.[151] The MOD assessed by the end of 1964 that the number of unidentified events of 4.0 mb and above occurring in the USSR was probably about 25 to 75 per year. Four mb represented the then current instrumental detection capability. British seismologists also believed that the number of unidentified

[149] TNA DEFE 69/501, B.M. Day to ACDS (Sigs) et al monitoring of Test-ban Treaty, 14 April 1965; and see also Detective Stations Tentative Communication Network; see also TNA AB 48/546, U.S./U.K. Agreement on Joint Test-ban Monitoring System, W.G. Penney to Sir Henry Hardman, PUS, MOD, 16 November 1965. The Australian Seismic Array Project was located at Tennant Creek and run by the Australian National University. The code name for the overall project was 'Crown Royal'; see D.F. Sim to Mr Hudspith, Crown Royal, 4 July 1965 see also TNA AB 48/546, Note for the File, Crown Royal Meeting on 14 July 1966.

[150] TNA DEFE 69/501, D.C. Fakley D Sc 3, MOD to N.J. Lewisohn, Shore Division (Navy), 4 January 1966; TNA CAB 148/21, O.P.D. (65) 87, Monitoring the Test-ban Treaty, Memorandum by the Secretary of State for Foreign Affairs, 26 May 1965.

[151] TNA AB 43/196, R.J. Andrew, MOD to E.C. Lester, HM Treasury, 22 August 1963; and E.C. Lester, H.M. Treasury, 10 September 1963.

events would be considerably smaller than their US counterparts believed.[152] Technically the complete elimination of the on-site inspection requirements for such events would not, or so it was thought, change the existing balance of power. As the Foreign Office pointed out, the main reason why a ban on underground tests had not been possible was that the means of discriminating between tests and earthquakes had been inadequate so that the level of on-site inspection required was such that the USSR could not accept. The research programme offered the best hope of a technical solution that might prove acceptable to the Soviet government, thus making a comprehensive test-ban treaty possible. This was still an important political objective and experience had shown that it was necessary for the UK to do its own research if technical improvements were to be translated effectively into political results.

By the end of 1965, the UKAEA's experimental work at Blacknest was concentrated on the refinement of array stations. The Authority hoped that this would result in improvement across a range of technologies, such as improving sensitivity by increasing the number of seismometers; refining data processing methods and improving instrumental techniques to reduce maintenance. Studies underway at that time offered some hope of improving the uncertainty of identification, but AWRE thought that it would be surprising if the number would increase above 95 per cent of events. There was a finite but small possibility that this small residue of unidentified events contained one or two very low underground nuclear explosions, or larger yield decoupled explosions. No means were available or foreseeable in the mid 1960s for recognising these as such, other than direct inspection at the site.[153] It is quite clear that the UK's position during the test-ban talks was based on a solid technical assessment of what was and was not possible and that a good deal of intellectual and practical effort had gone into finding ways of overcoming the identification and discrimination problems. UK support for a test ban was not just words; it was matched by practical deeds too. Moreover, much of the scientific expertise that had been involved in the recording of the UK's atmospheric tests now worked on test-ban treaty verification – an early example of swords into ploughshares. Work would continue for the rest of the period discussed in this book, but as we will see in Chapter 10, there would be no dramatic breakthrough.

---

152 TNA CAB 148/50, OPD (O) (D) (64) 5, A Guide to the Disarmament Negotiations, Note by the Secretaries. Document prepared by the Foreign Office, 17 December 1964.

153 Atomic Weapons Research Establishment, *The Detection and Recognition of Underground Explosions*, p. 50.

Chapter 7
The End of the Testing Moratorium 1961: UK Responses

Introduction

As we saw in Chapter 5, the UK had struggled to keep the US pursuing a test ban in spite of increasing internal pressure in Washington to resume atmospheric testing. Soviet attitudes to the test-ban negotiations had not helped, especially with positions that both the UK and US regarded as incompatible with effective verification. Although the UKAEA and MOD had some reservations about a test-ban's implications for the UK, there does not appear to have been any compelling need for further UK testing from autumn 1958 to spring 1960. There were however, plans for megaton tests in spring 1960 and kiloton tests in autumn 1960.[1] The UK was thus able to continue with its annual moratorium as Macmillan continued to work tirelessly for a comprehensive test-ban treaty. By mid 1961 the RAF was on its way to acquiring a thermonuclear capability with the production of Yellow Sun Mark II bombs with the Red Snow warheads. However, deliveries had fallen behind schedule and the Air Ministry and Ministry of Aviation had for some time been thinking about future requirements, such as a warhead for the Skybolt air launched ballistic missile and a new kiloton warhead to replace Red Beard. In fact, the Joint Naval and Air Staff Requirement GD 10/OR 1177 for what was to become WE177 appears to have been first set out in 1959.[2] With these requirements, could the UK have gone on with an indefinite testing moratorium? We cannot tell from the public record, but the provisional testing plans mentioned above suggests that the UK would at some point have been confronted with some unpleasant

---

1 TNA AVIA 65/870, Duncan Sandys to Prime Minister, Nuclear Tests, 9 January 1959.

2 The eventual decision was that there should be 90 Skybolt warheads, TNA AIR 6/151 and 153 Air Council Memoranda 1962, Annex to A.C. (62) 36, Warhead Programme for Vulcan/Skybolt Deterrent Force, 9 October 1962. The first Yellow Sun Mark IIs were not deployed with the RAF until early 1961, TNA AVIA 65/1792, Yellow Sun Mk II (Live) Air Ministry Requirement/Deliveries Quarter Ending September 1959–September 1963; for the date for WE177 origins see titles TNA AVIA 65/1862 ASR 1176 and 1177: nuclear weapon requirement 1959–63 and TNA AIR 2/17322 OR 1177 Improved Kiloton Bomb 1959–60 Closed documents, open descriptions. Retained under Section 3 (4) of the Public Records Act 1967. The OR however was only approved in 1963. TNA DEFE 19/103, The Navy's Requirement for WE177A, 19 May 1966.

decisions; either to continue with its self-imposed moratorium or become the first to resume atmospheric testing. Fortunately for the UK the Soviet decision to resume atmospheric testing in September 1961 on a large scale made it easier to start testing again. The question then became one of deciding what the UK should test rather than whether it should test. Keeping the test-ban negotiations going in such circumstances was going to be difficult and the US request to use Christmas Island for its own large-scale atmospheric testing programme put the UK in a decidedly awkward position. This chapter will see how Macmillan's government handled these conflicting issues and we will see how two of the most important UK tests, which were to set the basis for the UK deterrent for the rest of the decade and beyond, were decided upon and why.

USSR Breaks the Moratorium – September 1961: US-UK Reactions

Following the resumption of Soviet atmospheric testing on 1 September, the initial UK line was that the Soviet test created a new situation that required careful consideration.[3] For the UK at that time, the effect of the moratorium had been limited on its weapon R&D efforts, but if a moratorium had not been in place, some tests would probably have been conducted.[4] However, Arthur Schlesinger, Special Assistant to President Kennedy, reported at the end of July that he understood that some British scientists were unconvinced that a resumption of testing was necessary on security grounds.[5] Macmillan and Kennedy made a joint appeal on 3 September 1961 urging the Russians to refrain from further atmospheric testing.[6] Instead, Kennedy proposed an immediate agreement 'not to conduct nuclear tests which take place in the atmosphere and produce radioactive fallout'. The UK and US were prepared to rely upon existing means of detection to verify compliance with this as they believed them to be adequate for this purpose. Khrushchev rejected this offer on 9 September.[7] Both felt that the Geneva talks should go into recess until after the UNGA, unless the USSR made some constructive move. It did not, and after its 340th meeting on 9 September 1961 the Conference on the Discontinuance of Nuclear Weapon Tests went into recess. The talks in Geneva

---

3 TNA FO 371/157104, IAD 22/400, Foreign Office telegram no. 1166 to Geneva, 3 September 1961.

4 TNA AB 48/643, Harold Winterbourne, Ministry of Aviation to J.W. Russell, Foreign Office, 28 June 1961.

5 TNA FO 371/157076, IAD 21/10/G, Memorandum for the President, The Nuclear Test Ban, Issue 21 July 1961.

6 GEN/DNT/120, Conference on the Discontinuance of Nuclear Weapon Tests, 4 September 1961.

7 United States Arms Control and Disarmament Agency, *International Negotiations on Ending Nuclear Weapon Tests September 1961–September 1962*, Washington p. 9, October 1962.

did not resume until 28 November. HMG made it clear then that unless the three states concluded a test ban rapidly, it could not bind itself not to engage in or support the conduct of any further tests given the massive series of tests conducted by the Soviet government during the last three months.[8]

On 15 September 1961 the UK emphasised to the Russians that their rejection of an immediate ban on atmospheric testing as proposed by Macmillan and Kennedy was a missed opportunity, and that the UK regarded the US decision to resume underground testing as fully justified.[9] The MOD's assessment was that the USSR must have been planning the test series at least six months in advance, and if they were developing new designs, they would have needed twelve months, possibly more.[10] Both the UK and US rejected a new treaty proposed by the Russians since it amounted to an unverified ban on all tests. Moreover, when the UK and US had voluntarily refrained from testing, the USSR had used most of the negotiations in Geneva to make large-scale secret preparations for renewed testing. The Soviet draft also repudiated the principle of international supervision.[11]

Macmillan set out the formal UK position on the Soviet resumption in a statement on 31 October to the House of Commons. This was to shape UK thinking and public utterances for the next six months. It laid out very frankly the conflicting pressures facing HMG and Macmillan personally. On the one hand, it was important to respond to the concerns over the health impact of atmospheric testing. On the other, it was also essential to maintain the balance of power and effectiveness of the deterrent. Macmillan told the House that if tests had to be conducted for good military or scientific reasons, if possible, they would be held underground. He had in mind their need to ensure the safety in peace, and effective operation, of weapons either newly in service or under development or alternatively the study of new techniques, most probably of a defensive kind. The allusion to safety was probably a reference to the UK need to test a safe primary for the Skybolt warhead. However, the key paragraph, which became the anchor for UK policy, came right at the end: 'Certainly, we have no plans for such (atmospheric) tests in the near future. Nevertheless, I must be quite clear and frank. If I were to be convinced that a particular atmospheric test was necessary to maintain the balance and to preserve the freedom of the world, then our country would be bound to cooperate in or support their conduct.'[12] Its use of the word 'particular' was to come back and haunt Macmillan during the debates that followed soon after

---

8 TNA FO 371/157106, IAD 22/447, Foreign Office telegram no. 8597 to Washington, 23 November 1961.

9 TNA PREM 11/3592, Foreign Office telegram no. 6559 to Washington, 15 September 1961.

10 TNA FO 371/157124, IAD 26/153, Mr A. Potts, MOD to H.C. Hainworth, Foreign Office, 15 September 1961.

11 TNA CAB 134/2296, OD (63)21, Information Material on the Nuclear Test Negotiations Note by the Joint Secretaries, 13 August 1963.

12 Hansard, House of Commons, Oral Answers, Column 33, 31 October 1961.

the American request to use Christmas Island for its own atmospheric tests. The island had been put on a care and maintenance basis in 1958 at a cost of £1 million a year to keep open the possibility of megaton tests in the atmosphere at about 18 months' notice and to allow for a transfer of experimental work from Maralinga if politically necessary.[13]

By early December 1961 the Foreign Office was advising Ministers that the best strategy to adopt in Geneva would be to advocate a treaty banning tests in the atmosphere, with verification through national detection systems. Although, this would only entail a limited degree of international control through the creation of an organisation to register reports from national detection systems, it would enhance the UK and US positions in the UN. Officials recognised that it was unlikely the Russians would accept this, or that the US itself would be prepared to buy into this when it was considering its own needs for new atmospheric tests.[14] Even Lord Home himself thought that the US would refuse, but felt it was worth trying to persuade them of the political advantages for going down this particular road. However, the MOD was opposed to this in principle as it was against anything in this field or that of disarmament generally that claimed to be a verification control process, but was in reality far from it. They did not want to set a poor precedent.[15] Moreover, Ormsby-Gore, in Washington, advised strongly that London should not raise the proposals with the US until UK and US scientists had evaluated the significance of the Soviet tests. In his view, the Americans would certainly regard the proposal as something worth considering when the Prime Minister and the President met in Bermuda at the end of December 1961.[16] Given MOD views and the probable US reaction, the department was minded to accept Ormsby-Gore's advice. The Foreign Secretary agreed, and testing was added to the agenda of the Bermuda summit meeting on 22 December 1961. In the meantime, discussions continued on a number of possible options to be pursued in Geneva. Each one had its own pros and cons, but on balance the Foreign Office advised Macmillan that the best bet would be to merge the test-ban talks in the new general discussions on disarmament, which were scheduled to start in March or April 1962.[17]

---

[13] TNA CAB 21/6038, Nuclear Requirements for Defence Committee Meeting in Conference Room C Cabinet Office, Thursday 28 July 1960; see also Arnold and Smith, *Britain, Australia and the Bomb*, 2006.

[14] TNA FO 371/157115, IAD 24/84, Pemberton-Pigott to Minister of State/Secretary of State, 5 December 1961.

[15] TNA FO 371/157115, IAD 24/85, C.W. Wright, MOD to A.D. Wilson, Foreign Office, 6 December 1961.

[16] TNA FO 371/157115, IAD 24/86, A.D. Wilson to Private Secretary, 7 December 1961.

[17] TNA FO 371/1571101, IAD 22/506, Bermuda meeting: December 1961, Nuclear tests conference, a Geneva brief by the Foreign Office, undated. This new body was to become the Eighteen Nation Disarmament Committee – ENDC.

Macmillan and Kennedy agreed that it was now necessary, as a matter of prudent planning that preparations should be made for atmospheric testing to maintain the effectiveness of the deterrent.[18] UK and US experts also agreed at Bermuda that, notwithstanding resumption of large-scale Soviet atmospheric tests, the US still retained an overall technical advantage in warhead design, though it had been matched or surpassed in several areas. In the absence of new US tests, the USSR could gain an overall advantage in two or three years if it continued to test. Both sides were concerned about Soviet interest in anti-missile defence and possible Soviet progress in this area came to be a determining factor in the eventual UK agreement to the US request to conduct atmospheric tests at Christmas Island.[19]

US Resumes Underground Testing

The Soviet resumption of testing increased the pressure to test within the US and UK, and inside the Kennedy Administration the advocates of testing were emboldened. London too was aware that the Russians were working very hard on an anti-missile missile.[20] Just how should the UK and US respond? The Foreign Office sought to steer the Americans away from a statement that they would resume testing, towards one that merely regarded the US as free to do so. Officials hoped that the Americans would avoid accompanying this by any hint they would in fact start testing.[21] At Cabinet on 5 September, Macmillan proposed that a message should be sent to the US government suggesting that if the USSR did not refrain from atmospheric testing, consideration should be given to means of using this to allied advantage. He hoped that time would be allowed for full consideration of this question before the US itself decided to resume testing.[22] Kennedy, however, announced that day that the US would resume laboratory and underground testing. Macmillan was at least pleased that the President did not contemplate any early

---

[18] United States Arms Control and Disarmament Agency, *International Negotiations on Ending Nuclear Weapon Tests September 1961–September 1962*, Publication 9 released October 1962, Washington, p. 40.

[19] Outgoing Telegram Department of State Action Amembassy Paris 29 December 1961 at http://www2.gwu.edu/-~nsarhciv/NSAEBB/NSAEBB/tb38.pdf.

[20] Macmillan, *At the End of the Day*, p. 146 and p. 152.

[21] TNA FO 371/157120, IAD 26/77, Nuclear Tests – Joint Message from the Prime Minister and President Kennedy, C.T. Gandy to Mr A.D. Wilson, 4 September 1961.

[22] TNA CAB 128/35, CC (61) 49th Meeting, Item 2 Disarmament Nuclear Tests, 5 September 1961.

resumption of atmospheric tests.[23] Kennedy had made clear to the UK that he would consult closely before taking any such decision.[24]

Super Octopus and the US Request to use Christmas Island

In the UK, the pressures to test were of a different sort. AWRE had developed a new implosion principle called Super Octopus. At that point, the UK planned to follow their first operational thermonuclear bomb, Red Snow, with a quite different type of warhead for the Skybolt missile for the RAF V-Bombers. However, whereas the Red Snow warhead raised no problems for safety and reliability once anglicised, the existing proposal for the Skybolt warhead had revealed sharp differences of opinion between the British and American laboratories over safety aspects of the proposed primary device. Being unable to test, the UK could either adopt the proposed US primary, which had been tested but used more sensitive explosive than the UK Ordnance Board permitted, or use an untested British modification of it. Unfortunately, neither solution seemed likely to offer the same assurance of safety or reliability provided for Red Snow.

Super Octopus, however, offered a satisfactory solution to the problem, at least in the UK.[25] It utilised a new primary multi-point implosion system that could lead to a more rugged, reliable and lighter type of warhead than existing designs; it was also inherently single-point safe; and used the new EDC 11 HE. As such this would also go a long way to addressing the safety problem. The lighter weight would make it possible to incorporate the less sensitive explosive and a mechanical safeing device.[26] The principle was tested in 1958 and had attracted particular interest in the US, especially from Los Alamos.[27] A successful test of such a device would, the UKAEA thought, be a significant contribution to the UK/US collaborative effort. Furthermore, the concept could also be used in the Skybolt warhead. If proven, the design would be of immense value to the UK programme. It would open the way to significantly lighter and more efficient UK warheads with a reduction in size. If this was not tested it might well, in one worried official's mind, mean that the UK would be denied the opportunity to make significant advances in warhead design in future. This in turn could result in the flow of information from the US ceasing,

---

[23] TNA FO 371/157122, IAD 26/101/G, Foreign Office telegram no. 6285 to Washington, 7 September 1961.

[24] TNA FO 371/157122, IAD 26/115/G, Prime Minister's Personal Telegram T .496/61 Kennedy to Macmillan, 7 September 1961.

[25] TNA DEFE 19/92, Nuclear Requirements for Defence Committee, Memorandum by the Chairman, October 1961.

[26] TNA CAB 134/2238, Cabinet Nuclear Requirements for Defence Committee, N.D. (61) 1, Nuclear Tests, Note by the Secretaries, 16 October 1961.

[27] TNA PREM 11/ 3246, Note for the Record, Roger Makins, UKAEA, to Sir Norman Brook, Cabinet Office. 7 November 1961.

as there would be no ongoing UK contribution to UK-US collaborative research efforts.[28] Also without the successful validation of this design, there would be no safe or reliable military nuclear capability in the UK; military and technical rationales for this device were at one.

The problem was where to test it since the UK had no facilities for underground tests and finding a suitable location would be costly and time consuming.[29] Makins had mentioned Super Octopus to Glenn Seaborg (Chairman, USAEC) on 21 September 1961 and passed on details of the design. He asked whether it would be technically possible to fit a test of this concept into the current US programme; Seaborg promised to look into it.[30] Macmillan wrote to Kennedy on 3 November saying that, in the light of the Soviet resumption, London had given some thought to possible UK testing requirements.[31] This device promised to be safer and more efficient and in every way more satisfactory than the proposed alternative design for use in Skybolt. Macmillan therefore asked at the highest level if Super Octopus could be fitted into the ongoing US programme at Nevada before the end of the year as a suitable device could be made ready very quickly. He also indicated that there were plans for one or two other tests, but these would be in 1962. Kennedy readily agreed and promised to give the UKAEA technical data derived from the on-going US tests relevant to the weapons covered by the existing 1958 US-UK Mutual Defence Agreement.[32] He also suggested that if there were a leak of information, the visit of UK technicians to Nevada to prepare for the test would be explained by a reference to UK participation in the Vela Uniform programme, which was the US seismic research programme launched in October 1960 to improve detection and identification capabilities for underground events. It also examined detection of tests in outer space and atmosphere by means of satellites and the detection of tests in these environments using ground instruments.[33]

---

28 TNA DEFE 19/92, Copy of a minute from the Secretary of State for Aviation to the Ministers of Defence and Aviation, 28 September 1961.

29 AWRE started to look in a preliminary fashion at the possibility of finding suitable locations for underground tests; informal approaches were made to Australia in July 1958 and the search looked at British Commonwealth and colonial territories. Over 120 separate areas were listed and these were examined based on their geographic features. The most suitable sites for geological, topographical, and meteorological and access reasons were in the UK, Canada and New Zealand and East Africa. And even here there were problems. Political objections ruled out any UK site. In fact, the only really suitable candidate was in British Columbia. TNA ES 1/1349, Contained Nuclear Explosions Part II Suitable Sites, paper for Submission to NTPC, Sites for Underground Nuclear Tests, Note by D/AWRE, 22 April 1959.

30 TNA DEFE 19/92, Sir Roger Makins to Sir Robert Scott, 22 September 1961.

31 TNA PREM 11/ 3246, Prime Minister to President Kennedy, Foreign Office telegram no. 7977 to Washington, 3 November 1961.

32 TNA PREM 11/3706, Kennedy to Macmillan, 11 November 1961.

33 TNA DEFE 19/92, Foreign Office telegram no. 8579 to Washington, 22 November 1961; TNA PREM 11/3706, Washington telegram no. 3132 to Foreign Office, 20 November

Kennedy also explained the problems the US had in finding an alternative for atmospheric tests to Eniwetok Atoll. It was UN trust territory and the political problems in using such an area meant that Washington had to find somewhere else and Christmas Island fitted the bill.[34] It would help enormously, Kennedy wrote, if the US could have an early assurance that there would be no obstacles to US use of this site. The tests the US wanted to perform were weapons effects and development tests that could not be conducted underground. Kennedy had not yet decided to resume atmospheric testing, but wanted to start preparations to give the US the option should such a decision be required. Kennedy assured Macmillan that the US would not conduct any tests in the atmosphere unless they were a necessary part of a wider programme, the result of which would offer the possibility of improvements of substantial military significance; also that the results could not reasonably and effectively be obtained in any other way, and that the most rigorous steps were taken to avoid or reduce fall-out. Kennedy appeared here to be trying to satisfy the conditions that Macmillan had laid out in his 31 October statement to the House of Commons, in particular that Macmillan would have to be convinced that each test was necessary militarily.

Cabinet Deliberations

Kennedy's request to the UK to use Christmas Island for atmospheric tests put Macmillan in a quandary. On the one hand, he could hardly refuse his closest ally, but plans to resume testing went against the grain of his and the UK's commitment to a CTBT. Shortly before the formal request, Macmillan told his officials that if the Americans made a request for the use of Christmas Island, and made a case, the UK should sympathetically consider. The full Cabinet considered the US request on 14 November 1961. The Foreign Secretary argued that from the point of view of defence and the UK's relations with the US, the balance of advantage lay in making the facilities at Christmas Island available to the US. That said, this could only be on very strict conditions that would have to be justified to the UK government. He believed that the preparations for, and still more, the holding of atmospheric tests would give rise to severe domestic criticisms.[35] Macmillan was quite adamant about insisting that the UK had to be satisfied that the whole US programme fell within his definition of the types of test that would be defensible. Testing could not be open ended and there had to be a rough time limit, such as during 1962.[36] UK scientists would have to be satisfied that the proposed tests met

1961; see also Bates, Gaskell and Rice, *Geophysics in the Affairs of Man*, pp. 172–87.

34 TNA DEFE 19/92, Copy of a letter from President Kennedy to the Prime Minister, 11 November 1961.

35 TNA CAB 128/35, C.C. (61) 62nd Conclusions, 14 November 1961.

36 TNA PREM 11/3718, Prime Minister's Personal Minute Serial No. M 1/62 to Sir Norman Brook, 1962.

the strictest criteria of military necessity and this would involve the UK obtaining full information about the devices the US proposed to test.

From the MOD perspective, other things had to be safeguarded. The UK needed to obtain the maximum technical and scientific advantage out of these tests; it should be able to include atmospheric tests of its own in the US series if it were decided to do so; and the position of the Prime Minister and of HMG in general should be safeguarded.[37] (This alluded to Macmillan's 31 October statement to the House of Commons.) The Cabinet considered the US request again on 16 November 1961 having seen a memorandum circulated by the Foreign Secretary. This outlined the background and specifics of the US request.[38] The Americans had explained that the testing was necessary mainly to improve yield to weight ratios and thus reduce warhead weights. This was needed in order to carry more protection devices, as well as to develop counter-measures, and even counter counter-measures. Full information was promised by the US from the tests. Home wrote that the UK had already supported the US decision to resume underground testing back in September and had since stated (Macmillan's 31 October 1961 statement) that if the UK were convinced that a particular test was necessary to preserve the balance of the deterrent and preserve world freedom, then it would also support atmospheric testing too. Since the Americans were in real difficulty over finding a suitable location for atmospheric tests, this placed the UK in a position to withhold something that they badly needed. This gave the UK strong bargaining advantages. However, there was a down side as agreeing to the US request would generate world and domestic criticism, and both sides would come under strong pressure to halt the new programme.

On balance, Home thought there was advantage from both the foreign policy and defence perspectives in acceding to the US request, but only on very strict and precise conditions, which could be justified to the US on the grounds of the severe criticism that would follow directed at the UK government. The UK would need to be satisfied that tests met the strictest criteria of military necessity, and it could only be satisfied about this based on full scientific information, including design information about the devices to be tested. These conditions would have to apply to each individual test. Health and safety considerations would also be critically important. Home therefore recommended that his colleagues agree that officials should be instructed to look into the detailed conditions that should apply. Moreover, when the UK was ready to tell the US that it agreed in principle, officials should also make clear that there would need to be discussions on how both sides should present the test programme to the world in order to mitigate the severe criticism that would inevitably follow.

---

[37] TNA DEFE 19/92, Minute to Chief Scientific Adviser, Christmas Island, 10 November 1961.

[38] TNA CAB 129/107, C. (61) 179, Use of Christmas Island for United States Nuclear Tests, 10 November 1961.

Macmillan and Home went over the arguments with their colleagues and the minutes of the meeting make it clear that the Cabinet was not entirely comfortable with the idea.[39] Ministers expressed the view that it would be hard to justify to UK public opinion a series of atmospheric tests on the scale envisaged in the US programme, or to square it with the Prime Minister's own 31 October statement. The Cabinet was adamant that the UK would need to be given full information about the military and scientific justification for the US programme in order to evaluate it against the Prime Minister's statement.[40] Furthermore, it would be important to have this before any overt work began on preparing the facilities at Christmas Island, a fact that would no doubt become public knowledge. The US planned test programme, as a response to the Soviet breach of the moratorium, was excessive in the UK view. Macmillan's Cabinet colleagues were particularly anxious that any announcement on the use of Christmas Island for further tests should be coupled with a new disarmament effort.[41] Macmillan personally did not want to see atmospheric testing resume. As he and the government had to be persuaded that the tests that the US wished to conduct could be justified militarily, Penney and Zuckerman were despatched to Washington to hear in detail what the Americans proposed. Testing on barges or on the surface was unacceptable because of the fall-out that these would create.

Macmillan told his colleagues that he understood the deep anxieties that the US request had created. However, he suggested that he might reply that it would be necessary for the President and himself to be satisfied, on military and scientific grounds, that each test met the criteria laid down in their public statements (1 November and 31 October respectively), and could be justified to world opinion on that basis. Macmillan would make clear that there could be no certainty that he would always concur in the President's judgement that a particular test was required. In such circumstances, it might be best if the US were to begin to look to make some additional arrangements in case it proved impossible for the UK to agree to use Christmas Island. Meanwhile it would be good if the UK and US experts could jointly review the merits of the testing programme to help the President and Prime Minister decide whether each test was in fact necessary. For the present, a joint secret and small reconnaissance party could go to the island to examine what programme of work would be necessary to prepare the island for any resumption. This was agreed and the Cabinet decided to meet later that week to resume its discussions on the US request.

A few days later the Cabinet reverted to the Christmas Island issue and reviewed Macmillan's draft reply to Kennedy.[42] Once again, Macmillan emphasised the strict criteria that would be used to judge the necessity or otherwise of

---

39 TNA CAB 128/35, C.C. 62 (61) Item 3 Nuclear Tests, 1961.

40 For a detailed account of the tests that the US conducted at Christmas Island, see Hansen, *U.S. Nuclear Weapons*, pp. 83–7.

41 Macmillan, *At the End of the Day*, p. 153.

42 TNA CAB 128/36 C.C. 63 (61) Item 4 Nuclear Tests, 1961.

each individual test. The test that he had in mind was the direct contribution to preserving the balance of the deterrent, and especially against counter-measures or by developing anti-missile missiles. It would be essential to avoid prejudicing the moral predominance of the West by indulging in tests that were not absolutely essential. If Christmas Island were to be used for US tests, the UK would need a formal agreement to settle the financial and administrative terms of what in effect would be a joint effort enabling UK experts to have access to the information that the tests would yield.

The Cabinet agreed and instructed Sir Norman Brook to ensure that the actions flowing from the decision were put in hand expeditiously.[43] If a joint UK and US reconnaissance party went to the Island, secrecy was paramount. In fact, such was the sensitivity that the reconnaissance party had a cover story to disguise the real reason for the US visit. This was that the Americans were there to study a possible extension of the satellite-tracking programme.[44] A team would also go to Washington to find out what tests the Americans had in mind. Officials noted that marginal improvement in yield to weight ratios would not of itself justify tests in the atmosphere, whereas an improvement which enabled important new counter measures to be incorporated in the weapons, or provided some other new strategic application, might in some circumstances be judged necessary to preserve the balance of the deterrent.

Administrative and Technical Aspects of the Request

The Christmas Island issue emerged while the UK and US were still seeking agreement on the terms for use of the Holy Loch for Polaris submarines. There was still no written agreement covering liability in the event of an accident or radioactive leak. This was one reason why Brook and his colleagues felt the need for a written agreement setting out the conditions for the use of Christmas Island.[45] Another key issue was the protection of British sovereignty over the island. Such was the importance attached to these issues that the NRDC established a sub-committee to look into the details of a Memorandum of Understanding (MOU) governing the rights and responsibilities of the two sides. This committee had particular responsibility for safeguarding UK interests in relation to its right to receive technical information about the preparations for the US tests and their

[43] TNA DEFE 19/92, Nuclear Tests, Note of a Meeting in Sir Norman Brook's Room, 17 November 1961.

[44] TNA CAB 134/2235 Nuclear Co-ordinating Committee 'A': Meetings 1–3; Papers 1–8. 1961 Nov–1962 Jan, Nuclear Co-ordinating Committee 'A' 1961 Meeting, 24 November 1961.

[45] TNA CAB 134/2236, ND (61) 6 (Final), Cabinet Nuclear Requirements for Defence, Anglo-American Nuclear Tests: Brief for the Prime Minister, Note by the Secretaries, 15 December 1961.

results. This had to be included in the formal agreement to be negotiated with the US. NRDC members endorsed the view that advance US/UK agreement was necessary on the particular strategic and military significance of any proposed test or tests in the atmosphere.[46] UK officials certainly believed that it would not be difficult to make the case for a UK resumption as it wanted to meet a real military requirement; namely, the certification of warheads for UK use as safe.[47]

One of the factors in the US (and UK) decisions to resume atmospheric tests was whether the Russians had derived any significant advantage from their recent programme. Both the UK and US had conducted extensive radiochemical analysis of the debris with the aim of making some accurate assessments of the Soviet warhead designs being tested. Information exchange between the UK and US on their own tests and respective analyses, which dated from the early 1950s, gave both sides greater confidence in their diagnostic capabilities and therefore the accuracy of their respective assessments of Soviet tests. The Soviet test series included a number of delivery systems tests and at least four atmospheric weapons effects' tests; several also indicated advances in yield-to-weight ratios.[48] The high altitude tests created the greatest alarm since these seemed to suggest that the Russians were investigating possible ABM warheads. However, the UK assessment at the end of 1961 was that nothing learned by the Russians would actually shift the strategic balance of nuclear power in their favour.[49] AWRE's

[46] TNA DEFE 19/92, Use of Christmas Island for US Nuclear Tests, R. Press to Sir William Penney, UKAEA paper attached to letter, 22 November 1961.

[47] TNA CAB 134/2236, Cabinet Nuclear Requirements for Defence Committee, Nuclear Tests, Note by the Secretaries, ND (61(1) (Final), 16 October 1961.

[48] Thomas B. Cochran, William M. Arkin, Robert S. Norris and Jeffrey I. Sands, *Nuclear Weapons Databook Volume IV Soviet Nuclear Weapons* (New York, 1989), p. 352.

[49] TNA CAB 134/2236, ND (61) 6 (Final), 15 December 1961 Cabinet Nuclear Requirements for Defence Committee, Anglo-American Nuclear Tests: Brief for the Prime Minister. Note by the Secretaries. A report prepared for the British Nuclear Deterrent Study Group on 'ABM Defence systems' dated November 1962 noted that, 'The damage mechanisms resulting from the absorption of X-rays by the missiles skin are uncertain. Several mechanisms are conceivable, and estimates of lethal radii from a 1 MT bomb have ranged from insignificantly small values out to several miles. Although it may be certain materials are vulnerable at distances of several miles from a 1 MT bomb, we have no evidence to show that this order of vulnerability obtains for the complete range of permissible materials. In short, the damage effects of X-rays are insufficiently understood to allow us to consider them further here.' Personal correspondence. That said, the first time that the US tested one of its own warheads in an underground effects test the evidence appears to have been that none of the US or UK warheads would be effective against Moscow and other cities when ABM systems became available. This led to a requirement for systems that were immune to ionising and non-ionising radiation. Personal correspondence.

view, at least according to Zuckerman, was that there was always something to be learned in terms of technical understanding.[50]

Kennedy said much the same thing on 9 December following a USAEC press release on its preliminary analysis of the Soviet tests.[51] Yet, as analysis produced more results, it became clear to the US and UK that the Russians had in fact made significant progress.[52] Glenn Seaborg had told Roger Makins in mid November 1961 that the US assessment was that the Russians had now caught up in weapon science and in the sophistication, economy and efficiency of their weapon designs. Furthermore, the very large explosion on Novaya Zemlya in the Soviet Arctic (the 58 megaton Tsar Bomba) had attained a degree of cleanness, efficiency and economy that was probably beyond US capabilities at that time.[53] Penney warned Sir Norman Brook that the Russians might have made at least one impressive technical advance that could have deep military consequences.[54]

This fact was reported to the Cabinet as it debated what to do about the US request to use Christmas Island. There was a strong military case to resume testing if the US were not to fall behind the Russians.[55] Although these consequences were not spelled out, it is highly likely this referred to ABM capabilities. Furthermore, the UK assessed that the 58 megaton test showed an ability to obtain a high fusion to fission ratio, and UK analysis had picked up a faint suggestion that the fission yield of the total Russian series might have been less than originally supposed.[56] In the event of this evolving data, Penney and the NRDC recommended that the UK should discuss their assessments of the Soviet tests with the US before proceeding to the proposed Anglo-American discussions on the need to resume atmospheric tests.[57] Given this background, the pressures within both the US and UK to resume atmospheric tests were immense. Moreover, Kennedy continued to press for a response from Macmillan; in return Macmillan suggested that UK and US experts should meet to agree the terms that would be necessary for the US

---

50 Zuckerman, *Monkeys, Men and Missiles*, pp. 318–19.

51 Atomic Energy Commission Press Release 9 December, Preliminary Analysis of the United States Atomic Energy Commission of Soviet Nuclear Tests, December 1961 in United States Arms Control and Disarmament Agency, *Documents on Disarmament 1961*, p. 706.

52 TNA DEFE 19/115, Agreed record of Conversation with Mr A. Potts of the Joint Intelligence Bureau on Thursday 18 January (Zuckerman). UK assessment of Soviet tests depended almost entirely upon US data. The analysis carried out in the UK had been affected because of difficulties over filters. Many short-lived isotopes had decayed because UK filters had not been despatched to the laboratory until after a significant delay.

53 TNA PREM 11/3246, Washington telegram no. 3118 to Foreign Office, 18 November 1961.

54 TNA PREM 11/3246, W.G. Penney to Sir Norman Brook, 21 November 1961.

55 TNA CAB 128/36, C.C.1 (62) Item 4 Nuclear Testing, 3 January 1962.

56 TNA PREM 11/3246, Zuckerman to Minister of Defence, 24 November 1961.

57 TNA PREM 11/3246, Sir Norman Brook to Prime Minister, 22 November 1961.

to use Christmas Island. Macmillan still had to get the proposal past his Cabinet colleagues. Moreover, he did recognise that it would be unreasonable for the British government, should it approve the use of the island, to be in a position to object to tests that Kennedy thought necessary, even if the UK disagreed.[58]

Further Cabinet Deliberations

Macmillan met Kennedy in Bermuda on 21 and 22 December during which he discussed with Kennedy and the Secretary of State whether the US should resume atmospheric testing.[59] He and Home had made great efforts to persuade Kennedy that the UK and US should make a further major effort to halt the nuclear arms race.[60] Macmillan had told Kennedy that he was prepared to consider a private agreement with the US about the resumption of nuclear tests and the use of Christmas Island, although he would have to put this to the Cabinet. He hoped that it would be possible to make a public announcement of some new disarmament initiative. He agreed that it would be wrong for the UK to expect the US to waste time and money on preparation at Christmas Island if at the end of some months of work the UK vetoed the holding of tests. Macmillan would expect that if HMG decided to let tests begin they would also have to accept that, although the US would consult them about a decision to test, the UK could not at that stage exercise a veto. Kennedy was grateful for this statement.[61] Although Macmillan had suggested putting off tests for another year, Kennedy made clear that he felt that the West should now prepare to test in the late spring of 1962 unless there had been a change in the political situation before then.

After the Cabinet debated the issue on 3 January 1962 Macmillan was then able to write to Kennedy to say that in principle the UK could accept the US request to use Christmas Island.[62] However, in a covering note to the British ambassador Macmillan said that whilst he did not want to make a disarmament initiative a condition of granting facilities for US tests, he hoped very much that Kennedy would agree to act as he suggested. If he did not agree, Macmillan told Ormsby-Gore, that he would have to consult the Cabinet again. He might also have to think about what independent initiative the UK might then take on disarmament.[63]

[58] TNA DEFE 19/92, Macmillan to President Kennedy, 22 December 1961.

[59] TNA CAB 128/36, C.C.1 (62), Item 4 Nuclear Testing, 3 January 1962.

[60] Zuckerman, *Monkeys, Men and Missiles*, p. 321.

[61] TNA CAB 21/5557, Nuclear Tests, Record of a Meeting Held in Government House, Bermuda on Friday 22 December 1961.

[62] TNA PREM 11/3718, Macmillan to Kennedy, Prime Minister's Personal Telegram T5/62; TNA FO 371/163113, IAD 23/3, 1962.

[63] TNA PREM 11/3718, Prime Minister's Personal Telegram Serial T. 6/62, 5 January 1962. See also TNA CAB 195/20.

Ormsby-Gore saw the President on 10 January 1962. He explained that, although the Cabinet had agreed to put Christmas Island at US disposal, this was not conditional upon Kennedy accepting the precise proposal on disarmament set out in Macmillan's letter. The Cabinet had given its approval in the context of such an initiative and would wish to consider its decision if no agreement to act was reached.[64] Kennedy made clear his own reluctance to resume atmospheric testing, but said that in absence of any progress on East-West relations he would order resumption in the spring. Kennedy was also worried about limiting a disarmament initiative too closely to resumption of testing since that might well give the Russians the opportunity to string the US along for months.[65] Kennedy emphasised that the US was eager to work out a new joint initiative that might be a significant step forward. Whilst the US had some questions about the tactical aspects of Macmillan's suggestions, they had no difficulties at all on the importance and urgency of the need for progress.[66] He would ask Dean Rusk to work out the possible details with the Foreign Secretary. Macmillan still hoped that the UK could help by making the facilities on Christmas Island available and that the UK would not make use conditional on a new initiative toward disarmament. Kennedy would find it hard to accept conditions that might prevent him from meeting his responsibilities for military security. Working out the disarmament ideas would take time, but a decision on Christmas Island was needed within the next few days. And as a parting shot, Kennedy said that he would rather make preparations without Christmas Island than run the risk of putting that sort of strain on 'our old and vital friendship'.

There was indeed a risk. Macmillan pointed out in his reply to Kennedy the next day, that opinion in the UK, whilst accepting the decision to test was militarily justified, would not like an arrangement under which the UK gave the US the right to decide unilaterally whether or not actually to test from Christmas Island.[67] It had been for this reason that the UK wanted a joint statement from the President and Prime Minister that announced that they had jointly decided that tests were necessary rather than simply state that preparation on the island was underway and that no decision had been taken. However, Macmillan made clear that he did not think that testing should resume until a further attempt had been made to halt the nuclear arms race. It was by no means certain that the problems of anti ballistic missile defence could be overcome; however, neither side could afford to risk the other getting there first. He suggested that they jointly propose a fresh initiative in which the UK, France and US might take in the new Eighteen-

---

64 TNA FO 371/163113, IAD 23/4, Washington telegram no. 95 to Foreign Office, 10 January 1962.

65 TNA PREM 11/3718, Washington telegram no. 95 to Foreign Office, 10 January 1962.

66 TNA PREM 11/3718, Kennedy to Macmillan, 13 January 1962.

67 TNA PREM 11/3718, Prime Minister's Personal Telegram Serial No. T. 18/62, 14 January 1962.

Nation Disarmament Committee (ENDC) in Geneva when it held its first meeting in March 1962.

The Cabinet discussed the terms of Macmillan's draft message to Kennedy and is so doing expressed agreement that a determined effort should be made to prevent the opening of a new phase in the nuclear arms race and that the US request to use Christmas Island gave the UK a special opportunity of intervening to this end. Timing was a problem; the need to begin preparations on the island as soon as possible meant that once this work became public knowledge it would be assumed that the decision to resume had already been taken. (That was certainly Khrushchev's attitude later.) In any case, the Cabinet realised that if the US had decided that they must resume they would do so, with or without British help and any UK refusal would lose what influence London could bring to bear in Washington on the need for disarmament measures. Macmillan himself was quick to point out that if the UK tried to disassociate itself from US testing it would run the risk of grave damage to the Anglo-American relationship, and in particular to the bilateral cooperation agreements on nuclear defence. Macmillan suggested that the Cabinet might meet later in the day when he would arrange for Penney to be available to assist them in their consideration of the technical case for resuming testing. Macmillan had suspected at the end of December that there was a good deal of doubt amongst his Cabinet colleagues about going ahead with Christmas Island.[68]

Penney, now the Deputy Chairman of the UKAEA, attended the resumed Cabinet meeting later that day to give his assessment of the technical reasons that would justify the Western powers in holding a further series of atmospheric nuclear tests.[69] In his view, if no further tests were undertaken, the US would retain a slight advantage in capability. Penney did not think it could be assumed that the Russians would in fact refrain from further testing themselves once they absorbed the lessons from the last round. A study of these tests had shown that some of them were specifically directed towards the evolution of anti-missile warheads.[70] Although the perfection of such a system would be immensely costly and difficult, Penney could not see how the US could stand aside if the Russians continued to develop an ABM system. The work of the US scientists would be bound to lose momentum if they could not test the results of their research.

The new series of tests that the US wanted to pursue contained four broad categories. The first of these would consist of high altitude explosions for measuring ionisation and other effects and for studying the effect of a nuclear explosion on a missile at high altitude. Second, tests were needed of the warheads now in production for Polaris, Minuteman and Skybolt ballistic missiles, which had been conservatively designed and might well be reduced in weight in order

68 TNA FO 371/163113, IAD 23/2, Prime Minister's Personal Minute No. M. 401/61, 29 December 1961.

69 TNA CAB 128/32, C.C.2 (62) Item 1 Nuclear Tests, 1962.

70 It appears as if this was indeed the case: papers in Russian archives indicate that this was one of the purposes of the tests. Personal Correspondence.

to make room for safety devices to prevent accidental detention of the HE in the warhead.[71] In the third category, the Americans wanted to test new warheads that might increase very significantly the yield to weight ratio of warheads in the one kiloton and one megaton range. Finally, experiments were needed to verify certain completely new concepts, still in the theoretical stage, which, if successful, would produce a revolutionary improvement in yield to weight ratios. All of these tests would be important first steps towards devising a defence against missiles. Penney made clear that an effective ABM system could not be produced in less than ten years. Nevertheless, the consequences of falling behind were such, that if he had the responsibility of advising President Kennedy, he would feel obliged to recommend that testing should resume unless there were some guarantees that no further Russian tests would occur. He felt certain that US scientists would give the same advice, and at this point he left the meeting to give the Cabinet the chance to consider what he had told them. In fact, the UK assessment was that the Russians had made significant progress in improving yield to weight ratios, something of importance for ABMs because it would then become practicable to include various decoys and similar devices to protect the weapon.[72]

As the Cabinet Conclusions record, there was general agreement that, failing international agreement on a ban on testing, atmospheric testing in the first three categories described by Penney would indeed be justified. However, the Cabinet could not form an opinion on the fourth category until its nature and purpose could be more precisely defined. In these circumstances, the UK would be obliged to support the US decision to resume, if such a decision were taken, given the sound military case. It was therefore reasonable that UK facilities should be made available to the US.[73] Penney's intervention had clearly been decisive. Cabinet members thought that Kennedy would want to take some sort of disarmament initiative if he could and for this reason before finally committing the Western powers to resumption, a fresh initiative on disarmament should be taken. Macmillan therefore had secured agreement to go back to Kennedy with the affirmative UK response on the use of Christmas Island, albeit coupled with a disarmament condition. Messages were duly sent and responses received.

Lingering Doubts

In mid January 1962, however, Watkinson was still not convinced that there were compelling military reasons to resume atmospheric testing. He suggested to Macmillan that the UK should attempt once more to encourage the USSR to start

[71] It is not clear from the context whether this phase implies warhead hardening or warhead safety.

[72] TNA FO 371/163113, IAD 23/2, Prime Minister's Personal Minute No. M. 401/61, 29 December 1961.

[73] TNA CAB 128/32, C.C.2 (62) Item 1 Nuclear Tests, 1962.

talking again seriously on limitations in order to avoid another round of escalating tests.[74] Cabinet met again on 18 January to discuss the response from Washington, which was not entirely supportive of the UK's suggested line.[75] Kennedy, although welcoming the suggestion of a new initiative on disarmament, had urged the need for an independent decision on Christmas Island. Macmillan therefore had to propose a new tack and this time round a new public statement could be made which would put the onus on the USSR to respond constructively to Western proposals in the ENDC. There was a complicating factor. Kennedy also wished it to be clear that the final decision on testing rested in his hands. Politically he could not afford to have decisions on national security left contingent on a foreign government, even a friendly one. Macmillan's political problems were just the opposite. He could not put himself in the position of being seen to allow testing on British territory to taking place unless it were a matter of joint decision and in conformity with the 31 October 1961 statement. As before the arguments about meeting US requests were compelling, so too was the need to pursue disarmament options and to avoid placing the UK in an awkward position where it would attract criticism for associating itself with US policies. In any case, it looked as if US scientific opinion was by no means unanimous on the need to test in the atmosphere, which seemed to be the opposite of what London had originally been led to believe. It was certainly news to Penney, but in any case, he felt that London should not take these reports too seriously. There were conflicting currents of opinion in the US, but Sir William still felt that the one in favour of atmospheric testing would prevail.[76] He was right.

The Cabinet eventually agreed that the best course would be to suggest to Kennedy that the announcement should declare it to be the present view of both governments that the existing state of nuclear development, in which the Soviet tests were an important factor, would justify the West in making further tests as might be necessary for purely military reasons. The US and UK governments had therefore decided that preparations should be made in various places and as part of these the UK had made available the facilities at Christmas Island. Final wording on the draft statement would have to be decided when further progress had been made towards agreement on the details of new initiatives on disarmament. A further exchange of messages led to agreement on the terms of the announcement, but its release was delayed in the hope that the French might sign up to the disarmament elements. They did not.[77] Macmillan was worried about French attitudes and felt the need to keep de Gaulle in the loop.

---

74 TNA CAB 21/4755, Nuclear Tests in Atmosphere, Harold Watkinson to Prime Minister, 16 January 1962.

75 TNA CAB 138/36, C.C.6 (62) Item 2 Nuclear Tests, 18 January 1962.

76 TNA FO 371/163113, IAD 23/12, Resumption of Nuclear Testing, HAS Stephenson to Sir H. Caccia, 15 January 1962.

77 TNA CAB 138/36, C.C. 10 (62), Item 2 Nuclear Tests, 1 February 1962.

The Problem of Justification

Macmillan was anxious to avoid being pinned down on who had the final say on whether a particular test should proceed in the event that there was a disagreement between the UK and US. A satisfactory formula was eventually devised with Washington for dealing with this point. They had agreed that, if enquiries on this were pressed, each government would say that consultations would continue between them on both the scientific and political aspects. They would add that disagreement seemed unlikely, in view of their joint statements and continuing consultations. However, since the UK and US had both reached agreement in principle on the moral justification for making further tests, neither government would stand in the way of the other if, at the end of the day, one felt that further tests were militarily necessary.[78] London and Washington hoped to fudge this issue and that no one would press too closely. The Embassy in Washington reported that the Administration agreed to the UK's terms, but that they felt that there was still a risk that some one would indeed ask the pointed question on the final decision. However, the Ambassador felt that the UK could rely on Kennedy and Rusk sticking to the Prime Minister's formulation described above.[79]

Public Announcements

Macmillan made a public announcement on 8 February 1962; Kennedy made one at the same time, about the possible use of Christmas Island and disarmament. This made clear that both the UK and US felt that the recent Soviet programme had justified the West on resuming testing as might be necessary for purely military purposes. For this reason and as a precautionary step, preparations were underway in various places, including Christmas Island. In parallel, further attempts would continue on making progress at the ENDC, scheduled to open in Geneva on 14 March and that the Heads of Government of the UK, USSR and US should assume a direct and personal interest in these negotiations.[80] This was included in a statement to the House of Commons on the same day.[81] Kennedy however rejected the idea of a test-ban summit. In parallel, a committee was appointed, 'to advise the Prime Minister on the scientific aspects of nuclear tests'. This was the Nuclear Tests Advisory Committee (NTAC) chaired by Sir Roger Makins and included Penney and Zuckerman amongst its membership. Macmillan saw this

---

78 TNA CAB 138/36 C.C. 6 (62), Item 2 Nuclear Tests, 18 January 1962.

79 TNA CAB 21/4755, Washington telegram no. 198 to Foreign Office, 20 January 1962.

80 White House Statement on Anglo-American Arrangements for Nuclear Test Preparations at Christmas Island, 8 February 1962 in *Documents on Disarmament 1962 Volume I January–June, United States Arms Control and Disarmament Agency*, pp. 31–2.

81 Hansard, House of Commons, col. 630, 8 February 1962.

as working closely with its US counterpart that was to look amongst other things at reviewing the case for and against atmospheric testing and to consider a future programme of such tests.[82] There is little available in the public domain on the work of the NTAC.

UK-US MOU on Use of Christmas Island

An Anglo-American agreement for mutual testing cooperation was concluded on 14 February 1962.[83] This stated that Christmas Island could not be used indefinitely and the UK must be satisfied that the whole programme fell within Macmillan's definition of the type of test that would be defensible.[84] It also made clear that no test would take place unless the UK was satisfied that the safety precautions, which it considered necessary, were taken and that conditions were suitable for the proposed test. Macmillan told Kennedy that the exchange of letters on the MOU so quickly on the issue was very satisfactory. He was anxious to emphasise that the exchange of scientific information should be of mutual benefit and flow easily.

The Americans had given the UK a Statement of Principles for the use of the island on December 1961. Paragraph five stated that the US, in accordance with the 1958 Agreement, would furnish or otherwise make available to the UK detailed information concerning the tests making use of Christmas Island.[85] Macmillan hoped that the existing relevant agreements would be applied and interpreted on both sides in the broad spirit of collaboration.[86] This latter point was made at the request of Sir Roger Makins given UKAEA concerns that the US might try to interpret the MOU clause on exchange of scientific information too narrowly. Makins had reminded the Prime Minister that the Americans were inclined to be 'very lawyer ridden on these matters'.[87] In the event, the Americans permitted a small AWRE team (16 strong) to participate directly in the trials and to take their own measurements.[88] They were also there to provide safety advice to the US. In addition to the use of Christmas Island, the UK also agreed that the Americans could set up instrument stations in various places including Fiji,

[82] TNA FO 371/163091, IAD 21/6, Foreign Office telegram no. 1239 to Washington, 12 February 1962; Washington telegram no. 420 to Foreign Office, 12 February 1962.

[83] Hansen, *U.S. Nuclear Weapons*, p. 83; Clark, *Nuclear Diplomacy*, p. 220.

[84] Clark, *Nuclear Diplomacy*, p. 217.

[85] TNA CAB 134/2235, United States Statement of Principles handed to the Foreign Office Secretary on 10 December 1961.

[86] TNA CAB 21/4977, Prime Minister's Personal Telegram Serial No M68/62, Macmillan to Kennedy, 19 February 1962.

[87] TNA PREM 11/3719, Roger Makins to Mr de Zulueta, 15 February 1962; and Roger Makins to Mr de Zulueta, 19 February 1962.

[88] TNA CAB 21/4977, J.H. Williams for Director of Physical Research, Admiralty to DGAW, Ministry of Aviation, 9 March 1962.

Tonga, Canton, Gilbert and Ellice Islands and Washington Island to carry out air sampling during the test programme off Christmas Island and elsewhere in the Pacific.[89] An exchange of letters was agreed that permitted the use of the stations for one year – the same period as allowed for Christmas Island. Discussions with relevant local authorities were put in hand to secure agreement. Macmillan was to confirm to the House of Commons that all information from the tests would be interchangeable.[90]

UK Testing Responses: Options?

What was the UK to do in response to Soviet activities? A starting assumption seems to have been that it would be politically unacceptable to carry out tests in the atmosphere and that it would be necessary to think in terms of underground tests instead. However, the UK had no immediate test site of its own that could be used. Three years would be needed to prepare one at considerable cost; the only option was Nevada in the United States.[91] The British Nuclear Deterrent Study Group's Technical Sub-committee had noted earlier in 1961 that only marginal improvements could be foreseen in nuclear weapons and that without any testing, dramatic change was very unlikely.[92] So it would seem that there were no overwhelming strategic reason for UK testing, but this statement is at odds with what we now know about Super Octopus and its centrality to future UK warheads.

A UKAEA paper prepared for the NRDC outlined three options: the resumption of testing of British weapons at a British site; the resumption of British weapons tests at a US site as part of a joint Anglo-American programme; or finally some form of participation in a purely US programme.[93] There were immediate technical benefits from the second option. British tests would allow AWRE to produce and confirm stockpile versions with maximum economy, speed and safety – UK service designs were different from their US equivalents. Such tests would allow the UK to contribute to the joint US programme and increase the chances of information flow in the other direction. A resumption of British testing would

[89] TNA CAB 21/4977, Commonwealth Relations Office telegram no. 423 to Canberra, 17 March 1962; Nuclear Tests in the Pacific and Despatch No. 121 Foreign Office 29 March 1962.

[90] Hansard, House of Commons, Oral Answers col. 1512, 15 February 1962.

[91] TNA AVIA 65/1836, Draft Brief to the Minister on the Resumption of Nuclear Tests D.A.W.D., 3 October 1961.

[92] TNA DEFE 13/617, BND (TSC) (61) 15 (Final), 27 July 1961. British Controlled Contribution to the Nuclear Deterrent Report by the Technical Sub-Committee This was set up in 1959 to recommend what form the British strategic deterrent should take in future in light of improving Soviet defence.

[93] TNA DEFE 17/92, UKAEA Memorandum, sent to Sir Robert Scott under cover of a letter from Sir Roger Makins, 18 September 1961.

permit improvements to, and increase in the knowledge of, existing weapon types. It would also admit development of weapons of a new kind and advances in scientific knowledge. A joint test programme was, in the UKAEA's view, the only sound basis for close and continuing collaboration in the development of nuclear weapons and in making advances in scientific knowledge. If the UK were to test again there were four devices that could be included in a new programme: Super Octopus – a new implosion system (discussed above); the stockpile version of RO. 106 (Tony); the stockpile version of the British primary for the Skybolt warhead and a partial test of the complete Skybolt warhead.[94] All this was estimated to cost some £0.8 million for one test, or £2.75 million for a series of four tests as part of a joint programme with the US. Julian Amery, Secretary of State for Air, pointed out that if the UK did not test Super Octopus then the UK would be unable to make any significant advance in warhead design in future, since there would be a real danger that the flow of information from the US would cease without some continuing contribution from AWRE.[95]

Sir Robert Scott (Permanent Under Secretary, MOD and Chairman of the NRDC) wrote a memorandum for the Committee to set out the options for the UK in September 1961. Although this noted that one course for the UK would be to abstain from further testing, this would not go down too well in the US and ran the risk of cutting the UK off from the flow of atomic information. Scott came up with the classic Whitehall compromise. The UK should propose to the Americans that the UK participate in the US underground programme, but that the UK should try to dissuade the Americans from resuming atmospheric testing and that they make clear that the UK would have to dissociate itself from such a step. In addition, in a month's time and in light of consultations with the Americans, the UK would need to decide what to do about the facilities on Christmas Island. Once all Ministers of the departments and organisations represented at the NRDC were content, the

[94] TNA CAB 134/2238, ND (61) 1 Final, Nuclear Tests Note by the Secretaries Annex Test of United Kingdom Devices, 16 October 1961; the Tony warhead was also intended for use in Blue Water and Seaslug. It was one of the new sealed capsule types of warheads. TNA AVIA 65/784, Nuclear Weapons Safety Boards Minute 10, Blue Water Safety Organization. The DRPC's Atomic Energy Sub-Committee had agreed at the end of 1959 that there would be four warheads on which work would be done over the next four years: Red Snow for 1960–61; Tony for 1961–62; Low-yield Tony 1963–64 and MK 47 (a US warhead) for 1965–67 in TNA AVIA 65/1116, D.R.P. (A.E.S.)/M (59) 2 Meeting, Item 1 Warhead Development Programme, 15 December 1959. There was also a warhead called Gwen RO/106 that was similar to Tony but slightly smaller with a proposed yield of a half to two kilotons. It was also less economical than Tony in its use of fissile material. Tony was cylindrical and 34.5 inches long by 14.5 inches in diameter and weighed 175 lb; TNA WO 286/39, MOS ARDE, Fort Halstead to DREE, MOS, 10 February 1959.

[95] TNA DEFE 19/92, Copy of a minute from the Secretary of State for Air to the Ministers of Defence and Aviation, 28 September 1961.

Chairman proposed that the Minister of Defence and Foreign Secretary should minute Macmillan for approval for their planned course of action.[96]

It appears that Macmillan approved this. At the end of October 1962 he suggested to Kennedy that the US and UK should issue a Joint Declaration in which they would commit themselves to limit testing to underground only and that there were no military reasons to conduct atmospheric tests in the next six months.[97] Kennedy could not accept this as it would tie his hands and internal pressures were just too great to ignore. Moreover, at the same time there were indications coming out of Washington that if the UK wanted to test Super Octopus at Nevada, then the *quid pro quo* might be that the US might well ask to use Christmas Island.[98] Given the sensitivities over atmospheric testing, this was obviously disquieting if the UK were to ask formally for the use of Nevada and this request were linked directly to Christmas Island. In such circumstances, the UK would have to be careful.[99] Nonetheless as Dr Robert Press, MOD recognised, the UK would have to accept that the Christmas Island and Super Octopus issues were closely linked. For this reason London would have to realise that a tough attitude on the Christmas Island issue could endanger Super Octopus.[100]

Ministry of Aviation officials thought the same. Although Washington was not seeking a *quid pro quo*, there was reason to believe that a UK rejection of the US request to use Christmas Island would result in considerable delay, if not complete postponement, of the Super Octopus test.[101] The timing of the Super Octopus test gave Macmillan some concern. It was, as of the end of January 1962, fixed for March 15 – the day after the new ENDC opened. Given the embarrassing conflict, Macmillan urged Watkinson to either advance the date or postpone.[102] The date was brought forward to February, but it is not clear that the embarrassment factor was the sole motivating factor in any event plans reverted to the original schedule.

---

[96] TNA DEFE 19/92, Nuclear Requirements for Defence Committee, Testing Memorandum by the Chairman, September 1961.

[97] TNA DEFE 19/92, Foreign Office telegram no. 7630 to Washington, 24 October 1961.

[98] TNA DEFE 19/92, Sir Roger Makins, UKAEA to Sir Robert Scott, MOD, 24 October 1961.

[99] TNA DEFE 19/92, Sir Robert Scott, MOD to Sir Roger Makins, UKAEA 1 November 1961.

[100] TNA DEFE 19/92, Brief on Use of Christmas Island for US Nuclear Tests, 11 November 1961.

[101] TNA AVIA 65/1836, Use of Christmas Island for United States Tests, Memorandum by the Secretary of State for Foreign Affairs, A.S. /A.W., 13 November 1961.

[102] TNA FO 371/163113, IAD 23/22, Evelyn Shuckburgh to Mr Pemberton Pigott, 25 January 1962.

UK Nuclear Tests and Warheads: 1962–1964

Super Octopus was tested successfully in March 1962 – the Pampas shot – and produced the expected yield of about ten kilotons.[103] The device appears to have used tritium supplied by the US in the US, the amount being counted against the quantity due to the UK under the barter agreement.[104] Macmillan had earlier told the House of Commons that if successful, the test would establish a significant advance in UK weapon technology.[105] Pampas showed that the device could be used if modified, either by increasing the fissile material content, in which case it would be necessary to incorporate a mechanical safeing arrangement with some loss of reliability and ease of handling; or by increasing the HE content, in which case it would be necessary to test the device again. The latter solution was the preferred option.[106] The device would have a larger charge of EDC 11 explosive, which would permit the restriction of the amount of fissile material to a level that would ensure single point safety but without the incorporation of mechanical safeing means.[107] This was all connected to the Skybolt warhead; the outstanding problem was to select an effective design for the primary stage of the Skybolt warhead. Pampas was a successful step in this direction.

Officials agreed that a further test (Pampas II later changed to Tendrac) would also be required. The Americans had told the UKAEA that the necessary facilities could be made available in November 1962; as the MOD pointed out to the Prime Minister, if this window were missed another might not occur in the immediate

[103] TNA PREM 11/3706, Roger Makins to Prime Minister, 2 March 1962. According to a US Department of Energy Openness Press Conference Fact Sheet 27 June 1994 p. 179, the yield of PAMPAS was 9.5 kilotons and that it released radioactivity detected off site; TNA AB 49/14, Roger Makins to Lord President, CH (62) 22, 12 March 1962.

[104] TNA AB 49/14. It also seems that the UK device did not require a UK courier when shipped on a US transport plane from Greenham Common, although the device was transported in component form. Draft letter from Chairman, UKAEA to Chairman USAEC, Proposed Underground Nuclear Test of a UK device at the Nevada Test Site. CH (61) 166.

[105] Hansard, House of Commons, Oral Answers col. 1512, 15 February 1962.

[106] TNA CAB 134/2239, Nuclear Requirements for Defence Committee, Item 1 Nuclear Tests N.D. (62) 2nd Meeting, 17 July 1962.

[107] TNA CAB 134/2239, Nuclear Requirements for Defence Committee, Item 1 Nuclear Tests N.D. (62) 2nd Meeting; and TNA PREM 11/3706, The Skybolt Warhead, MOD, 10 August 1962. The US equivalent used a primary Tsetse, originally developed as a kiloton warhead. The British RO 106 warhead was based on this design, but in developing RO 106 some explosive power was deliberately sacrificed in the interests of using an explosive (EDC 11) that was less likely to detonate in the event of an accident, than the PBX 9404 used in the US Tsetse design. As result, the yield of the RO 106 was less than Tsetse and could not be used as the UK Skybolt primary. TNA PREM 11/3706, The Skybolt Warhead, MOD, 10 August 1962.

future.[108] As before, US laboratories were interested in the results because they were developing a similar implosion system.[109] Super Octopus was clearly seen then as a likely basis for future British warheads to follow 1965. It became the basis for the implosion system used in both the WE177 (high and low-yield variants with necessary safety and other experiment on the design being carried out at Maralinga in 1963[110]) and Polaris warhead primaries.[111] (At each successive modification the dimensions were reduced so that mechanical safeing could be included to ensure single point safety.)[112] This test took place successfully on 7 December 1962 with a satisfactory order of yield. Makins told Macmillan that the UKAEA had almost certainly achieved the test objective, namely to provide an inherently safe design of trigger bomb using British HE that would be suitable for Skybolt. He also had reason to believe that the same trigger could be used in a warhead for Polaris, though this could not be confirmed until more information was available on the Polaris missile system.[113]

---

[108] TNA PREM 11/3706 Underground Testing of UK weapons in US 1961–62, Peter Thorneycroft to Prime Minister 23 August 1962

[109] Robert Norris, Andrew S. Burrows, Richard W. Fieldhouse, *Nuclear Weapons Databook Volume V British, French and Chinese Nuclear Weapons* (Oxford, 1994), pp. 402 and 404 footnote 63.

[110] TNA CAB 134/2240, Nuclear Requirements for Defence Committee, Future of United Kingdom Nuclear Weapon Tests and Experiments Memorandum by the Ministry of Aviation, N.D.(63) 8, 15 October 1963.

[111] Hawkings, *Keeping the Peace*, pp. 58–9; TNA AIR 6/152, Appendix to A.C.(63) 14 22 March 1963 noted that ASR 1176, development of the ten kiloton tactical warhead, continued satisfactorily; a device of similar design was successfully tested (Tendrac) at the NTS last autumn. Tendrac comprised a developed version of the Pampas 'multipoint implosion system' that was fired in March 1962. The device contained rather more explosive of the EDC 11 type and was designed to be inherently 'one point safe'. The expected yield was ten kilotons. TNA AB 48/306, Roger Makins to Dr Glenn T. Seaborg, USAEC, 5 September 1962. The first reference we see to a multipoint initiation systems comes in a letter from William Cook to E.S. Jackson on 2 December 1957 when he notes that AWRE, 'have under consideration a system of multipoint initiation to eliminate lenses and reduce weight for a given yield. This will find adaptation to triggers, but if it is of more general application it would of course benefit all kiloton weapons'. TNA AVIA 65/1217, Future trials from 1956–59: tentative programme 1956–59. The prior announcement of this test created a bit of a political storm and much was made domestically of the Soviet statement about its possible effect on test-ban negotiations. MOD ministers became most anxious to accelerate the date of the test if at all possible. MOD London to BDS Washington, to Rowlands from Zuckerman, personal message to General Betts, 13 November 1962. AWRE were instructed to take every possible step towards completing *Tendrac* one week in advance of the currently planned date. MOD London to BD Washington, 14 November 1962. In the end, the US was not able to help. MOD to BDS Washington, 16 November 1962.

[112] Personal Correspondence.

[113] TNA PREM 11/3706, Roger Makins to Prime Minister, 21 December 1962.

Whilst it was within UKAEA capabilities to design and produce almost any of the possible warheads being considered at that time for Polaris, the choice to be made depended upon further examination and discussion with the US.[114] Skybolt was in the event cancelled by the Kennedy Administration and Macmillan had managed to convince Kennedy at Bermuda in 1962 to let the UK have the Polaris missile instead.[115] A UKAEA team planned to go to the US in mid February 1963 and final decisions would depend on the final report from this meeting. The size and economy of fissile material and efficiency of the implosion system meant that there could be a considerable reduction in the diameter of the warhead compared to the first and second generation of UK nuclear weapons. Zuckerman informed the Prime Minister in mid May 1964 that the Polaris warhead was being developed from the results of previous tests.[116] The US equivalent had been tested in 1962 and the UK had a choice of either copying the US design as closely as possible or of modifying it by substitution of a UK designed primary to be in closer accord with UK practices. The secondary was to be the same in both cases.[117] The UK primary was similar to, but smaller than that used in WE177. The smaller size required the introduction of 'mechanical safeing', which AWRE had been developing for several years, in anticipation of its need in small primaries. High reliability needed to be demonstrated too. There had been no nuclear tests of UK devices of the reduced size primary to date, and Aldermaston had based its design on the results of US tests of smaller devices. The two underground tests in 1964, the first of which was Cormorant, were not specifically directed towards Polaris warhead developments, but it was just possible that the results could influence the detailed design of the UK warhead.[118] Cormorant was the first test in the UKAEA's three-year research programme. AWRE had access to all US experience in this respect. Aldermaston had full confidence in the adopted design, based on both US and UK test information – it is possible that some of this may have come from the US Christmas Island tests, which included a full system test of a Polaris missile and warhead. Arrangements were made with the US to ensure complete compatibility of the British warhead with the US A. 3 missile. Polaris warhead manufacture although scheduled to begin in November 1966, did not in fact get underway until

[114] WO 32/17069, Army Council Secretariat Extract from minutes of 1st (63) Meeting of the Nuclear Requirements fro Defence Committee, 7 February 1963.

[115] See Richard Moore, *Nuclear Illusion. Nuclear Reality Britain, the United States and Nuclear Weapons, 1958–64*, London, 2010.

[116] TNA PREM 11/4737, Zuckerman, MOD to Prime Minister, Polaris Warheads, 15 May 1964.

[117] The warhead, a version of the UK Mark 58 warhead incorporating a British designed primary, would be inside the basic Mark 2 Mod 0 Re-entry systems. In April 1964 the Secretary of State authorised a study on a penetration aid programme for Polaris. Personal correspondence

[118] TNA AVIA 65/1836, J.S.W. Henderson to U.S. (S.A.R.), 6 May 1964.

December.[119] It seems that this weapon may have been intended to have a yield of 500 kilotons and weigh 300 lb.[120] The warhead itself appears to have combined UK and US concepts in its overall design.[121]

In June 1964, the Ministry of Aviation wanted to carry out a Vixen B type experiment to prove the safety of the British warhead for Polaris sometime in January 1966.[122] However, this requirement was disputed and once reviewed by the MOD's Nuclear Weapons Safety Committee, it emerged that the need could not be entirely ruled out. If required, such an experiment could take place at Nevada; there were no technical reasons to maintain funding to keep Maralinga on a care and maintenance basis for this purpose.[123] There had been no experiments at Maralinga since 1963. However, it seems that subsequently both AWRE and the Ministry of Aviation agreed that a Vixen B experiment was not likely to be necessary so far as technical considerations were concerned since the then present design was fitted with a mechanical safeing device.[124] However, it seems that officials decided in March 1964 that the UK would adopt the Polaris Mk. 3 Mod. 0 re-entry system, a version of the US Mark 58 warhead incorporating a British designed primary. Officials also decided that a Vixen Type B experiment would be necessary to prove the safety of the UK warhead for Polaris after the warhead design was frozen. Such experiments would be best held sometime between 1 January and March/April 1966, shortly before the Maralinga agreement ran out.[125] It never was. The decision appears to have been that such tests were not needed after all.

Future UK Tests

Makins told the NRDC in the summer of 1962 that the AWRE design teams had to be able to look forward to a substantial programme of further development work

---

[119] TNA PREM 13/1388, D. Wilson, Ministry of Aviation to P. Le Cheminat, 5 October 1965; and Lt Col R. Rubens to M.J.O. Wright, 5 October 1965; TNA CAB 164/931. Minister of Aviation (F.W.M.) to Secretary of State for Defence, 23 November 1966; DEFE 13/700, Healey to Minister of Aviation, 25 November 1966.

[120] TNA AVIA 65/1771, Brief for meeting of Nuclear Requirements for Defence Committee on 30 October 1963, 28 October 1963.

[121] TNA PREM 13/3129, Zuckerman, MOD to Prime Minister, 16 December 1964.

[122] TNA CAB 134/2241, Nuclear Requirements for Defence Committee, Item 5 Future of Maralinga Range, N.D. (64) 5th Meeting, 25 June 1964.

[123] TNA CAB 21/6039, Nuclear Requirements for Defence Committee, The Future of the Maralinga Range, Draft NRDC paper, November 1965.

[124] TNA DEFE 7/2364, Draft Minute from CSA to PUS, Future of Maralinga, R. Press, ACSA (N) to CSA, 14 July 1964.

[125] TNA CAB 134/2241, Nuclear Requirements for Defence Committee, Future of Maralinga Range, Note by the Ministry of Aviation, ND (64) 9, 23 June 1964.

if they were to be kept together and available for unforeseen tasks. This would include underground and atmospheric tests. It was also necessary that some such programme should be authorised if the present satisfactory level of collaboration with the US was to continue. If no new development programme were approved, the human resources would begin to run down rapidly between 1964 and 1967, and this process would be virtually irreversible.[126] Therefore, there was an urgent need to proceed as quickly as possible to the point of seeking government decisions on the question of future warhead development. In the meantime, it was acceptable for the UKAEA to ask the USAEC unofficially whether an associated programme of underground testing could be accommodated within an American test series. The question of possible atmospheric tests was not to be mentioned given the current diplomatic exchanges in Moscow about an early test-ban treaty.

The NRDC finally concluded that there was now no justification for retention of Maralinga even as a contingency basis for underground tests. The balance of technical and financial advantage was always likely to favour Nevada. Although as of early 1966 there was no approved programme for underground tests, the NRDC felt that if one were approved, this would be likely to reduce the risk that the US would refuse the use of its facilities at the NTS. In these circumstances, there was no real case for providing underground test facilities at Maralinga. A preliminary study of Maralinga as a possible underground test site had been undertaken in 1958 and this concluded that owing to underground water supplies, the use of a hill would be essential; but there was no suitable site within 100 to 200 miles.[127] Accordingly, the NRDC recommended that the UK should seek agreement of the Australian government to fulfil the UK's security and safety obligations under the existing agreement for the use of the range, but otherwise to relinquish use of Maralinga on expiry of the agreement on March 1966. The Defence Secretary, Dennis Healey, endorsed these recommendations and so too did the Prime Minister thus closing another chapter in British nuclear testing history.[128]

Kennedy Decides to Resume Atmospheric Testing

Kennedy decided at the end of February 1962 to resume atmospheric testing and proposed to announce this publicly. Kennedy had been obliged to conclude that it was indeed desirable on military grounds to resume atmospheric testing on 15 April. The Americans simply could not allow the USSR to make unmatched

---

[126] TNA CAB 134/2239, Nuclear Requirements for Defence Committee Meeting, N.D. (62) 2nd Meeting, 17 July 1962.

[127] TNA DEFE 16/432, Brundrett to Penney, 2 September 1958.

[128] TNA PREM 13/722, Healey to Prime Minister, The Future of the Maralinga Range (Australia), 24 January 1966.

progress on anti-missile systems.[129] Kennedy had come under intense pressure from the USAEC and Joint Chiefs of Staff (JCS).[130] He also concluded that it would be preferable to announce this before the ENDC opened and planned to do so on 1 March unless an international agreement for the cessation of nuclear tests had been concluded. Macmillan was troubled by this. He immediately wrote to Kennedy to plead for a delay in any announcements and suggested that the US postpone the test series until 1 June or at least until 3 May 1962 in order to give the ENDC sometime to produce positive results.[131]

The Cabinet was told that the Prime Minister, Foreign Secretary, the Minister of Defence and UKAEA scientific advisers were satisfied that the proposed series was justified within the terms of his statement to the House of Commons on 31 October 1961. As such, the Cabinet accepted that the UK should support the US decision, but were a little concerned about the timing of the announcement. On the one hand, it would be better to make it before the ENDC began, but equally it would be preferable not to give the Russians an ultimatum. Macmillan, in light of the Cabinet discussion, urged Kennedy to delay a few days in order to avoid causing undue problems for him in parliament and to inform close allies, a request that Kennedy was happy to accept.[132] He also assured Macmillan that the US genuinely wanted a decent treaty if one could be had and that he would be instructing US experts to work full steam ahead on this in order to develop a good proposal for Geneva. In the UK view that since there had been already three years worth of negotiations on a test ban, if the Russians wanted one it could be concluded in a matter of weeks. However, Macmillan hoped that Kennedy would avoid giving the impression of an ultimatum in the US announcement and that the US statement should emphasise the military necessity of further tests.

Officials continued to worry away at the justifications for the US tests; a revised list of proposed US tests now included two full systems tests for Polaris and Thor. Macmillan had established the NTAC to review the US plans and this body confirmed that the US tests, apart from the systems test, met the conditions previously set down for UK support.[133] As for the system tests, if these involved detonations of nuclear warheads, this would be better done at a pre-determined safe point in the trajectory, and then these too would be acceptable. Makins suggested

[129] TNA PREM 11/3718, Washington telegram no. 631 to Foreign Office, 27 February 1962.

[130] Robert A. Divine, 'The Early Record on test moratoriums', *Bulletin of Atomic Scientists*, May, 1986, p. 26.

[131] TNA CAB 128/36, C.C. (18) 62, Item 3 Nuclear Tests, 1 March 1962; TNA PREM 11/3718, Foreign Office telegram no. 1798, Prime Minister's Personal Telegram Serial No T. 89/62, 1962.

[132] TNA CAB 21/4755, Washington telegram no. 631 to Foreign Office, 27 February 1962; Foreign Office telegram no. 178 to Washington, 28 February 1962; President Kennedy to Prime Minister, 28 February 1962.

[133] TNA PREM 11/3718, Makins to Prime Minister, 2 March 1962.

that the Prime Minister might wish to make this point to Kennedy. The NTAC also recommended that the Prime Minister seek further information about these tests.[134] A message went to Kennedy on 11 April once the UK-US dispute over the length of the warning notice to mariners and aircraft for Christmas Island had been resolved to argue that such tests should not be conducted; the British technical view was that dummy warheads with instrumentation ought to be able to do the job. Watkinson felt that what the Americans proposed was indefensible.[135] Macmillan was worried too about accidents and giving encouragement to the Russians to fire nuclear-armed missiles into the Pacific, thereby opening up a dangerous new area of superpower competition.[136]

Kennedy replied that there were other considerations that seemed decisive to him and promised to discuss these when Macmillan was next in Washington.[137] The MOD provided a technical brief that made the case against such tests for Macmillan to use in his April meeting with the President.[138] In the end, UK anxieties made little difference as the US tested the Polaris missile at full range with the warhead detonating above Christmas Island.[139] The Administration cancelled the Atlas test; and although it had tried to reduce the number of tests as far as it could, it still wanted to make sure that the safety device in the Polaris warhead was effective. Macmillan told his Cabinet colleagues on 30 March 1962 that it was now clear that the US would not be prepared to revive its earlier offer to conclude a test ban in the atmosphere only. Attitudes in Congress had hardened and it now seemed likely that the Administration would go ahead with more tests.[140]

Once the UK accepted the general nature and purpose of the nuclear test programme, both sides agreed that the scientific data arising from the series of tests to be conducted should be transmitted to the UK under the 1958 Agreement as amended in 1959. At the time the UK, or to be more precise the Air Ministry had been thinking about the need for a British multi-megaton bomb of its own. AWRE's view was that it was very doubtful whether there would in fact be an advantage in

---

134 TNA PREM 11/370 Use of Christmas Island Part 4, Washington Talks April 1962 Polaris and Atlas (Nuclear) System Tests Brief by the Ministry of Defence, 17 April 1962.

135 TNA PREM 11/3720, Godfrey to P.F. de Zulueta, 28 March 1962.

136 TNA PREM 11/370, Prime Minister's Personal Telegram Serial No. T. 199/62 Foreign Office telegram no. 2937 to Washington, 11 April 1962.

137 TNA PREM 11/370, Prime Minister's Personal Telegram Serial No. T. 200/62, From the President, the White House to Prime Minister Macmillan, 12 April 1962.

138 TNA PREM 11/370, Polaris and Atlas (Nuclear) System Tests, Brief by the Ministry of Defence, 17 April 1962.

139 Thomas B. Cochran, William M. Arkin, Robert S. Norris, Milton M. Hoenig, *Nuclear Weapons Databook Volume II US Nuclear Warhead Production* (Cambridge, Massachusetts, 1987), p. 161.

140 TNA CAB 128/36, C.C. (25) 62, 30 March 1962; TNA PREM 11/370, Record of a conversation in HM Embassy, The PM, Mr McNamara and Mr de Zulueta, Washington, 20 April 1962.

developing such a system, as it would require development of a suitable delivery vehicle capable of carrying a very large payload. The lower limit below which the weight of warhead of this capacity could not be reduced was of the order of six or seven tons.[141] In Penney's view there was no important technical advantage to the UK in carrying out atmospheric tests at that time. The requirement would however have to be reviewed in the light of the results of the forthcoming US tests at Christmas Island.

British expertise and instrumentation techniques, which had been developed in the 1950s UK test programmes, would be made available if of use to the US.[142] Indeed, it seems that some of these techniques were of interest to the USAEC.[143] These included measurements of neutron multiplication rates for airdrops and balloons, radio flash measurements at various distances and techniques for radiochemical diagnostics. Seaborg himself noted that the USAEC made valuable use of AWRE expertise from the Christmas Island tests and would want to take up AWRE expertise in radio flash measurements and welcomed independent analysis of radiochemical samples.[144] Subject to formal USAEC agreement, AWRE planned to measure radio flash and atmospheric fluorescence by stationing three small groups at Johnson Island, Palmyra Island and Maui (Hawaii).[145] The first in a series of 24 tests took place at or near Christmas Island on 25 April 1962 with the last test being on 11 July.[146]

During the test series there was some anxiety in London over US plans for one of the high altitude shots. It seemed that the Americans were now suggesting that there were was no military necessity for this test, which was embarrassing for Macmillan because he had been defending the highest altitude tests in exactly this manner.[147] There had been concerns over possible adverse effects on the Van Allen

---

141 TNA AIR 2/18270, Secretary of State meeting with Sir William Penney, 19 April 1962, J. Roberts' Private Secretary to Secretary of State; see also Roberts' record of Secretary of State meeting on 22 March 1962.

142 TNA DEFE 19/92, Scientific Collaboration in Proposed Tests Based on Christmas Island.

143 TNA CAB 21/4977, Roger Makins, UKAEA to Dr Glenn T. Seaborg, Chairman, USAEC, 19 January 1962.

144 TNA CAB 21/4977, Glenn Seaborg to Roger Makins, Chairman, UKAEA, 12 March 1962.

145 TNA CAB 21/4977, Foreign Office telegram no. 2365 to Washington, 22 March 1962.

146 TNA AB 49/13 NRDC; correspondence 1959–1966 Roger Makins to Sir Robert Scott, MOD, 9 July 1962. AWRE clearly had detailed results of these tests, which had implications for revised plans for testing – the planned programme of research tests. TNA AB 49/13 NRDC; correspondence 1959–66 Roger Makins to Sir Robert Scott, MOD, 9 July 1962.

147 TNA PREM 11/370, Prime Minister's Personal Telegram Serial No. T. 276/62, Foreign Office telegram no. 3942 to Washington, 27 May 1962.

Belts. The Americans asked for UK agreement for an additional shot of some three to five megatons, but it would not result in an increase in the total fission yield of the whole series as originally planned as one of the earlier devices had failed. The US pointed out that this additional test fell within the first category that they had provided in February as well as in the class of advance concepts. As a result, they considered that it was within the criteria of military need that had been previously agreed.[148] Sir Roger Makin's NTAC agreed with the US assessment and recommended that Macmillan agree; he did.[149]

Parallel Efforts on a Test Ban

The efforts in the ENDC to persuade the Russians to consider international verification ran into the sands by late March 1962. Macmillan reluctantly had to recognise that once the USSR rejected Western overtures on this key issue, then the UK would have to accept that the US tests at Christmas Island would have to take place.[150] Macmillan asked the Foreign Secretary at the Cabinet meeting on 22 March 1962 to look into the possibility of proposing a treaty banning atmospheric tests only in which both detection and verification would be provided by national systems. Foreign Office officials did not think that this was at all practical politics as the US was bound to disagree given the extent of its own preparations for its nuclear test programme at Christmas Island.[151]

Macmillan and Home reviewed the issue on 24 March. The Foreign Secretary said that he would discuss the various possibilities in Geneva after consulting his officials first. The more that Macmillan and Home thought about the issue, the less they liked the prospect of resuming nuclear tests. In their view whatever the Russians may have gained from their last series, the West could not hope to match these advances by the proposed US tests. UK advisers agreed that the West would be better off militarily if there were no more tests at all. For this reason, the Foreign Secretary felt that they should be trying much harder to end the whole business.[152] In addition, the US justification seemed to Home to be thin. It was only with some difficulty that the UK had managed to bring them within the definition of necessary tests made by the President and the Prime Minister the previous autumn.

---

148 TNA PREM 11/370, Prime Minister's Personal Telegram Serial No. T. 238/62, Washington telegram no. 1713 to Foreign Office, 2 July 1962.

149 TNA PREM 11/370, Roger Makins to P.F. de Zulueta, 5 July 1962 and P. de Zulueta to Prime Minister, 5 July 1962.

150 TNA PREM 11/370, Prime Minister's Personal Telegram Serial No. T. 154/62, Foreign Office telegram no. 2338 to Washington, 21 March 1962.

151 TNA FO 371/163091, IAD 21/11, J.O. Wright to P.F. de Zulueta, Admiralty House, 1961.

152 TNA PREM 11/370, Foreign Office telegram no. 2535 to Washington, 28 March 1962.

The strong advice from the embassy in Washington was that there was no chance of persuading the President to back off from the planned resumption of atmospheric testing; it was just not practical politics in view of Soviet intransigence and Congressional opinion to take a different tack. The justification for the American test series that had influenced Kennedy was not because the Russians had gained a lead from their last test series, but that if they tested again while the US sat on its hands, they could get ahead. The latest intelligence showed that preparations for another test series were well advanced.[153] On 4 April 1962 the warning notices to mariners and airmen were issued about the possible conduct of nuclear tests on 10 April, Macmillan and Kennedy issued a joint declaration announcing the resumption of tests.[154] There had been much debate about the drafting of this statement and exchanges between Kennedy and the Prime Minister, who had written on 30 March suggesting that a last appeal might be made to the Soviet leader.[155]

The President's reply gave Macmillan just enough encouragement to send a further message to Khrushchev.[156] Macmillan had been insistent that the joint UK-US statement explicitly left the door open for Khrushchev to accept the principle of verification. However, Kennedy told Ormsby-Gore that some vague acceptance of the principle of on-site inspection by the Russians would not be sufficient for him to hold up the US test series; a clear desire to conclude as rapidly as possible a treaty embodying effective verification would be necessary.[157] Moreover, by so doing a test-ban treaty might be speedily achieved.[158] This was an extremely important point presentationally for the British. On 10 April, Macmillan's letter to Khrushchev was handed over in Moscow and this urged him to reconsider his position and to give the most earnest consideration to the Western proposal on the need for verification.[159] However, it seems that the Foreign Office felt that such initiatives were unrealistic.[160] Indeed they were because all that Macmillan received in reply was a long and argumentative reply that officials thought 'was so churlish in tone that the most dignified course would be to let the correspondence

153 TNA PREM 11/370, Washington telegram no. 959 to Foreign Office, 29 March 1962.

154 TNA PREM 11/370, Foreign Office telegram no. 2727 to Washington, 4 April 1962.

155 TNA FO 371/163091, IAD 21/32, Kennedy to Macmillan, 3 April 1962.

156 TNA FO 371/163091, IAD 21/33/G, Prime Minister's Personal Minute M 92/62 to Foreign Secretary, 4 April 1962.

157 TNA FO 371/163092, IAD 21/27, Washington telegram no. 10765 to Foreign Office, 7 April 1962.

158 TNA PREM 11/370, Prime Minister's Personal Telegram Serial No. T. 164/62, Foreign Office telegram no. 2412 to Washington, 23 March 1962.

159 TNA FO 371/163092, IAD 21/28/G, Foreign Office telegram no. 955 to Moscow, 7 April 1962.

160 Schrafstetter and Twigge, *Avoiding Armageddon*, p. 111.

rest there'.[161] As a result, the reply was 'perfunctory and rather boring', which Home and the Foreign Office Permanent Under Secretary thought the best option. It was quite plain from this that the Russians were not willing to accept the principle of international verification and were retreating from their earlier readiness to accept international control. In these circumstances he did not think it right to urge the President to postpone the series of tests on Christmas Island.[162]

Kennedy and Macmillan discussed the testing problem during their meeting at the White House on 28 April. Kennedy underlined that the current US testing programme would take about two to three months and he assumed that the Russians would test in either parallel or shortly thereafter. Analysis of the test results would be needed and in light of this, it could be decided that it might be worth making another offer to the Russians, either on an atmospheric treaty only or a revision of the current draft treaty now on the table. He suggested that the UK and US might begin to consider what sort of proposal could be made, subject of course to the results of the current test series at Christmas Island.[163] Macmillan welcomed this and thought that British and American scientists should meet to consider the next UK-US offer. When he returned to London the Foreign Office and MOD were given their marching orders to study the possible options. Moreover, Macmillan wanted to be in a position to agree on a plan of discussion to put to Kennedy within the week.[164]

Foreign Office officials promptly produced a memorandum following discussion with their MOD colleagues, the Office of the Minister of Science and the UKAEA.[165] Home's minute recommended that the UK should only propose at that stage a non-technical examination of various possible courses of action and their implications, with the clear understanding that any conclusions would be subject to the evaluation of recent US and Soviet tests. There were at that point three possible courses of action: a new treaty combining an unverified ban on atmospheric tests, under little or no international control, with effective control over underground testing; declarations on a date to be decided by the US, UK and USSR that they intended to conduct no further tests in any environment, provided they were satisfied that the other side were not doing so; and a comparable declaration limited to atmospheric tests. There were of course problems with all of these, not least their likely unacceptability to either or both the USSR and US. In the case of declarations, the UK would have to say explicitly that in the present

161 TNA FO 371/163091, IAD 21/40, Pemberton-Pigott to Secretary of State, 15 April 1962.

162 TNA PREM 11/370, Foreign Office telegram no. 3193 to Washington, 23 April 1962.

163 TNA FO 371/163096, IAD 21/102/G, Extract from record of meeting at White House, 28 April 1962.

164 TNA FO 371/163095, IAD 21/100/G, Prime Minister's Personal Minute to Foreign Secretary M. 115/62, 3 May 1962.

165 TNA FO 371/163096, IAD 21/107/G, A.C.I. Samuel to P.F .de Zulueta, 11 May 1962.

state of weapons development, the UK was ready to accept a deliberate risk of some cheating by the Russians. Home believed that UK scientific advice was that such a risk was probably acceptable. If there were to be a radical breakthrough in the ABM field that necessitated some further testing, then tests for these would have to be conducted in the atmosphere. Whatever advantages were to be gained from further underground tests, they were not of the same order of importance. It was clear to the UK that whether the US would accept this would depend on the results of the evaluation of the most recent tests. There was no future in trying to revive the Western draft treaty based on the 1958 Experts' report; the aforementioned possibility should be discussed with the US, along with a range of possible variations. Home advised that these should be raised by Ormsby-Gore in secret and that he should emphasise that the UK recognised the need for any new initiative, for reasons of domestic US politics, appearing to come from the American side.

Although the UKAEA argued subsequently that any discussion ought to be preceded by scientific exchanges, the Foreign Office felt that Macmillan's requirement for immediate action precluded any such thing, although it recognised that science had to underpin any concrete proposal that might eventually be put to the Russians.[166] In general, the UK objective was to set the Americans thinking sooner rather than later about the kind of treaty that might eventually be agreed. Whether this was done immediately, and at political level or scientific level, the problem of tactics in Geneva was urgent. Ormsby-Gore needed instructions to work out with the Americans a line that Western delegations might follow in Geneva until the US was able to present something radically new. Macmillan approved the Foreign Secretary's 11 May memorandum and Ormsby-Gore was duly instructed, after a further round of internal Whitehall consultations. [167] Ormsby-Gore discussed all of this with Foster (ACDA) on 12 June and he promised to provide a draft treaty text for UK comment.[168] The Americans eventually came up with a revised draft treaty proposing a quota of inspections, but no international control posts in the USSR. This was roughly the same as UK views at that time.[169] ACDA's draft was a welcome simplification over the April 1961 effort, but neither the Foreign Office nor Ormsby-Gore thought that its provisions on inspections (the quota limit was

[166] TNA FO 371/163097, IAD 21/126, Nuclear Tests Ban Treaty Future Tactics, Brief for the Secretary of State, 18 May 1962; TNA PREM 11/4046, P. De Zulueta to Prime Minister, 25 May 1962. Sir Roger Makins had detected a good deal of irritation with the UK in scientific circles in the US during his recent visits to nuclear laboratories.

[167] TNA FO 371/163097, IAD 21/132/G, Foreign Office letter for the Secretary of State to Ormsby-Gore, Washington, 7 June 1962.

[168] TNA FO 371/163098, IAD 21/159, The New US Draft Treaty on Nuclear Tests, H.B. Shepherd to Mr Pemberton-Pigott, 25 June 1962.

[169] TNA FO 371/163098, IAD 21/152, A.D.F. Pemberton-Pigott to Mr A.D. Wilson, 14 June 1962.

too high and inspections could be requested by 'the other side'), withdrawal and the grounds for denunciation of the treaty were very satisfactory.

UK-US talks took place on 5 July 1962 in Washington in which the Americans explained the technical considerations of an international detection system with integrated national stations. The US handed over a draft treaty for a ban on tests in the atmosphere, outer space and in the oceans. There was some unexpected embarrassment when Michael Wright suggested that there should be a detailed discussion of the draft comprehensive treaty that ACDA had shared with London in June. It transpired that the DOD and USAEC had caused problems for ACDA during the drafting of what they thought was an internal US working draft and were disturbed to find that it had been shared with the UK before they had a chance to comment in detail.[170] Nor were the US convinced that the ENDC neutral delegations verification proposal would offer sufficient degree of detection and identification for reliance to be placed on such a system; in other words one without seismic stations on Soviet territory. This would have to be studied. Wright urged the Americans to see the wisdom of the West being able to put forward a comprehensive test-ban treaty in Geneva on 16 July, when the ENDC resumed, based on or inspired by the neutral CTBT paper.[171] There was risk that the Russians might do so and if there were no Western position, then this would be to its political disadvantage.

However, whilst the US delegation agreed, it was impossible for Washington to take a decision for several weeks. The US was also solidly against tabling a draft atmospheric treaty as long as there was a hope of tabling a CTBT first or simultaneously.[172] Wright felt that the US preference for getting through two or three weeks of ENDC deliberations without tabling any treaty in the hope that it might be possible to table a comprehensive text; or if this were not possible, the text of an atmospheric text could then be tabled instead, was on balance right. He recommended that London should agree, however reluctantly.[173] London accepted that it would be better to table a CTBT before an atmospheric treaty; it would be important to oblige the Russians to refuse a reasonable offer based on the neutrals ENDC proposal before falling back on the offer of an atmospheric test ban.[174] London did, however, think that it would be essential for the US to table the new data

[170] TNA FO 371/163099, IAD 21/168, Washington telegram no. 1738 to Foreign Office, 5 July 1962.

[171] Conference of the Eighteen-Nation Committee on Disarmament, ENDC/28 16 April 1962, Joint Memorandum, Brazil, Burma, Ethiopia, India, Mexico, Nigeria, Sweden and United Arab Republic.

[172] TNA FO 371/163099, IAD 21/170, Washington telegram no. 1740 to Foreign Office, 5 July 1962.

[173] TNA FO 371/163099, IAD 21/171, Washington telegram no. 1741 to Foreign Office, July 1962; IAD 21/170 Foreign Office telegram no. 4906 to Washington, 9 July 1962.

[174] TNA FO 371/163099, IAD 21/171, Foreign Office telegram no. 4857 to Washington, July 1962.

on their recent scientific findings about the detection of seismic events. This would satisfy the demand for some movement on the Western part, without a commitment to the crucial details of a treaty. UK-US seismic research had made some progress following study of the seismic signatures of the recent underground nuclear tests at Nevada. New data had been acquired which modified the requirements for seismic stations by reducing considerably the number required compared to the original Experts' 1958 report – about 170. A treaty would however be the next step. Home suggested that Macmillan make these points to the President.[175]

It was now clear to the UK that the new data indicated that there was no technical justification for the US insistence of a neutral presence at the seismic control posts. London took the view that this was logically indefensible and had no scientific justification.[176] The UK had thought that scientific discussions between the experts in Washington had accepted that there was adequate machinery for detecting and inspecting underground tests since a national system of detection posts reporting to an International Control Commission could provide this. Home urged Dean Rusk to think again about this in a message despatched to Washington on 5 August.[177] If he could not do so, then at least the only thing for Geneva to do would be to avoid discussion of it. In response, Rusk made a plea for the UK that they should not magnify Anglo-American differences when the essential difference was with the Russians. They should dispense with argument between themselves on this or that point in US proposals until the point in dispute had some actuality in negotiations with the Russians. That would only be when the Russians had reacted and indicated that they were prepared to accept the principle of on-site inspection.[178]

Late summer 1962 was thus clearly a problematic period in UK-US test-ban debates and this did not make it any easier to move the Western position forward in Geneva. Congressional attitudes made it very difficult for the Administration to show any flexibility on the question of international control. Foreign Office officials certainly felt that the fault lay with the Americans, but they remained anxious to avoid revealing their differences in public at Geneva. The Minister of State, Joseph Godber, was particularly incensed at the manner in which the Americans had proceeded. In his view, the UK had not been fully informed of the changes made that appeared to overturn without warning earlier agreements.[179] At

---

175 TNA FO 371/163100, IAD 21/199/G, Home to Prime Minister, 31 July 1962.

176 TNA FO 371/163102, IAD 21/222/G, H.B. Shepherd to Mr A.D. Wilson, 8 August 1962.

177 TNA FO 371/163104, IAD 21/276, Norwich to Shepherd, 20 August 1962; TNA FO 371/163108, IAD 22/362/G, Foreign Office telegram no. 5469 to Washington, 5 August 1962.

178 TNA FO 371/163100, IAD 21/207, Washington telegram no. 1949 to Foreign Office, 3 August 1962.

179 TNA FO 371/163104, IAD 21/276, Godber to Secretary of State, 16 August 1962.

best, officials had hoped that these could be fudged, at least in the short term.[180] By mid August 1962 Wright and Fisher agreed that the best tactics were for the UK and US to table both a draft CTBT and a draft atmospheric treaty before the ENDC went into recess in early September.[181] This idea came from the UK and was in fact the outcome in that both a draft CTBT and a draft atmospheric test ban were tabled in Geneva on 27 August, even though the UK had some reservations about the details. A joint statement from Kennedy and Macmillan accompanied this event and this noted that the latest scientific findings from their seismic research programmes underpinned the texts.[182] Wright felt that they scored a tactical success in the ENDC and although there was little indication that the Russians would accept either of the drafts, the West had put itself in the best possible position with the uncommitted nations and would hopefully help the UK and US at the UNGA in the autumn.[183]

As ever the public relations aspects of the testing debate were never far from view. There was, however, no weakening of the US insistence on the staffing of control posts with Western experts; Rusk told Viscount Hood in Washington that to do so would result in Administration heads being on a platter before Congress.[184] Russian intransigence and incredibly clumsy tactics made it easier for the British to conceal the extent of their differences with the Americans, much to their relief.[185] Between the tabling and early November 1962, the ENDC sub-committee on nuclear testing remained in session. Khrushchev had indicated in early September privately to the President that he was prepared to negotiate a treaty banning nuclear tests in the atmosphere, in outer space and under water in the immediate future. Kennedy suggested that the US, USSR and UK delegations should work toward the earliest possible agreement on a text before 1 January 1963.[186] However, this was short lived. The Russians changed their minds at the end of November, no

---

[180] TNA FO 371/163102, IAD 21/222, H.B. Shepherd to A.D. Wilson, 8 August 1962.

[181] TNA FO 371/163102, IAD 21/229, Geneva telegram no. 509 to Foreign Office, August 1962.

[182] TNA FO 371/163104, IAD 21/273, Joint Statement by the PM and President of the US 1100 BST, 27 August 1962.

[183] TNA FO 371/163102, IAD 21/235, Washington telegram no. 2056 to Foreign Office, 20 August 1962; TNA FO 371/163103, IAD 21/254, Geneva telegram no. 529 to Foreign Office, 27 August 1962.

[184] TNA FO 371/163102, IAD 21/236, Washington telegram no. 2062 to Foreign Office, 20 August 1962.

[185] TNA FO 371/163104, IAD 21/276, Godber to Secretary of State, Divergence between the British and Americans on the Question of a Nuclear Test-ban Treaty, 16 August 1962.

[186] TNA FO 371/163105, IAD 21/287/G, Kennedy to Macmillan, 11 September 1962; Macmillan to Kennedy 11 September 1962.

doubt influenced by the Cuban crisis, and in a letter to Macmillan Khrushchev insisted that they pursue a CTBT instead.[187]

Future Use of Christmas Island: Extension of MOU

In late summer 1962, London was worried that there might be another approach from the US to extend their use of Christmas Island. Contingency work was underway and officials began to review the issues and costs. Kennedy in fact expressly raised the possibility with Macmillan on 27 July; he wanted to consider with Macmillan the possibility of maintaining Christmas Island on a purely stand-by basis as part of a UK-US posture under a safeguarded test ban. This would retain a capability to respond if the Russians cheated. Continuing Soviet refusal to contemplate inspection of possible test preparations meant that the logical alternative was to sustain the West's own test readiness.[188] Kennedy also suggested that the UK and US might offer two options to the Russians: an atmospheric ban only or a comprehensive test ban with a small number of inspections. This appealed to Macmillan; as for Christmas Island he did not think that the UK would find it too difficult to keep the installation on a care and maintenance basis, and if the treaties were tabled the UK would certainly consider doing as Kennedy suggested.[189] Macmillan instructed the Minister of Aviation to look into the position of costs and associated matters on 2 August.[190] In his response the Minister's Parliamentary Secretary pointed out that the current MOU only related to the recently completed US tests, it would not be appropriate to use the current agreement for a reserve capability, in any case this expired in mid February 1963. Therefore, the arrangements could only be a stopgap until future policies could be decided in light of developments in the CTBT negotiations.[191] Although the US financial support for the Island was welcome, Macmillan was worried that undue dependence on US money would inhibit British freedom of action in future. This was an issue that would need to be looked at carefully.[192]

A draft MOU was prepared and agreed at the Nuclear Coordinating Committee on 3 October 1963, which built on the initial agreement, but with changes and additions to cover four separate points. The first of these was Macmillan's anxiety

---

187 TNA FO 371/163107, IAD 21/330/G, Foreign Office telegram no. 994 to Moscow, 30 November 1962.

188 TNA PREM 11/370, Kennedy to Macmillan, 27 July 1962.

189 TNA PREM 11/370, Prime Minister's Personal Telegram Serial No. T. 377/62, 1962.

190 TNA PREM 11/370, Macmillan to Minister of Aviation, Christmas Island, Prime Minister's Personal Minute Serial No. M 202/62, 2 August 1962.

191 TNA PREM 11/370, Minister of Aviation to Prime Minister, 9 August 1962.

192 TNA PREM 11/370, P.F. de Zulueta to A.C. Russell, Ministry of Aviation, 13 August 1962.

that the UK should retain its freedom of action with regard to the use of the island for nuclear tests. Second, there was still a need, albeit remote, to retain the right to hold UK tests. Third, it was still essential to make clear that the UK was the senior partner in the enterprise, at least in theory; and finally it was important to make it clearer still that US activity on the island would not affect the question of sovereignty.[193] The Americans made their formal request for an extension on 31 October 1962.[194] A Cabinet Office *ad hoc* committee looked into the details of the US request on 1 November.[195] This noted *inter alia* that if the US request related to completion of the 1962 test series, on the supposition that it was unfinished, this would be politically very awkward. Under the terms of the original MOU, the only veto open to the UK would have to be on safety grounds. Ministerial approval would have to be required before there could be any negotiations on a new MOU. In particular, it would be necessary to have agreement in principle that the US should be allowed to continue to maintain base facilities and personnel for the purposes of conducting nuclear tests; and that the US should contribute financially and possibly in some other way to the maintenance of all facilities on the island.

Home minuted the Prime Minister on 19 November 1962 recommending that the UK agree to the US request.[196] By allowing the use of the island, the UK retained the ultimate decision whether tests could take place there and the UK thereby was in a good position to influence the US. In the defence field Home also understood that the use of the facilities was helpful in ensuring that the UK received information from the US tests. Macmillan wondered whether there was likely to be a need for any more UK atmospheric tests.[197] In response, Peter Thorneycroft, Minister of Defence, noted that there was no current requirement, but in the absence of a treaty banning such tests, it was not possible to say that the UK would never have such a requirement. He made clear that if only for tactical reasons, the UK should not voluntarily deny itself the capability.[198] This alone justified the current care and maintenance policy.

However, it seems that the US found the UK imposed conditions on the use of Christmas Island too restrictive. When, in 1963, the US started to think about what steps would be needed for test readiness, the USAEC commented that Johnston

[193] TNA CAB 21/4977, Outline of a Possible Memorandum of Understanding (MOU) with the United States Government concerning the maintenance and use of nuclear testing facilities on Christmas Island, 17 September 1962.

[194] TNA CAB 21/6040, P. Wilkinson, Washington to A.D.F. Pemberton-Pigott, Atomic Energy and Disarmament Department, Foreign Office, 31 October 1962.

[195] TNA CAB 130/189, GEN.777/1st meeting, Facilities at Christmas Island, 1 November 1962.

[196] TNA CAB 21/60440, Home to Prime Minister PM/62/148, Use of Christmas Island, 19 November 1962.

[197] TNA CAB 21/60440, Prime Minister's Personal Minute Serial No. M316/62, Macmillan to Foreign Secretary, 20 November 1962.

[198] TNA CAB 21/60440, Thorneycroft to Prime Minister, 6 December 1962.

Island and Christmas Island would be needed, but with fewer restrictions in the former case.[199] In fact the USAEC preferred Christmas Island to Johnston Island since it was vastly superior for this purpose, but again the USAEC did not want to accept the kind of conditions that the US were likely to have to accept if Christmas Island was to be used.[200] As we have seen, the UK would not permit barge or surface bursts in order to avoid radioactive contamination of the island; Aldermaston insisted on strict procedures to ensure the safe conduct of tests with air dropped or balloon supported nuclear devices.[201] These had been worked out and agreed with the Americans during General A.D. Starbird's visit to London on 12 February 1962. London also wanted reassurance that each test was militarily justified.

Macmillan had asked that the talks about Christmas Island and its future should be kept going for the present and that no agreement should be reached until the UK had a chance to see how the Polaris Sales Agreement negotiations progressed.[202] Home did not agree since he thought it would be out of keeping with Macmillan's relations with the President to stall over Christmas Island. Instead, the UK should move ahead quickly and initial the agreement rather than sign until the Polaris negotiations were complete.[203] Peter Thorneycroft agreed with the Foreign Secretary and that the UK should move ahead quickly on this issue, including the proposal that the UK and US share costs in more or less equal shares.[204] A further revised MOU was prepared for Ministerial consideration in January 1963. This preserved the provisions from the first MOU on safety, claims and scientific collaboration. It also established clearly the requirement for a separate political decision before the US could either conduct a further series of tests near Christmas Island, or use the facilities on it for the support of tests in other areas.[205] Macmillan succumbed to the advice of his Ministers.[206]

Both DGAW in the Ministry of Aviation and AWRE had made clear that the UK might need to carry out experiments on the lines of the Maralinga Experimental

---

[199] United States Atomic Energy Commission Washington 25 D.C. Memorandum from Mr Bundy, Maintenance of a Condition of Readiness Under a Test-ban Treaty 14 February 1963 at http://www2.gwu.edu/~nsarchiv/NSAEBB/-NSAEBB/tb46.pdf.

[200] United States Arms Control and Disarmament Agency Memorandum of Conversation Nuclear Test-ban Treaty Meeting of Committee of Principals 17 April 1963 at http://www2.gwu.edu/~nsarchiv/NSAEBB/NSAEBB94/-tb49.pdf.

[201] TNA CAB 21/4977, Christmas Island Testing – Safety Arrangements, Note by DAWRE, 2 February 1962.

[202] TNA CAB 21/60440, P.F. Zulueta to A.P. Hockaday, MOD, 8 January 1963.

[203] TNA CAB 21/60440, J.O. Wright, Foreign Office to P.F. de Zulueta, Admiralty House, 14 January 1963.

[204] TNA CAB 21/60440, Thorneycroft to Prime Minister, 25 January 1963.

[205] TNA CAB 21/60440, J.M. Wilson, Ministry of Aviation to Sir Robert Scott, MOD, 22 January 1962.

[206] TNA CAB 21/60440, P. de Zulueta, Admiralty House to A.P. Hockaday, MOD, 29 January 1963.

Programme and wanted to be sure that the new MOU would not involve the UK in consultation with the US before such experiments could be initiated. However, others felt that the MOU as drafted in no way prevented the island from being used for such purposes.[207] The Foreign Office instructed the Embassy in Washington to approach the Americans to begin negotiations as soon as possible on the revised MOU. In many places the text was identical to the initial agreement, including the paragraphs on the exchanges of scientific information under the 1958 Agreement. The new draft however separated the provisions to be in force permanently from those that would apply in the event of the UK agreeing to a US test series based at Christmas Island.[208] Although the US had asked for an extension for a minimum period of two years, the new MOU offered three. From the UK perspective, this was a compromise between the undesirability of committing HMG to an agreement of long duration and the need not to make it so short that expenditure on maintaining facilities would become speculative. The new draft was given to the US on 18 February 1963, but as of 10 May, they still had not opened negotiations on the text. London had indications that the US requirements for the island would be much reduced, which meant that Washington would only pay a less substantial share of the costs than the British had first hoped.[209]

Between 16 February and August 1963, the Americans were granted two three-month extensions of the original MOU, although these extensions did not amount to agreement to any testing. These were to allow more time to negotiate the new MOU. The US had asked for a third extension, but with the agreement on the PTBT in July 1963, the Americans had now indicated their intention to withdraw from Christmas Island and to rely on Johnston Island should further atmospheric testing be needed as result of any Soviet breach of the PTBT. The additional three months was needed to give the Americans the necessary time to withdraw from the island.

What would the UK do now? On the one hand, it would be justifiable for the UK to maintain its own capability to resume tests if required. However, on the other hand the idea of a Grapple-like series had to be considered unlikely at the time since such tests that the UK might wish to conduct could be undertaken in the US. Moreover, there was considerable political advantage in down-playing any reserve testing capability. In fact, by that point, all UK special nuclear test facilities had already been withdrawn. Other military uses for the Island would have to be considered separately as the Treasury had only agreed that money should be spent to maintain the test capability. Once the US made their withdrawal announcement in August 1963, the UK followed suit and made clear that the future of the airfield and other facilities was under review. The Secretary of Defence agreed to the Chiefs of Staff recommendation that the staging facilities on Christmas Island should be

[207] TNA CAB 134/2237, Nuclear Coordinating Committee 'A' 1963 Meeting, 17 January 1963.

[208] TNA CAB 134/2237, Instructions Sent by the Foreign Office to the British Embassy, Washington, 13 February 1963.

[209] TNA CAB 21/60440, Home to Prime Minister, 10 May 1963.

closed down and the garrison withdrawn.[210] Washington was informed on 25 May and it was hoped to have all service personnel withdrawn by 30 June 1964.[211] This marked the closing of the final chapter on UK atmospheric nuclear testing.

Close of 1962: The Test Ban and Future Testing

By mid December 1962 the UK view could be summarised as follows. First, London accepted that there was very little prospect of the Russians budging on the principle of on-site inspection, or accepting a partial treaty. Second, there was equally little prospect of the US moving from the current Western stand on the principle of on-site inspection. Third, a continued requirement for on-site inspection was likely to lead the UK into increasing difficulties from the scientific point of view in light of further progress in detection capabilities. However, for the present the line would need to be held that given present knowledge some on-site inspection remained necessary.[212] Given Russian inflexibility it was back to the drawing board again: what offers could be put on the table to move the process along? The Minister of State sent some ideas from Geneva for consideration ahead of the Prime Minister's meeting with Kennedy later in December. These ranged from standing pat on the current drafts, or offering to negotiate on a variety of detailed points. These points were: the size of the quota, how it might be divided, numbers of nationally operated detection posts, or a three-month moratorium on underground tests provided that the Russians agreed that their scientists would meet their western counterparts to thrash out the differences on national detection stations for example.[213]

As before, there were many cons in these options, not least the need for a change in US positions. Moreover, there was little room for manoeuvre here largely because the UK had put all it could into the two draft treaties at the end of August 1962. In light of this Home advised Macmillan that all the UK could do for the present was to stand on the two August treaties as the US would find it difficult to do anything else.[214] On-site inspection was still necessary and the UK and the US should indicate that they remained ready to exchange scientific data about

---

210 TNA CAB 21/60440, J.H. Lapsley, Secretary Chiefs of Staff Committee to Chief of the Naval Staff, Chief of the General Staff and Chief of the Air Staff, 20 May 1964.

211 TNA CAB 21/60440, Foreign Office telegram no. 6782 to Washington, 25 May 1964.

212 TNA FO 371/163108, IAD 21/344/G, Foreign Office telegram no. 3315 to Paris, 11 December 1962.

213 TNA FO 371/163108, IAD 21/355/G, Godber to Secretary of State, 6 December 1962.

214 TNA FO 371/163108, IAD 21/356/G, J. Thomas to Secretary of State, 13 December 1962; A.D. Wilson to Mr Blaker, 14 December 1962; Home to Prime Minister, 14 December 1962.

means of detection, including the role of 'black boxes' (remote seismic stations). This concept had first been raised at an unofficial Pugwash meeting by American and Russian scientists at a conference in the UK in September 1962.[215]

As for continuing military utility of tests, Penney advised the Foreign Office that a low-yield test of up to ten kilotons could be important for one of several reasons: in its own right as a very light weight, robust, 'safe' device for an old or new operational requirement, or as a trigger for a thermonuclear warhead where the demand for bigger yields from less weight would always exist. If ABM work continued, the cost of the system was more or less proportional to the weight of the warhead. Since the cost was astronomical, the pressure for the lightest possible warhead would last for years. There would also be a need for such a device to check some ideas as well as was possible in the circumstances where large yield tests were not allowed. To illustrate these points Penney cited the two low-yield UK tests in Nevada held earlier that year. Their purpose had been to check and prove a new idea (Super Octopus) that made quite a difference to the Skybolt warhead; the Americans knew about this and developed it on their own account and tried several underground shots.[216]

Non-nuclear Explosive Tests

Ground initiation and flight trials of the WE177 tactical nuclear bomb with natural or depleted uranium instead of fissile material were held in 1964; these were essential if the weapon was to be available for TSR 2, the planned next generation RAF bomber. It seems that the mechanisms of this bomb were extremely complicated and unless they were fully tested, there could be no assurance that it would work. Douglas-Home, now Prime Minister, was a little worried about such tests in the run up to a General Election; he also worried that others might see such trials as a breach of the PTBT, which of course they were not. The Prime Minister agreed that such tests could proceed, but there were to be no parachute release trials or any other dropping or firing before the October 1964 General Election.[217]

The new Foreign Secretary also had anxieties; he too was cautious when it came to nuclear weapon-related testing activities. AWRE wanted to fire assemblies of natural uranium at the Foulness range in 1964.[218] They acquired the necessary

[215] TNA CAB 134/2296, Information Material on the Nuclear Test Negotiations Note by the Joint Secretaries, Official Committee on Disarmament OD (63) 21, 13 August 1963.

[216] TNA FO 371/163108, IAD 22/359, W.G. Penney, UKAEA to H.B. Shepherd, Foreign Office, 21 December 1962.

[217] TNA PREM 11/4719, Note for the record of PM meeting with Minister of Aviation, 9 June 1964.

[218] TNA FCO 66/371, E.F. Newley, Director AWRE to V.H.B. Macklen, DCA (PN) MOD, 5 June 1972.

authorisations from other relevant government departments, including the Foreign Office. Legal advice from the Foreign Office confirmed that such tests would not be a breach of the PTBT. However, the Foreign Secretary took a different view. He considered that the government might be exposed to criticism at home since the technical justification for the experiment in relation to the PTBT would not be easily understood and that the UK position could become difficult during disarmament discussions either in Geneva or New York. The matter should be deferred until after the General Election; it was, and the issue did not arise again until the middle of 1972.

Home, when Foreign Secretary in the previous year, had noted in response to an Admiralty request for a system test of Red Beard, that it had never been the UK's intention that the test-ban treaty should prohibit the conduct of a trial of this nature. A Red Beard surveillance round would be dropped off the Aberporth sea range to test that the weapon systems as a whole worked as intended, but there would be no fissile material in the device. This would be an extension of earlier flight and handling trials. The Admiralty was keen to conduct this trial in order to establish confidence in the weapon in normal service conditions. Natural uranium would be used instead, but this would have caused atmospheric contamination when the HE component was detonated. This proposal was rejected since in 'current circumstances', presumably the test-ban talks, the UK could not contemplate any action that would be construed as causing atmospheric contamination, albeit of a very minor nature.[219] Instead, the natural uranium would be replaced by steel components; the trial would entail detonation of the weapon's supercharge – 1000 lb of HE. (Super Octopus used about half this amount if the details of the Porton trials described in Chapter 4 are any guide.) However, since the test-ban talks in Moscow were underway in the summer of 1963, the Foreign Secretary thought it best to reserve the UK position. There were no objections to preparations being made, but the question of the drop would be brought up again some weeks before the date of the planned trial.[220] MOD duly obliged and the Foreign Secretary's view was sought in August 1964 for a test of the Red Beard Mark 3, on 29 September.[221] He had no objections. But the Prime Minister felt that the danger of misrepresentation was not acceptable in the run up to the October 1964 General Election. He did not believe it could not be done later in October despite Ministry recommendations that it could not be given availability of the necessary aircraft carrier prior to a Far Eastern deployment.[222] A 1000 lb HE detonation on the Irish Sea would probably not go unnoticed. As far as we can see from the archives, this test does not appear to have been carried out.

---

219 TNA PREM 11/4720, A.P. Hockaday to P.J. Woodfield, Admiralty House, 27 August 1963.

220 TNA PREM 11/4720, Home to Minister of Defence, 19 July 1963.

221 TNA PREM 11/4720, H. Godfrey to the Hon T.E. Bridges, Foreign Office, 14 August 1964.

222 TNA PREM 11/4720, J.O. Wright to A.P. Hockaday, MOD, 28 August 1964.

Conclusions: The End of the Moratorium 1961: UK Responses

Events in 1961 to 1963 showed quite clearly the conflicting pressure on UK policies on arms control and its own weapon requirements as well as the need to maintain effective working and diplomatic relations with the US if its own defence and diplomatic strategies were to be successful. This period also marked a transitional phase in warhead development as the UK was beginning to move towards in effect just two systems – the WE177 family of warheads and the single strategic system, Polaris. The underground tests conducted in Nevada in this period were of crucial importance for the future shape of the stockpile. Persistent attempts to secure a test ban was a hallmark of Macmillan's policy in this period too. The Soviet resumption of atmospheric testing had provided a cover for UK tests, but the extent of the US requirement for its own tests in response and the request for use of Christmas Island put Macmillan in all sorts of difficulties. He had tried to square the circle of keeping the test-ban option alive whilst satisfying his American ally's key defence requirements. British testing requirements were clearly much less than those of the US, but it is very clear that the UK benefited considerably under the rubric of the 1958 Agreement for Cooperation on the Uses of Atomic Energy for Mutual Defence Purposes from the information that the Americans obtained from their atmospheric tests at Christmas Island. We do not know exactly what was learned however as the specific details are retained in closed files in the National Archives. We can see nevertheless that AWRE and the UKAEA were in this period starting to think about future testing programmes and how this would play a key role in sustaining close ties with the US through the exchange of warhead design information. We will have a closer look at UK plans for future testing and warhead requirements in Chapter 9. Meanwhile we need to look in detail at the UK role in the 1963 PTBT, as this was to prove the only tangible outcome from all the efforts in pursuit of a CTBT since 1958.

Chapter 8
The UK and the Partial Test-Ban Treaty (PTBT): 1963

Introduction

Kendrick Oliver has ably chronicled Britain's role in the PTBT negotiations using papers in The National Archives at Kew.[1] For the purposes of this study we need only address the extent to which the UK made particular efforts to ensure a successful outcome and the extent to which military considerations played a part in HMG's decision making on test-ban policy. It is important to ascertain whether UK policies were based on technical and scientific factors, such as lessons from the UK's own nuclear weapons programme and knowledge arising from its own seismic research programme, which we reviewed in Chapter 6, or whether these were downplayed by political imperatives, namely the need to make progress on, or towards a comprehensive test-ban treaty. Or were both considerations – technical and political – at play? Looking at the main events and internal UK deliberations that led up to the Moscow talks in July 1963 should help us answer these questions.

The Conference on the Discontinuance of Nuclear Weapon Tests: Suspension 1962

The Russians made it impossible for the UK and the US to move forward towards the sort of nuclear test ban they wanted to see – one with effective verification provisions. The Russians insisted that their 28 November 1961 proposal, which rejected the core principles that the UK thought had been accepted back in 1958, meant that there was little point in continuing with the Conference. Neither the UK nor US could accept the November 1961 Soviet proposal for a halt to nuclear weapons tests because it amounted to an unverifiable pledge.[2] In January 1962 the UK still wanted to make clear to the Russians that it wished most earnestly to negotiate a nuclear test ban that accepted the principle of reasonable, international

---

1 Oliver, Kendrick, Kennedy, Macmillan and the Nuclear Test-Ban Debate, 1961–63 (Basingstoke, 1997).

2 TNA FO 371/163109, IAD 22/35, Washington telegram no. 247 to Foreign Office, 26 January 1962.

and impartial verification.[3] However, the UK did not want to close the door on this forever, partly for presentational reasons it would not do to be seen walking out from such an internationally important conference; there would be too much negative publicity, which would give Soviet propaganda a field day. Instead, the UK and US proposed that the Conference should adjourn until a common basis for negotiation could be re-established either at the Eighteen Nation Disarmament Committee (ENDC), through diplomatic channels, or through informal contacts among delegations at Geneva.[4] Therefore, the conference while not officially terminated went into indefinite recess after three and half years and 353 meetings. Tsarapkin chaired the final meeting and walked off with the gavel as a souvenir.

Michael Wright noted that the Soviet representatives had been careful to leave themselves a loophole for reverting at any time to acceptance of the principle of international control, which had been the basis for the Geneva negotiations for the previous three and half years. They used phrases such 'in the present international situation' or 'given the present international tension'. If there were a thaw in East–West relations, the draft Western proposals from April 1961 remained on the table. They could serve as a possible blueprint for a future system of international control over nuclear tests. Some two-thirds of the draft treaty and two of the three Annexes contained material to which at one time the Soviet government had agreed.[5] Successful arms control and disarmament was therefore dependent on a favourable international environment; it could not create improved East-West relations on its own, there were too many Cold War exigencies to make that possible, certainly in the late 1950s and early 1960s. Indeed, in one of his earlier despatches on the negotiations, Wright's principal point had been that the main decisions on a nuclear test ban were likely to be political rather than technical.[6]

Impact of Cuban Missile Crisis

The mutual fright created by the Cuban Missile Crisis of October 1962, however, did act as a catalyst for what became the PTBT, which the UK, US and USSR negotiated over a ten-day period in July 1963. Following this unsettling crisis the US and USSR made renewed attempts to move things forward on testing issues. This resulted in an exchange of letters between Kennedy and Khrushchev on testing and agreement on private bilateral talks following Soviet agreement at the end of December 1962. The USSR now indicated that it was ready to accept

---

3 TNA FO 371/163109, IAD 22/1 Nuclear Tests Conference Report on Proceedings 16/22, December 1961, CRO telegram to Ottawa et al., 1 January 1962.

4 TNA FO 371/16311, IAD 22/46, Michael Wright, Despatch No. 3 United Kingdom Delegation to the Nuclear Tests Conference, 3 February 1962.

5 TNA FO 371/16311, IAD 22/46, Michael Wright, Despatch No. 3 United Kingdom Delegation to the Nuclear Tests Conference, Conclusion paragraph 20, 3 February 1962.

6 TNA FO 371/163112, IAD 22/63, A.D. Wilson to Sir H. Caccia, 15 March 1962.

three inspections per year and two to three 'black boxes' in agreed seismic areas. Knowledge of this exchange was on very close hold in Washington, with only seven officials aware of it. Macmillan welcomed this development when informed and gave his blessing in a personal message to Kennedy on 6 January 1963.[7] The UK believed that the UK and US should aim for agreement on three unmanned seismic arrays in the USSR and a small quota of obligatory on-site inspections.[8] Zuckerman led a small team of UK experts (Press from MOD, Whiteway and Carpenter from AWRE) to Washington to discuss the details of what might be pursued in renewed testing talks with the Russians, particularly in relation to the location of unmanned seismic stations. This resulted in a submission by the US experts Drs Long and Wiesner to Kennedy, which made clear that given the developments in detection and discrimination, an annual quota of five inspections would be entirely acceptable and that unmanned seismic stations were not necessary on Soviet territory.[9] According to Zuckerman, Kennedy had accepted this position. Congress was of course another matter; Congressional opinion was still wedded to 20 inspections and after the Cuban Missile Crisis was not in any mood to trust in Soviet intentions.

London also thought that HMG might propose to the Americans that the UK might take part in the bilateral talks. Ormsby-Gore had written from Washington to make the case; he had two reasons. First, the US team leader William Foster (Director, ACDA) was not in his view an ideal person for conducting these talks. He was rather pernickety in his approach to these matters and extremely nervous of Congressional opinion. Ormsby-Gore told the Foreign Secretary that he thought that Foster would be more concerned to ensure that Congress could not accuse him of giving away too much to the Russians other than that a treaty should be finally agreed. Second, although he had encouraged the idea of the US and USSR meeting to settle their differences, he gathered that public and political opinion in the UK at that time was extremely touchy on the subject of US leadership that paid insufficient attention to the opinion of America's allies. In such circumstances, Ormsby-Gore believed that it would be better if the UK did not contract out of an extremely vital time in the test-ban negotiations after having played such a key part to date.[10] If Home and Macmillan agreed, it ought to be possible to suggest to the President that the Americans and Russians should invite the UK to participate in the private talks scheduled for New York (and later Washington).

---

7 TNA FO 371/171214, IAD 1114/2/G, Foreign Office telegram no. 192 to Washington, 1962.

8 TNA FO 371/171214, IAD 1114/3/G, H.B. Shepherd to Mr Pemberton-Pigott, 7 January 1963.

9 TNA FO 371/171214, IAD 1114/14/G, Notes on Test-Ban Discussions Washington DC, 9–12 January 1963, S. Zuckerman to J.B. Godber, 17 January 1963.

10 TNA FO 371/171214, IAD 1114/6/G, D. Ormsby-Gore, Washington, 9 January 1963.

Macmillan, taking up the Ambassador's advice, wrote to the President on 13 January 1963, noting that the fact of the talks had now become public. He suggested that, if it were agreeable, the UK might be associated with the talks, possibly after the first meeting. The UK and US had of course worked together on a test ban so closely that it ought to be quite natural for Anglo-American cooperation to continue as before.[11] Kennedy agreed.[12]

A New Initiative: January 1963

Ormsby-Gore would act for the UK and the US proposed British participation to the Russians at the first meeting on 14 January. However, the Russians did not seem too keen on the idea that the UK should become involved, but undertook to refer the question to Moscow.[13] In the end, there was no Russian objection, but they were discouraged by the first week of talks as, in their view, there were three problems complicating discussions: differences over the number of black boxes and on-site inspections and the US announcement of further underground tests.[14] Gromyko told the British Ambassador in Moscow, Sir Humphrey Trevelyan, that the Russians had gone as far as they were prepared to go. The three issues together could well finish the talks, although they continued for the time being with the Russians still questioning the need for more than three inspections and the West insisting that any agreement must be rooted in scientific considerations, which required more than three inspections.

Kennedy did however ask the USAEC to postpone its planned underground tests to help keep the atmospherics of the meeting sweet. However, by 30 January 1963 Tsarapkin was even more emphatic than he had been to date; the USSR would make no more concessions on their figures for on-site inspections and automatic seismic stations. If the US could not accept this, then the talks would be fruitless and it would be the West's fault.[15] Despite this the US wanted to make a new offer to the Russians and hoped that the UK could support it as a joint initiative. In essence, if certain technical conditions were met the US would be ready to accept a quota of seven on-site inspections and possibly six automatic seismic stations – the present figures on the table were eight and seven respectively. Since this

---

11 TNA FO 371/171214, IAD 1114/7/G, Foreign Office telegram no. 481 to Washington, 13 January 1963.

12 TNA FO 371/171214, IAD 1114/7/G, Foreign Office telegram no. 486 to Washington, 13 January 1963.

13 TNA FO 371/171214, IAD 1114/9/G, Washington telegram no. 147 to Foreign Office, 15 January 1963.

14 TNA FO 371/171214, IAD 1114/15, Moscow telegram no. 105 to Foreign Office, 21 January 1963.

15 TNA FO 371/171215, IAD 1114/36/G, New York telegram no. 126 to Foreign Office, 30 January 1963.

was a clear advance in the direction in which Joseph Godber, the Minister of State had been advocating, the Foreign Office authorised the UK delegation to support the US offer.[16] Although this was seen as 'the worst example so far of a tactical bounce', A.D. Wilson (Assistant Under Secretary, Foreign Office) told Sir Harold Caccia (Permanent Under Secretary) that it was not right to talk of 'following blindly' when the US had taken an important step in the direction previously urged by the UK. That said, the Minister of State himself doubted that the US initiative was an 'important step'. It did not matter either way; the Russians broke off the private talks on testing issues in early February 1963 even though they had only been underway for a few weeks. Two days later the Russians pulled the plug and suggested that the subject revert to the ENDC. They claimed that the West was demanding far too many inspections and seismic stations, when they had earlier indicated that far fewer would be required (two to three) something that the US denied emphatically. Soviet insistence on two or three inspections was unbending. This was where a 'compromise' figure of five came into the reckoning. Godber thought that this might be a reasonable middle ground and if only Khrushchev would seize the chance, then a test-ban treaty could be there for the taking.[17] Having spoken to Dean Rusk and George Ball in Washington on 6 February 1963, Godber had the clear impression that the Americans still wanted an agreement and were prepared to 'horse trade' provided the Russians showed some flexibility too. He reaffirmed the importance of securing sufficient detection and inspection capability to satisfy US public opinion that a treaty could be effectively verified.[18]

A New UK Initiative: February 1963

Godber suggested to the Secretary of State on 13 February 1963 that the UK should now make a direct initiative in Geneva in favour of a compromise on a test-ban agreement.[19] He had tried to persuade the US to be specific on the figures that they would be prepared to accept both for on-site inspections and automatic recording detection posts in the USSR. There were indications that these might be ten and seven respectively, whereas the Russians wanted three for both. Therefore, it seemed to Godber that there were opportunities for a UK political initiative to break the log jam by offering compromise figures of five and five. This would be

[16] TNA FO 371/171230, IAD 1115/24/G, A.D.F. Pemberton-Piggott to Mr A.D. Wilson, January 1963.

[17] TNA FO 371/171230, IAD 1115/33 G, Nuclear Tests, Joseph Godber to Secretary of State.

[18] TNA FO 371/171216, IAD 1114/48, J. Godber to Sir Harold Caccia, Record of Conversations with Mr Dean Rusk and Mr George Ball in Washington on 6 February 1963.

[19] TNA FO 371/171235, IAD 1117/5/G, J. Godber to Secretary of State Nuclear Tests, 13 February 1963.

a break from previous attitudes where the UK had consistently supported the US in public, while at times disagreeing with them in private, but such a proposal might not be so repugnant to Kennedy and if presented properly might even ease his difficulties with Congress in making such a concession. There might be a risk to UK-US relations even though the points that divided the delegations in Geneva were so small as to merit taking some risks in order to make a breakthrough.

Ormsby-Gore, now Ambassador in Washington, however disagreed strongly. It was already down to the US that the negotiations had moved close to final agreement. The Americans had already made a lot of movement towards Russian positions. Moscow needed to show some reciprocity. In such circumstances, a UK initiative along the lines suggested would probably be regarded by the Americans as destroying their negotiating position and playing straight into Russian hands. A unilateral British initiative would thus not be at all welcome in Washington.[20] Given such advice Home saw clearly that there was no need for a new UK proposal at that time, and in any case the US were about to receive new instructions in Geneva; it would be better to await these before deciding on the next tactical steps.[21] Home therefore agreed that there should be no question of the UK acting without either consulting or at least informing the US in advance.[22] For the immediate future British policy was best focused on working with the US in trying to convince the Russians to move from their rigid position and in doing all that was possible to assist the Administration to secure a reasonable treaty that would be accepted by Congress.[23] Further consideration of this idea was overtaken by new US proposals put privately to the head of the Soviet delegation on 19 February. The key point in these proposals was the offer to reduce the on-site inspection quota from eight to ten to seven.

Military Utility of Testing and Impact on UK Thinking: March 1963

Whilst this debate was going on, Penney's advice on the military value of tests, as AWRE then understood the problem in early 1963, was that a single test had little military value. Given the present state of weapons technology in the US and USSR, the UKAEA did not think that anything less than series of three or more tests would have any significance. Referring to the recent UK underground tests that had produced valuable military information, these two tests had 'rounded

[20] TNA FO 371/171235, IAD 1117/4/G, Washington telegram no. 518 to Foreign Office, 15 February 1963.

[21] TNA FO 371/171235, IAD 1117/4/G, A.D. Wilson to Secretary of State, 16 February 1963.

[22] TNA FO 371/171235, IAD 1117/12/G, Scope for an independent UK initiative to settle the nuclear tests problem, H.B. Shepherd, 26 February 1963.

[23] TNA FO 371/171235, IAD 1117/12/G, A.D.F. Pemberton-Pigott to Mr Wilson, 26 February 1963.

off' a lot of prior work. If testing stopped at that point, then in a few years time there would be new ideas, but a single test of a device of 20 kilotons or so would not have major significance. It was conceivable that a series of three or more might have, but a series was easier to discover than single isolated test.[24] This assessment helps us understand the context in which, by March 1963, Macmillan and Home were becoming very much concerned about the state of the nuclear test-ban talks in Geneva.[25] There were important reasons for pursuing a test ban. The most vital reason, even more important than stopping contamination of the atmosphere, was stopping the spread of nuclear weapons, especially to Germany. Consequently, Macmillan proposed to suggest to Kennedy that the three leaders should meet to resolve the remaining issues in the space of two to three days. He asked Ormsby-Gore for his views on this idea; both he and Home felt that the matter could not drag on. The Ambassador agreed and saw every advantage in the Prime Minister making an approach along the lines he had suggested; timing would always be a problem, but the perfect moment was always an elusive article.[26] Congressional problems were considerable, and Kennedy would need time to work on the doubters there. A key to any meeting would be some prior assurance from the Russians that they would be prepared to move; an inflexible Khrushchev would make it most awkward for Kennedy to report back in the face of such intransigence. Macmillan therefore wrote to the President on 16 March to outline his ideas for a summit meeting on the test ban. He also included some other possibilities for breaking the log jam, including personal messages, or sending an emissary such as Averell Harriman or Robert Kennedy to Moscow to clear up any misunderstandings and find out whether there was a chance of settling around about a quota of five inspections.[27]

Unfortunately for Macmillan, the President was not overly enthused by any of these ideas, but asked for his views on a possible draft message that might be sent instead.[28] This proposed that agreement should be sought on objective criteria for the conduct of an on-site inspection and on inspection team composition for instance. Ormsby-Gore was not much impressed with the suggestions and neither was Trevelyan in Moscow. In Ormsby-Gore's view, this represented the kind of tentative and bureaucratic approach focusing on detail that would make

---

24 TNA PREM 11/4555, Summary of Scientific Position Regarding Nuclear Test Control, W.G. Penney, 8 March 1963.

25 TNA FO 371/171235, IAD 1117/13/G, Foreign Office telegram no. 2549 to Washington, 12 March 1963.

26 TNA FO 371/171235, IAD 1117/13/G, Washington telegram no. 798 to Foreign Office, 14 March 1963.

27 TNA FO 371/171235, IAD 1117/20/G, UK Policy towards NTB negotiations: discussion with US and tripartite talks in Moscow between USSR, UK and US, 1963.

28 TNA FO 371/171216, IAD 1114/50/G, Foreign Office telegram no. 3050 to Washington, 28 March 1963.

no progress.[29] Instead, it would be better to make a direct, personal and private approach to the Soviet leader in which the full extent of UK/US interests in a test ban were made clearly manifest in a letter and reinforced by sending an emissary. Trevelyan suggested this too; it was the most hopeful way of proceeding. Home also agreed with the idea of an emissary: Harriman was his choice.[30] Some hint might be given in the letter, reaffirmed by an emissary, of a top-level meeting to reach final agreement on the inspection quota and any other main problem that remained and was ripe for solution. Moreover, such a letter would also need to hold out the clear hope and intention of reaching agreement. We see here therefore the genesis of the strategy that was to lead to the PTBT later that summer; Ormsby-Gore's role must therefore receive due credit. Macmillan did not need too much persuading and sent off another message to Kennedy, reflecting the ideas from Ormsby-Gore.

Origins of PTBT: April–May 1963

Macmillan wrote to Kennedy on 3 April and enclosed a draft letter that they might send to Khrushchev jointly.[31] This suggested that there might be a secret meeting in Geneva to look at the remaining problems and to see if the gaps could be narrowed, then the three leaders themselves might meet to complete the process of concluding a test-ban treaty. Meanwhile, in Geneva the UK and US presented a memorandum to the ENDC on on-site inspections; this indicated that subject to the conditions governing the satisfactory conduct of inspections, they would be prepared to accept an annual quota of seven inspections.[32] However, the idea of a summit meeting caused the US some heartburn; they did not want to face a situation where Khrushchev made no concessions before the meeting.[33] The Americans were taking their time in responding to this, but when Kennedy spoke to Macmillan on April 11, the latter emphasised that there would be problems in getting all the unresolved issues resolved at level lower than heads of government;

[29] TNA FO 371/171216, IAD 1114/50/G, Washington telegram no. 969 to Foreign Office, 28 March 1963.

[30] TNA FO 371/171214, IAD 1114/50/G, Tokyo telegram no. 159 to Foreign Office, 2 April 1963.

[31] TNA FO 371/171216, IAD 1114/50/G, Foreign Office telegram no. 3236 to Washington, 3 April 1963.

[32] TNA CAB 134/2295, Official Committee on Disarmament, Nuclear Tests Note by the Secretaries, OD (63) 13 April 1963. This body lasted until 17 November 1964 when it was dissolved, with its functions transferred to the Defence and Oversea Policy (Official) Committee, Sub-Committee on Disarmament.

[33] TNA FO 371/171216, IAD 1114/51/G, Washington telegram no. 1121 to Foreign Office, 10 April 1963.

that was where an emissary might help. Kennedy said he would certainly consider the suggestion of Harriman.[34]

After further exchanges, the UK and US agreed a draft letter for the Soviet leader in mid April.[35] Kennedy was able to accept nearly all of Macmillan's draft.[36] However, the draft downplayed the reference to the role of a summit to reflect US concerns as the State Department was emphatically opposed to any mention of a possible summit meeting. The letter revived the emissary role. Special emissaries might go to Moscow after Khrushchev had received the letter and had it explained to him by UK and US ambassadors. The core of the proposal was that the UK and US would be prepared to arrange a private tripartite discussion in whatever seemed the most practical way. Macmillan and Kennedy hoped that these discussions would bring the test-ban treaty issue close enough to a final decision so that it might then be proper to think in terms of a meeting at which a definite agreement could be reached. Alternatively, they might send very senior representatives empowered to speak with Khrushchev direct. This letter also included a tentative proposal on a floating quota of inspections running over several years. Sir Humphrey Trevelyan received instructions on 15 April to be ready to deliver the message in concert with his US counterpart on 16 April.[37] Macmillan made clear that he was very anxious to keep the pressure up both on the Russians and the US in order to reach agreement and to lead if possible to a summit meeting. Trevelyan was left to interpret his instructions flexibly to ensure that this outcome was achieved. For the US, the priority was to ensure that the Soviet leader received the message personally and Macmillan himself suggested that if Khrushchev was not available for several days, then Trevelyan and the US ambassador should ask to go to see him.[38] An interview was arranged for 24 April. Now, although Khrushchev was not unfriendly, he was completely negative throughout the interview with the UK and US ambassadors. Once he found out that they were not there to agree to three inspections only, it was quite a job to persuade him to have the letter read to him.[39] He saw no basis for agreement in the letter although it would be studied and a reply sent in due course. He added that it was clear to him now that even offering

---

34 TNA FO 371/171216, IAD 1114/52/G, Foreign Office telegram no. 3557 to Washington, 11 April 1963; TNA PREM 11/4566, Record of a conversation between the Prime Minister and President Kennedy, 11 April 1963.

35 TNA FO 371/171216, IAD 1114/58/G, 1963.

36 TNA PREM 11/4566, Prime Minister's Personal Telegram T. 171/63 From the President, the White House to the Prime Minister, 11 April 1963.

37 TNA FO 371/171216, IAD 1114/57/G, Foreign Office telegram no. 793 to Moscow, 15 April 1963.

38 TNA FO 371/171216, IAD 1114/59/G, telegram from Admiralty to House to Foreign Office for despatch to Moscow, 16 April 1963.

39 TNA FO 371/171217, IAD 1114/63/G, Moscow telegram no. 746 to Foreign Office, 24 April 1963.

three inspections was a mistake and might now need to be withdrawn. Any idea of a floating quota was also unacceptable.

Despite the unpromising reaction from Khrushchev on 24 April, the formal Soviet reply came on 8 May. This accepted that Moscow would be happy to consider a proposed visit of emissaries; it was important that they be empowered to negotiate on the question of ending tests on a realistic and equitable basis without inspection 'by spies'.[40] Khrushchev however denied acceptance of the principle of on-site verification and came near to withdrawing his earlier agreement to two to three inspections. Macmillan quickly jumped on what he saw as the 'helpful' elements in the reply, especially the Soviet leader's willingness to contemplate a meeting of emissaries in Moscow; it would be important to explore the possibilities and to remove certain misconceptions that Khrushchev had on the scope and purpose of inspections.[41]

Risks of a Test Ban: Military Opportunities and Limitations: May 1963

In parallel Macmillan asked Penney and Zuckerman for the best analysis that they could produce of the minimum number of annual inspections of Soviet territory for which the West could safely ask.[42] This analysis considered the military risks from further testing; the only significant development that could emerge from continued testing was in the anti-missile field. In such developments refinements of the warheads would play a comparatively small part, the real and more demanding difficulties in developing and deploying an ABM system occurred in the radar, guidance systems and propellants for the interceptor missiles. Decisive progress in ABMs did not therefore depend on underground nuclear tests alone. Underground tests might allow some advances in developing a very clean bomb, but even this would require testing in the atmosphere eventually before a final decision on devoting immense resources to the creation of a system of defence based upon the use of such a warhead. Any advantages from clandestine testing thus would not put the cheat at a crucial advantage. The real conclusion to be drawn from this argument was that underground tests were not of crucial importance (but it was conceded that they were of political importance) and that a treaty to ban all tests

[40] TNA FO 371/171218, IAD 1114/80/G, Moscow telegram 866 to Foreign Office, April 1963.

[41] TNA FO 371/171218, IAD 1114/83/G, Foreign Office telegram no. 4457 to Washington, 9 May 1963; TNA CAB 128/37, C.C. (63) 34th meeting, Item 3 Nuclear Tests, 23 May 1963.

[42] TNA PREM 11/4557, Macmillan to Sir William Penney, Nuclear Tests, 9 May 1963.

could be signed immediately without any provision for inspections even though cheating did take place.[43]

Underground tests, however, would be of assistance in improving yield to weight ratios of warheads that might be suitable for ABMs. If a clandestine test of a few tens of kilotons took place in a large cavity, the chance of detection and identification by seismic means was small. In such circumstances a violator could at great cost and effort make marginal improvements in nuclear devices of a few tens of kilotons. Penney was firmly of the view, having thought about the problem extensively after his meeting with Macmillan, that the development of an ABM would demand nuclear tests. A test-ban treaty would therefore add considerably to the technical problems of an ABM defence – problems that to him gave the appearance of being insoluble, even with US and Russian resources.[44] To be technically meaningful clandestine underground nuclear tests would have to be above the detectable level; this was the MOD view. They also held the view that were a series of tests conducted, they could hardly fail to be recognised.

Macmillan hosted a brainstorming meeting at Chequers on 18 May 1963 to review the options for a test ban. Only a handful of Ministers and officials were present: Godber, Ormsby-Gore, Penney, Zuckerman, Wilson and de Zulueta.[45] In light of the technical advice, this meeting concluded that underground tests were not of crucial importance and that they could agree a treaty to ban all actual tests immediately without any provision for inspections for underground events, even though cheating might take place. It is quite clear from this, and earlier research by Penney and Zuckerman on the scientific arguments in favour of quotas, that the UK assessment rested as much on technical considerations as opposed to political ones whereby a test-ban treaty should be pursued regardless of costs. Perhaps the main risk was clandestine testing of ABM warheads, but if this were to be a ‘clean warhead’, then it would be necessary for a yield of between ten and 100 kilotons with a fusion to fission ratio of 99 per cent to 1 per cent. Although it might be possible to make progress with such a warhead by underground tests, tests involving several tens of kilotons would be easy to conceal. Furthermore, even if an ABM warhead’s performance could theoretically be perfected by underground tests, its importance would be so great that it would certainly have to be tested in the upper atmosphere at some point.[46] Macmillan accepted Penney’s advice that

---

[43] In 1959, the advice appears to have been that even a single explosion might be enough to give valuable data to a country such as the USSR that already made substantial progress in weapon development. TNA CAB 130/130, GEN.604/9, Quota System of Inspection to Police a Suspension of Nuclear Tests, 7 May 1959.

[44] TNA PREM 11/4557, W.G. Penney, UKAEA to P. de. Zulueta, Admiralty House, 16 May 1963.

[45] TNA CAB 21/6033, Summary record of meeting at Chequers on Saturday, 18 May 1963.

[46] TNA PREM 11/4557, Nuclear Tests: the basic Argument. This was prepared after discussion with Penney at Chequers on 11 May, P. de Zulueta to J.O. Wright, Foreign Office,

the ABM question went beyond just warheads; interception problems required solution and he told Kennedy that perhaps too much significance was devoted to underground tests in isolation and their identification as man-made events rather than natural seismic events.[47]

Zuckerman and Penney (but in practice primarily the former) prepared a detailed memorandum setting out these arguments in detail for Macmillan to send to Kennedy.[48] This stemmed from discussions between Ormsby-Gore and Kennedy on 22 May in which the President had agreed with the suggestion that it might be useful if Macmillan sent him a memorandum prepared by UK scientists on the significance of underground tests in the then present circumstances – early summer 1963.[49] This memorandum concluded that, whether one considered tactical, strategic missiles or ABM warhead requirements, underground tests were not a consideration that should prevent agreement on a test ban. The UK and US ought to be able to settle readily for three on-site inspections without endangering national security.[50] Ormsby-Gore however did not like the version of the paper that he had been asked to hand to the President. In the form presented it was just an argument in favour of accepting the present Soviet position on on-site inspections as opposed the US one, which until that moment the UK had supported. As such it would not convince the US; it was well known that the President could not accept three inspections and the flat statement in the concluding paragraph on three inspections would only cause great irritation. Moreover, there was no urgent need for this paper either since the US had agreed to the UK's immediate objective of sending a representative to Moscow and the Russians had agreed to receive him.[51] Home agreed; the draft was too political and he hoped that a hybrid version drawing on Penney's earlier May minute for the Chequers meeting could be produced instead.[52] Consequently, the paper was re-written and the offending

---

12 May 1963. Macmillan wrote to Penney to thank him for his help and noted that 'it may be that we shall never succeed in getting a test-ban agreement. What encourages me and inspires me to go on is that you have made it so clear that it is worth trying', Macmillan to Penney, 12 May 1963.

47 47 Macmillan, *At the End of the Day*, pp. 470–71.

48 TNA FO 371/171236, IAD 1117/31/G, Memorandum about the present situation concerning nuclear tests, 12 June 1963.

49 TNA FO 371/171236, IAD 1117/32/G, Macmillan to The President of the USA, 21 June 1963.

50 PREM 11/4557, Zuckerman to the PM, 23 May 1963. This had been prepared with the approval of Penney's staff, 'dealing with the arguments to prove that further underground nuclear tests will not be of significance', P. de Zulueta to Prime Minister, 24 May 1963.

51 TNA FO 371/171236, IAD 1117/31/G, Washington telegram no. 1802 to Foreign Office, 14 June 1963.

52 TNA FO 371/171236, IAD 1117/32/G, A.D. Wilson, Foreign Office to P.F. de Zulueta, Admiralty House.

passages deleted along with some other improvements. The new version went to Kennedy on 21 June.

Russian Attitudes: On-site Inspections

There was now some doubt over Soviet readiness to accept on-site inspections as Khrushchev had said privately, but not officially to either the UK or the US, that he had withdrawn his previous offer of two or three inspections. In fact this much was clear from reports from the UK Ambassador in Moscow.[53] For this reason the British and Americans realised that they might have to fall back on a partial treaty banning tests in all environments except underground; Khrushchev himself had indicated that this might be possible in a speech in Berlin on 2 July. This appeared to be a new Soviet position since it was the first time that there was no mention of the usual condition that such an agreement would have to be accompanied by an unverified moratorium on underground tests. This was a very welcome development. A partial treaty would be an important step in the right direction if a comprehensive test-ban treaty proved unattainable.

Announcement of the Emissaries' Trip to Moscow: Preparations June 1963

The respective governments announced publicly on 10 June to the world that Averell Harriman for Kennedy and Lord Hailsham for Macmillan would travel to Moscow for a meeting on test-ban issues in mid July. As preparations for the Moscow meeting in London and Washington began, the possibility of a fall-back position on a limited test-ban treaty was recognised by both Ormsby-Gore and William Foster, (ACDA Director), even though the Russians might insist that it should be accompanied by an unverified agreement banning underground tests too.[54] Nevertheless, the first aim was still a comprehensive test ban with adequate verification of uncertain seismic events underground.[55] National detection would be insufficient; a proportion of on-site inspections would still be needed. Inspections could be carried out under strict safeguards against the possibility of espionage. If the Russians refused to entertain a comprehensive test ban the UK and US might explore the possibility of some intermediate position between a comprehensive and partial test ban; for example, this might entail offering an atmospheric ban

---

[53] TNA FO 371/171236, IAD 1117/35/G, R.H. Mason to Mr Pemberton-Piggott, Briefs for Prime Minister's Talks with President Kennedy, 20 June 1963.

[54] TNA FO 371/171220, IAD 1114/121, Washington telegram no. 1892 to Foreign Office, 20 June 1963.

[55] TNA FO 371/171220, IAD 1114/127/G, Talking Points, Nuclear Tests and Other Related Measures Brief by the Foreign Office, Brief No.1 June 26, 1963, Talks with President Kennedy, 29 June 1963.

with some annual limit on the number of underground tests. Both Home and Rusk, reflecting Kennedy's view, felt that the Moscow talks would probably be the last chance to secure a test treaty-ban.[56]

Macmillan and Kennedy met at Birch Grove, Macmillan's country residence, to finalise Anglo-American positions ahead of the Moscow trip. They devoted a substantial part of their meeting to reviewing the instructions for Harriman and Hailsham for the Moscow talks, which were to start on 15 July.[57] Both agreed that the aim was a comprehensive test ban with adequate verification of uncertain seismic events underground; for this reason the UK and US required a quota of inspections. The Western position was flexible. Kennedy, for instance, told Macmillan that he was very doubtful that anything worthwhile had come out of the series of US tests in the Pacific, which he had authorised after their Bermuda meeting and that he would like to have certain technical evaluations to put against the views of the Joint Chiefs of Staff who were opposed to a treaty. The opposition in the US to a test ban was nearly as great as the opposition in the UK to a mixed-manned NATO force (the Multilateral Force – MLF).[58] Macmillan rehearsed all the arguments about the utility of further testing supplied earlier by Zuckerman and Penney.[59]

John T. McNaughton, General Counsel of the Department of Defense, and Zuckerman were tasked to address Kennedy's questions. What was the maximum size of underground nuclear tests which the USSR could repeatedly carry out without significant fear of detection? What could be learned technically from such small underground tests and what would the military significance of the increased knowledge be? Would atmospheric tests be required to prove out developments made by underground testing?[60] For the first question, they concluded that the Russians could test up to three kilotons without decoupling and risk only a 10 per cent chance of detection. For the second point, increases in several fold in the yield to weight ratio could be accomplished for weapons in the low kiloton range between three to ten kilotons. The fraction of fissionable material in thermonuclear weapons could be reduced and some weapons effects tests could be conducted, specifically the effect of radiation on warhead components. Studies of this nature would permit development of a wide range of low-yield tactical nuclear weapons.

---

56 TNA FO 371/171216, IAD 1114/131/G, Record of Conversation Between the Foreign Secretary and the United States Secretary of State in the Foreign Office on Thursday, 27 June, 1963; TNA FO 371/171222, IAD 1114/170, A.D. Wilson to Private Secretary, 28 June 1963.

57 TNA PREM 11/4558, Zuckerman note on US JCS paper against a CTBT to Kennedy shown to the Prime Minister at Birch Grove, 2 July 1963.

58 TNA FO 800/951, Secretary of State's Tentative Comments on the ANF Minute, 26 December 1964.

59 TNA PREM 11/4558, Nuclear Tests, Record of a conversation at Birch Grove House, 29 June 1963.

60 TNA FO 371/171221, IAD 1114/144/G, Record of a meeting between the Prime Minister and the President at Birch Grove at 1200 noon on Sunday, 30 June 1963.

Underground tests could also lead to improvements, by reducing the proportion of fissionable material in the warheads of ABMs. However, the warhead was only one component in the necessarily complex ABM system and the offensive forces had such means of counter-improvement at their disposal that the possible improvements for ABM warheads did not appear to be of great military importance. As for the third question, McNaughton said that atmospheric tests were not needed for tactical weapons nor for improving some components of strategic weapons, but a significant amount of such testing would be required for meaningful changes to strategic weapons, to test operational effectiveness of anti-ballistic missile systems and for a range of significant weapons effects tests.

Moscow: July 1963

After Birch Grove Macmillan and Home agreed on 5 July that the right tactic for Moscow would be to accept the Soviet offer of a test-ban treaty in the three environments, subject to there not being an unverified moratorium on underground tests; to discuss whether and how a comprehensive test ban could be achieved; to discuss how other countries might adhere to a test-ban treaty and possibly on how a non-proliferation treaty might be organised. Such a treaty however would not be at the expense of a test-ban agreement.[61] When Harriman and Hailsham had their first meeting on 12 July they noted that their instructions were to seek a comprehensive test-ban treaty with a partial test ban in the three environments as second best solution.[62] Sure enough at the first meeting with Khrushchev on 15 July the Soviet leader made it clear that he would not look at a comprehensive test-ban treaty which involved inspections and that the offer of a quota of two to three on-site inspections per year was now withdrawn. He did however express himself ready to agree immediately on a test ban covering environments other than underground and tabled a new draft of such a treaty.[63] Both the UK and US were of course ready to sign such an agreement at once. Once it became clear during the Moscow talks that this was all that was on offer Macmillan was pleased enough to report to the Cabinet that this would make a start of a gradual reduction in tension between East and West.[64] After three days of discussions, it was quite clear too that there was going to be no need for detailed technical discussions on verification, so Penney came home early as there was nothing for him to do. The Russians

61 TNA FO 371/171221, IAD 1114/145/G, Prime Minister's Personal Minute No. M 238/63, 5 July 1963.

62 TNA FO 371/171222, IAD 1114/178/G, Note of Discussion between Governor Harriman and Lord Hailsham and their advisers, 12 July 1963.

63 TNA FO 371/171222, IAD 1114/1720/G, Moscow telegram no. 1457 to Foreign Office, 15 July 1963.

64 TNA CAB 128/37, C.C. (63) 47th meeting, Item 2 Nuclear Tests, 18 July 1963.

had resolutely refused to establish any technical discussions on seismic research between Western and Soviet scientists.[65]

The UK and US agreed that there was no point in trying for some understanding about continuing discussion on underground tests. Harriman was particularly opposed, as it would only complicate matters with Congress.[66] Once the problems over the withdrawal clause (the Russians did not want one and the Americans did), the role of peaceful nuclear explosions (the US wanted a reference to this and the Russians were against) and depositary powers (problems over acceptance of instruments of ratification from states that were not recognised) had been overcome, the way was clear to reach agreement on an initialled text. Hailsham, Harriman and Gromyko agreed the text of a Treaty Banning Nuclear Weapon Tests in the Atmosphere, in Outer Space and Under Water (PTBT) on 23 July after only 11 days of negotiations. The Foreign Ministers signed the treaty in Moscow on 5 August. Kennedy wrote to Macmillan on 29 July and noted that what no one could doubt was the importance in all of this of Macmillan's own persistent pursuit of a solution. He had never given up for a minute, and more than once had produced the initiative that had started things moving again. As such, his indispensable contribution was well understood and highly valued.[67]

Whither a Comprehensive Nuclear Test-Ban Treaty: October 1963

The Minister of State and Fischer, Director of ACDA, discussed on 22 October what the UK and US should do if there was pressure in the ENDC in Geneva from the neutral and non-aligned states for more discussions for a comprehensive test-ban treaty. The answer was that the UK and US would propose instead further scientific talks on underground detection. If the Russians refused, they would be seen as the ones responsible for lack of progress.[68] This was to be the basis for UK-US action in Geneva for much of the remainder of our period. Nothing had happened to suggest that the Russians would actually agree on further scientific research, but London considered at the end of January 1964 whether to put forward a formal UK proposal for the establishment of a suitable scientific body at the ENDC. Its composition would be open ended and its terms of reference would be to study and report on the scientific aspects of the detection and identification problem. As for the practical arrangements of such a gathering, the model in mind was that of the technical working groups that

65 TNA FO 371/171223, IAD 1114/195, Moscow telegram no. 1486 to Foreign Office, 19 July 1963.

66 TNA FO 371/171216, IAD 1114/203/G, Moscow telegram no. 1502 to Foreign Office, 19 July 1963.

67 TNA FO 371/171226, IAD 1114/255, Message from President to Prime Minister on 26 July by direct wire, Foreign Office telegram no. 7165 to Washington, 29 July 1963.

68 TNA FO 371/171237, IAD 1117/62, H.B. Shepherd, Foreign Office to Dr R. Press, MOD, 14 November 1963.

had met during the Conference on the Discontinuance of Nuclear Weapon Tests. The UK wanted a short concentrated effort to reach scientific agreement and not some kind of semi-permanent sub-committee.[69]

Ministers had made clear in Parliament that such technical and scientific talks were the only possible way to make progress towards a comprehensive test-ban treaty. The delegation in Geneva however was not convinced. In Sir Paul Mason's (UK Ambassador at the ENDC) view, there was little chance of this making much impact. So long as the Russians refused to play ball, there was not much running that could be made with it.[70] More to the point, the delegation had heard from their US colleagues that the difficulties within the US Administration between the pro and anti test lobbies over the desirability of talks such as proposed by the UK, were unlikely to be solved in less than two months. Washington therefore instructed the US delegation to take no initiative. For this reason such a proposal might not be most appropriate coming from a British Minister either. Despite this the UK delegation in Geneva thought that the UK should at least say something about this in order to make it clear that the UK was not on the defensive on this subject. So once again, arguments over verification requirements and internal US problems put a spoke in the wheels of immediate practical progress towards a test ban.

Conclusions: The Lead Up To, and UK Participation in the Partial Test-Ban Treaty

Throughout the test-ban negotiations, not just the eleven days in Moscow, the British delegation and more importantly, Ministers from the Prime Minister downwards maintained a continuing commitment to finding a solution that would result in limitations on nuclear tests. However, there is also evidence that at times UK pressure in Washington came close to being counter-productive.[71] At least one occasion a USAEC Commissioner complained about UK attitudes on test-ban issues and noted that the USAEC might make things difficult over the transfer of nuclear weapons information.[72] This was always a risk that upsetting the Americans over test-ban issues might harm the harmonious and productive operation of the 1958 Agreement, and this was a prospect that would become much more of a preoccupation in British minds by the end of the decade, but we will see why this was so in Chapter 11.

[69] TNA FO 371/176437, IAN 1016/1, Foreign Office telegram no. 1136 to Washington, 24 January 1964.

[70] TNA FO 371/176437, Suggestion that UK might propose a scientific confrontation, Geneva telegram 15 to Foreign Office, 24 January 1964.

[71] Schrafstetter and Twigge, Avoiding Armageddon, p. 102.

[72] Schrafstetter and Twigge, Avoiding Armageddon, p. 102.

UK policies were never entirely divorced from technical realities since they were also key considerations, such as the belief that a few low-yield clandestine tests would not alter the strategic balance. Indeed as Penney had pointed out, the UK's two underground tests in Nevada had adequately addressed immediate UK nuclear defence requirements.[73] In 1963 the UK would have probably settled for three on-site inspections per year without any fear of endangering national security whether considering Soviet tactical, strategic, or ABM warhead developments. However, unless there was a right to carry out such on-site inspections, given the then current seismic detection capabilities and state of knowledge achieved by AWRE, there was no way the uncertainty associated with unidentified seismic events could be removed. Moreover, in political terms this could adversely affect the East-West confidence-building aspects of a comprehensive test-ban treaty and would create a poor precedent for other measures of disarmament.[74] The UK was always mindful of the precedents that might be set for the wider arms control and disarmament agenda from any settlements over verification in a test-ban treaty. Effective international control allied to a careful assessment of the potential military risks of any cheating remained a key feature of UK thinking and negotiating style.

There were no prospects at the end of 1964 that the US would accept a lower quota of inspections, or that the Russians would accept any at all. Western states had consistently urged the Russians to agree to official discussion between scientists to try to reconcile these differences, but the Russians had always refused. This was the basic position for the rest of our period. Even if the Russians had changed their minds, the US Administration would have found it exceptionally difficult to persuade the Senate to ratify a comprehensive test-ban treaty. So the PTBT was the high water mark in efforts to achieve a comprehensive test-ban treaty and never again in our period would the UK play such a prominent role in its pursuit. To understand why this was so we must look both at US and USSR attitudes in the subsequent years and the growing questions on the UK's own nuclear deterrent's long-term effectiveness against emerging Soviet ABM capabilities. These questions were to generate requirements for new UK tests to sustain the British deterrent into the late 1970s and beyond. Chapters 10 and 11 will address these issues.

---

[73] Pampas and Tendrac, March and November 1962.

[74] TNA CAB 148/50, Arms Control and Disarmament Policy, Ministry of Defence Review, Note by the Ministry of Defence, Defence and Oversea Policy (Official) Committee Sub-Committee on Disarmament OPD (O) (D) (64) 3, 24 November 1964.

Chapter 9
UK Testing and Future Plans 1963–1965

Introduction

As we saw the first UK nuclear test for four years took place in 1962, quickly followed by a further test in the same year, but what of other plans and requirements? It has always been unclear, from the public record at least, why the UK stopped testing again in 1965: what were the reasons for this and what impact did it have, if any, in the short term on the UK weapons programme? Arms control pressures were much less acute in this period following conclusion of the PTBT, so neither the Macmillan, Home or Wilson governments of these years had to worry unduly about domestic and international imperatives comparable to those of the late 1950s. However, were there any vestiges of arms control and disarmament considerations in testing policy in this period?

An Independent Underground UK Test Site

Some exploratory modelling work on the options for possible locations and costings for an underground UK test site took place as early as the late 1950s. This was re-examined in the very early 1960s. The model assumed that a UK series would consist of four shots with a maximum individual yield of 100 kilotons and the optimum at 20; the UK knew that the US had been successful with measurements from radiochemical sampling, gamma time-intensity measurements, pinhole camera studies, hotspot techniques and earth shock measurement.[1] Preliminary examination of alternative sites suggested that Fraser Valley in British Colombia was the most suitable technically and probably politically. This was certainly an underestimate given Canadian problems with seismic verification research calibration tests (see discussion in Chapter 6). Alternatively, the Nevada Test Site (NTS) offered the capability of firing weapons up to 100 kilotons in the Lawrence Radiation Laboratory (LRL) tunnel sites and LRL had indicated without prejudice to political approval that they would welcome UK tests there if the UK desired this. There was a major advantage in this offer: many of the support facilities would already be at hand and a much closer collaboration on weapons programmes and testing would be possible.[2]

---

[1] TNA ES 1/1351, E.R. Drake Seager, Planning Field Experiments Branch, 6 November 1959, reprinted 26 June 1961.

[2] TNA ES 1/1351, Underground Nuclear Testing, I. Maddock HFE, 28 June 1961.

AWRE undertook some studies at Maralinga in 1965 to see whether the location would be suitable for underground nuclear tests; this was a precautionary measure since the UK would, it hoped, still have access to the NTS. Topographical information was available from earlier borings made for water supplies. This study led to the conclusion that, although some technical difficulties were likely to arise from the presence of sub-surface water, underground nuclear tests would be feasible in the Maralinga area. This was at odds with the outcome of the earlier studies. However, this assessment still needed to be confirmed with Australian geologists and probably by trial borings too. Costs for preparation of the site over two years followed by three years' testing at the rate of three tests per year (range support during tests and their preparation) would amount to some £7.1 million.[3] Officials were nevertheless very conscious that access to NTS was only implicit in the 1958 Agreement for Cooperation on the uses of Atomic Energy for Mutual Defence Purposes and might not necessarily always be open to the UK, especially since it was scheduled to expire unless renewed in 1969.[4]

Costs, planning and preparations were also investigated. As long as the NTS facilities continued to be available at reasonable cost, the balance of financial advantage lay in continuing use of Nevada for future UK tests and for such safety tests as may be necessary. Maralinga would only be more advantageous if an assured programme of several tests for a number of years could be guaranteed. There was also a political dimension; the Australians would no doubt insist on being taken into greater confidence about nuclear weapons matters if underground testing were to take place in future. Any request for information would have to be reviewed in relation to UK undertakings to the US under the 1958 Agreement. There was, however, no justification for keeping the site on a care and maintenance basis on the then envisaged level of proposed safety tests that would be required. Ministers had decided in early 1966 to close the Maralinga range, but there was a continuing requirement for surface experiments but these were not of a scale to justify its maintenance.[5] Small-scale experiments with natural uranium and HE could take place at Aldermaston's 'bomb' chamber.[6]

Future Testing Requirements: Courser and Charcoal

Following the resumption of UK testing with the Pampas shot in 1962, the Maralinga Experimental Programme for 1962 was scaled back considerably and

[3] TNA CAB 134/2241, Nuclear Requirements for Defence Committee, The Future of the Maralinga Range, Note by the Ministry of Aviation, N.D. (65) 8, 16 November 1965.

[4] TNA CAB 21/6039, Cabinet Nuclear Requirements for Defence Committee, 1965.

[5] TNA DEFE 7/2364, CRO telegram no. 454 to Canberra, 8 February 1966; and J.B. Bourn, PS/PUS to D. Sc 3 (Mr Fakley), 17 February 1966; see also Arnold and Smith, *Britain, Australia and the Bomb*, Chapter 12.

[6] TNA DEFE 7/2364, R. Press, ACSA (N), to PUS, 14 December 1965.

there were no trials at all in that year. Pampas was successfully carried through with the result that the technical picture had changed considerably. Results of this trial and US data meant that the AWRE scientific departments were hard pressed to develop any new programmes.[7] The UKAEA submitted detailed proposals to the Official Committee on Nuclear Requirements for Defence on 8 January 1963; these built on the agreement in principle in November 1961 that the UK would continue underground nuclear tests. The UKAEA wanted a research programme involving nine tests over a three-year period at an estimated new money cost of £4.8 million plus an estimated £1.1 million for fissile material. This programme was not aimed primarily at the development of specific weapons. Instead it sought to maintain UK nuclear weapons technology and close cooperation with the US.[8] The Committee recommended to the Prime Minister that approval should be given for firm planning to proceed for two low kiloton tests in 1964. The test programme for 1965 and 1966 would not be planned in detail until the end of 1964, as it would be dependent on further developments and on the outcome of the two tests in 1964.[9] The primary aim of the first of these would be to prove a mechanical method of ensuring single point safety in nuclear weapons and as a means for further reducing the size and weight of low kiloton yield devices. The improvements then achieved could be applied to kiloton weapons and would be a very useful step towards smaller and lighter megaton weapons. In addition, this initial programme could well throw some further light on the possibility of achieving a much higher fusion to fission ratio. One of the consequences of the PTBT was that in future tests of the Vixen B type would have to be conducted underground to avoid any accusations of cheating.[10]

The UKAEA readily accepted in September 1963 that there was no foreseeable need for tests in respect of the nuclear weapons already approved and being developed – the WE177 and Polaris – for service use.[11] No tests were required for Polaris warheads at this time since these were to be exact copies of the US model.[12] The NRDC decided in early 1964 that nuclear testing facilities should not be retained at Christmas Island and the remaining RAF personnel finally pulled

---

[7] TNA DEFE 7/2364, CRO telegram no. 550 to Canberra, 11 April 1962; CRO telegram no. 800 to Canberra, 7 June 1962.

[8] TNA DEFE 24/291, British Underground Nuclear Tests, Minister of Defence to Prime Minister, 20 January 1963.

[9] TNA CAB 134/2241, Nuclear Requirements for Defence Committee, A British Programme of Underground Nuclear Tests, N.D. (64) 1st Meeting, 1964.

[10] TNA CAB 134/2240, Nuclear Requirements for Defence Committee, N.D. (63) 4th Meeting, 30 October 1963.

[11] TNA PREM 11/5173, A. Hockaday, MOD to P.F. de Zulueta, Admiralty House, 13 September 1963.

[12] TNA CAB 134/2240, Nuclear Requirements for Defence Committee, N.D. (63) 4th Meeting, 30 October 1963.

out in 1964.[13] Testing was needed to concentrate on certain research problems as part of a programme required to maintain an efficient nuclear weapons technology in the UK. Macmillan wondered what this meant.[14]

There were four main reasons for testing in the UKAEA's view. First, so long as UK nuclear weapons were in service, a post-design capability had to be maintained to deal with problems of manufacture, transport, storage, handling, maintenance and modifications. Second, studies were also required on the effects of nuclear explosions on men, materials, structures, communications and radar. Third, the previous reasons called for maintenance of research on nuclear warhead physics, materials and technology at a level necessary to retain personnel skills and an ability to undertake a new warhead project, if it were required, and to make a useful contribution to Anglo-American collaboration in the nuclear defence field so long as the UK required and desired that collaboration. Fourth, proposals to fulfil these objectives did not entail tests for weapons already in service and being developed, but they would almost certainly entail some such tests for new concepts such as the use of new materials, effects tests and possibly research into PNEs.[15] The US would want to know what UK intentions were as they planned their own testing schedule; the UKAEA, therefore, wanted to be able to tell Seaborg (Chairman, USAEC) if he asked that they were preparing to submit proposals to the government and if these were accepted, might well lead to a request to use the facilities at Nevada. Makins told Seaborg that he had concluded that it would be in the interests of AWRE as a laboratory that its future programme should be based on the possibility of conducting underground nuclear tests.[16]

Makins made clear to the NRDC in October 1963 that if British scientists were to keep their knowledge up-to-date and continue their present close cooperation with the US, there would need to be agreement in principle to the continuation of underground testing. Meanwhile the absence of a firm programme was having an unsettling effect on UKAEA staff. Areas of greatest interest for further experimental work included advanced warheads, and nearly clean kiloton devices with only a few per cent of the yield coming from fission. In the discussion at the NRDC participants thought that from the foreign political point of view there seemed to be no overriding objection to the continuation of UK underground nuclear testing.[17] Furthermore, the Ministry of Aviation believed that so long as the UK undertook or contemplated any development work on nuclear warheads

13 TNA AIR 20/9807, W. Kyle, VCAS to Secretary of State, 19 February 1964 ref to NRDC ND (63) 4th meeting (Item 1); J. Roberts for Private Secretary to Secretary of State to VCAS, The Future of Christmas Island, 21 February 1964.

14 TNA PREM 11/5173, T.J. Bligh to A.P. Hockaday, MOD, 16 September 1963.

15 TNA PREM 11/5173, Arthur Hockaday, MOD to T.J. Bligh, Admiralty House, 17 September 1963.

16 TNA PREM 11/5173, Arthur Hockaday, MOD to T.J. Bligh, 3 October 1963.

17 WO 32/17069, Extract from minutes of 4th (63) Meeting of the Nuclear Requirements for Defence Committee, 30 October 1963.

there would also be a continuing, if intermittent, need for a programme of Vixen B and/or Tims-type experiments.[18]

There were problems however in conducting such experiments in the US. Officials thought it politically difficult, if not impossible, to conduct experiments in a US facility to confirm the safety of US designed warheads intended for British service use. The US would have already conducted its own safety tests, in which the UK might not have full confidence; it was thus essential to conduct such experiments independently. Any UK test would have to fit in with a US programme, which might mean delays thereby upsetting UK timings. It was essential for the UK to bring independent information to the collaborative programme, this was compromised by the need to share all details of UK proposed experiments with the US, who in turn made the measurements. Maralinga was therefore the best location for such experiments. Moreover, if Vixen B and Tims trials were conducted in boreholes, it could then be stated unequivocally that such experiments were not prohibited by the PTBT.

Officials sought guidance on future policy on testing from the new Prime Minister, Sir Alex Douglas Home on 26 November 1963; the main argument for continuing to test was that such a programme was essential if UKAEA scientists were to keep their knowledge up to date and therefore ensure continuance of the current close cooperation with the US. Without such a programme, AWRE would have little purpose for their work. Home agreed.[19] The UKAEA's detailed plans, endorsed by the NRDC, were submitted to the Prime Minister on 20 January 1964. The basic plan involved a three-year research programme involving nine underground tests at an estimated cost of £5.9 million, £1.1 million of which was for fissile material. Two kiloton range tests were envisaged for 1964. The expenditure for these tests was included in the 1964–65 defence budget. The NRDC recommended that the research programme and tests should continue after 1964, that there should be a review after the first two tests, and that a further report would be submitted to Ministers for the timetable for the 1964 tests.[20]

The 1964 tests were mainly directed towards improving the safety and reducing the size and weight of low kiloton devices, in the latter case with applications both to kiloton and megaton weapons. The programme was also expected to throw further light on the possibility of achieving a much higher fusion to fission ratio in nuclear weapons. Home agreed with the proposals provided Ministers were consulted again, after discussions with the US had taken place. He also agreed with the Chief Secretary to the Treasury's view that Ministers would need to look again at the situation after the test in the light of the AWRE run down and the

[18] TNA CAB 134/2240, Nuclear Requirements for Defence Committee, Future of United Kingdom Nuclear Weapon Tests and Experiments Memorandum by the Ministry of Aviation, (63) 8, 15 October 1963.

[19] TNA PREM 11/5173, Thorneycroft, Minister of Defence to Prime Minister, British Underground Nuclear Tests, 26 November 1963.

[20] TNA PREM 11/5173, Minister of Defence to Prime Minister, 20 January 1964.

Polaris programme.[21] Penney wrote to Seaborg on 21 July 1964 concerning plans for a forthcoming UK test in Nevada in November of that year. The original plan had been an experiment related to multistage warheads. However, the UK now wanted to see whether it would be possible to reduce the usage of plutonium in UK Polaris warheads.[22] Shortage of fissile material was a major concern at this time. There were few doubts that savings could be made in the primary design since it contained sufficient fissile material to ensure that it could be put into service without a test.[23] Performance could not be guaranteed without one or possibly two tests. NRDC members felt that it might well arouse political criticism if the UK Polaris warhead was put into service without a test. This would save about 100 kg of plutonium amounting to some £2.5 million.[24] Earlier indications had been that savings would amount to 130 kg in respect of 250 warheads, although much would depend on whether one or two tests were undertaken: if two, the saving might be as large as 200 kg.[25] Since Ministerial approval already existed for two tests, one of the research tests could be cancelled to accommodate the economy test. There was also some urgency in this driven by concerns over the possibility of an international agreement on the cut-off in the production of fissile material, which was a key consideration at that time.[26] This would enable the UK to carry out somewhat earlier any disarmament agreement involving a cut-off.[27] In fact, these concerns about the impact on supplies had led AWRE to re-examine possibilities of warhead economy in the use of fissile material.[28]

It seems that Penney was not convinced, initially at least with the case for savings of a quarter of a kilogram of plutonium per warhead – a total of 100 kg.[29] Costs were no doubt a factor too given the expense of fissile material.

21 TNA PREM 11/5173, J.O. Wright to A.P. Hockaday, MOD, 29 January 1964.

22 TNA DEFE 24/291, W.G. Penney to Dr G.T. Seaborg, USAEC, 21 July 1964; TNA AB 49/0015 British Underground Test in Nevada.

23 TNA CAB 134/2241 Nuclear Requirements for Defence Committee, Item 4 Economy in Weapons Use of Fissile Material, N.D. (64) 5th Meeting, 25 June 1964.

24 TNA CAB 21/6037, folio 151, Polaris Economy Test.

25 TNA CAB 134/2241, Nuclear Requirements for Defence Committee, Item 4 Economy in Weapons Use of Fissile Material, N.D. (64) 5th Meeting, 25 June 1964; and N.D. (64), 8 Economy in Weapons Use of Fissile Material, Note by the Ministry of Defence, 22 June 1964.

26 TNA DEFE 24/291, Brief for Permanent Under Secretary on Nuclear Weapons Policy, 12 January 1965.

27 TNA CAB 21/6037, Draft paper to Cabinet, Polaris Economy, Tests Memorandum by Secretary of State for Defence.

28 TNA CAB 134/2241, Nuclear Requirements for Defence Committee, Economy in Weapons Use of Fissile Material, Note by the Ministry of Defence, N.D. (64) 8, 22 June 1964.

29 TNA AB 49/0015, W.G. Penney to Dr N. Levin, AWRE, Aldermaston, 14 May 1964. AWRE later managed to cut the quantity by 0.45 kg per warhead.

There is also a suggestion from some Prime Ministerial statements made in the House of Commons that the warhead also had to be re-shaped to fit the missile.[30] The Americans acceded to Penney's request and the Courser shot was fired on 25 September 1964. This was a replica of the existing design for the primary of the UK Polaris warhead except that the amount of plutonium was reduced by some 0.45 kg. There was no nuclear yield, which caused considerable concern. However, the post-shot diagnostics revealed that all the actual warhead components had functioned correctly; the failure was down to a fault in the external neutron generators.[31] These were US components and it turned out to be a fundamental flaw in the US design; a US test had failed in October for the same reason. Therefore, the test had to be repeated; and that was proposed to Ministers.

The new Labour government elected in October 1964 was nervous about continued testing for political reasons given its opposition to future UK nuclear weapons, but accepted the need to repeat the economy test in view of the potentially very significant savings that could result from a successful test.[32] However, there was to be no publicity. Contingency lines were prepared in the event of a leak of the story before or after the test.[33] Penney wrote to Seaborg on 9 April 1965 to make the formal request to begin the joint UKAEA and USAEC preparations; the estimated maximum possible yield of the device was less than 40 kilotons.[34] This test would bring nearer the time by which certain UKAEA production resources at Calder Hall and Chapelcross could be switched from military work to the civil side in accordance with UK disarmament plans in the field of production of nuclear material for military purposes. This was an allusion to the cut-off.[35] Charcoal, the test code name, took place successfully on 10 September 1965.[36] Wilson told the House of Commons on 18 November that the test was conducted purely on the grounds of economy of expensive materials. After the test, Healey told Wilson that they had been, 'very lucky to have got away with the test'. UK seismic stations had not detected the event and, as Healey no doubt wryly observed, if they had not registered the test, this would make nonsense of the UK position in the disarmament

30 TNA DEFE 24/291, G.H. Green, Head of D.S. 3 to AUS (R &D), Prime Minister's Statements on Testing of Polaris Warhead, 18 August 1965; R. Press, ACSA (N) to Head of D.S. 3, PQS on Nuclear Weapons, 17 August 1965.

31 TNA DEFE 24/291, Cabinet Nuclear Requirements for Defence Committee, British Programme of Underground Nuclear Tests, Note by UKAEA.

32 TNA PREM 13/123, Healey to Prime Minister, Nuclear Weapons Policy and Underground Tests 1965/66, 27 January 1965.

33 TNA PREM 13/123, J.A. Peduzie, MOD to J.O. Wright, No. 10 Downing Street, 3 September 1965.

34 TNA AB 49/17, Penney to Seaborg, Chairman, USAEC, 9 April 1965.

35 TNA AB 49/17, W.G. Penney to Minister, 22 July 1965.

36 TNA FO 371/181436, IAM 1013/19, D.C. Fakley, D Sc 3, MOD to M. Kendall, Foreign Office, 28 October 1965; and TNA AB 49/17, W.G. Penney to Dr Glenn Seaborg, USAEC, 30 September 1965.

negotiations about the detectability of underground tests.[37] Presumably he was alluding to the UKAEA position that new seismic techniques had improved detection capabilities – see Chapter 10. However, it was early days yet and all the seismic data at UK and other stations had not been processed and analysed yet. Healey, as events proved, was mistaken; Eskdalemuir and the covert US stations as well as the overt ones had all identified the event and no doubt located it to Nevada.[38]

Testing Plans 1965 and Beyond

In December 1964 the UKAEA once again proposed a full quota of three tests for 1965 and 1966, the first of which was proposed for July 1965. Its objectives were the development of smaller, lighter low kiloton yield devices and improvements of yield to weight ratios of lightweight megaton devices. The weapon development programmes at that point had changed considerably from that outlined in 1962. Three warheads were now in the pipeline for the RAF and Royal Navy:

1. Ministry of Aviation Reference PT 176: a nuclear warhead in the kiloton range for the general purpose bomb WE117A to meet requirement GDA 16/ASR1176; in-service date December 1966;
2. Ministry of Aviation Reference ZA 297: nuclear warhead for general purpose bomb WE117B to meet ASR 1195; in-service date June 1966; and,
3. Ministry of Aviation Reference ET 317: Polaris missile system to meet Staff Requirement GD 209; in-service date June 1968.[39]

---

[37] TNA PREM 13/123, Note for the Record, 16 September 1965.

[38] TNA PREM 13/123, Zuckerman to Secretary of State, 16 September 1965.

[39] TNA CAB 134/2241, Nuclear Requirements for Defence Committee, Draft British Programme of Underground Nuclear Tests 1965/66, Note by the UK Atomic Energy Authority, AWRE, 9 December 1964. The WE177 B was not delivered until September 1966, eight months behind the target date. The final delay was caused by warhead modifications, a safety requirement and the need to obtain the Prime Minister's approval for road movement. TNA AIR 6/169, Annex to AFB (66) 35, 25 October 1966. There were also delays in component supplies, which was the primary cause of the slip. TNA AIR 6/167, Air Force Board Memoranda, Annex to No. 14 1965, Quarterly Progress Report to the Air Force Board on New Aircraft and Weapon Systems, British Nuclear Weapons, 19 May 1965. However, the Air Force Board was informed in 13 January 1966 that the slip was due to failure of the thermal batteries for the warhead and fuzing system and of the RF filter to survive environmental tests. The weapon was now six months behind the target date. TNA AIR 6/169, Quarterly Progress Report to the Air Force Board on New Aircraft and Weapon Systems, British Nuclear Weapons, 13 January 1966.

The UKAEA, while very conscious of the possibility of heavy financial pressure in 1965 and 1966 did regard its own programme of tests as of considerable importance. Testing would maintain adequate levels of collaboration with the US. For this reason, the Authority searched for means to achieve reductions of expenditure in that year by means other than deferring the test programme.[40] A reasonable practical research programme was, it seems, essential to enable AWRE to operate as a genuinely viable unit; and that some work on underground tests, so long as it was politically permissible, was a most important part of such a practical programme. The MOD's scientific staff no doubt felt the same; the absence of testing could cause great damage to be done to continued UK-US collaboration. UKAEA officials still wished, at least at the end of 1964, to carry out the full quota of three research tests; the first of which was planned for July 1965 and the other two at roughly four-monthly intervals; namely, one in November 1965 and the remaining one in February 1966.[41] Objectives were more or less the same: the development of smaller, lighter, low kiloton devices and the improvement of the yield to weight ratio of lightweight megaton devices.

The Defence Secretary had minuted the Prime Minister in early January 1965 on the question of what do about the second stage UKAEA's three year underground test programme that would require three or four tests in 1965 and 1966.[42] This programme was not only to be conducted for its own purposes, but also in order to maintain UK scientific collaboration with the Americans in nuclear weapons development. There was no requirement for underground tests in connection with the approved warhead development programme. However, the proposals for research tests were a different matter. Healey noted that until defence expenditure had been reviewed, additional commitments had to be kept to a minimum and policy decisions unprejudiced. Just to complicate matters, the Americans wanted to know if the UK intended to continue with a worthwhile research contribution. If the decision were taken to abandon testing, the UK research programme would wither on the vine with a consequential decline in scientific capability. Testing was thus important because it would help maintain research and development teams at AWRE.

The NRDC decided in January 1965 to defer the earlier programme of underground research tests pending the outcome of the defence review initiated by the new Labour government. The NRDC recommended that HMG should keep its options open with regard to further experimental tests, but refrain from entering into any commitments at that point.[43] In so doing, it recognised that these

---

40 TNA AB 49/0015, W.G. Penney to Sir Henry Hardman, PUS, MOD, 10 June 1964.

41 TNA CAB 134/2241, Nuclear Requirements for Defence Committee, Draft British Programme of Underground Nuclear Tests 1965/66, Note by the UK Atomic Energy Authority, AWRE, 9 December 1964.

42 TNA PREM 13/123, Polaris Economy Test, 1965.

43 TNA CAB 134/2241, Nuclear Requirements for Defence Committee N.D. (65) 1st Meeting, 22 January 1965.

tests would be of considerable value to the UK R&D programme, but there were political considerations that favoured deferment of these tests and these were reinforced by the fact that the defence review was still in progress.[44] Further tests should be postponed, since it should be assumed that a decision would be taken to continue to develop and produce nuclear weapons, but that the government had not yet taken that decision.[45] However, the committee agreed that to maintain the UK's close collaboration with the US, it would be important to assure the US that the UK's continued interest on a research programme, including tests, and of British determination to collaborate effectively.[46] Given approval to repeat the economy test – the first had failed – a final Ministerial decision on the question of the research tests could be deferred until June or July, but continued uncertainty would only increase tension in collaboration with the Americans.[47]

The MOD estimated that the three underground research tests would cost £2.4 million, including the fissile material. The real charge on resources would be about £1.7 million. Zuckerman's instructions, however, for a forthcoming trip to the US to discuss UK future contributions were to make no commitments about likely UK intentions on testing. Wilson decided to refer the matter to the Defence and Overseas Policy Committee, which eventually decided against further tests for the foreseeable future in part for economic reasons.[48] By August the MOD told the UKAEA that among the firm assumptions that it was taking was that no provision for underground tests would be made for 1966 and 1967. MOD officials were working to a clear direction that defence expenditure for the next year was not to exceed the current level in real terms.[49] This appears to have derived from Healey's paper to the Defence Council on the Defence Estimates target for 1966–67.

By early 1965 AWRE's view was that it would now prefer to conduct safety experiments at Nevada so long as the UK conducted underground tests there. This course avoided the difficulties that the UK had experienced in being unable to inform the Australians fully about the nature and purpose of the safety experiments.[50] On the need for tests generally, the UKAEA's position by August 1965 had been made

---

44 TNA CAB 134/2241, Nuclear Requirements for Defence Committee, Item 1 Nuclear Policy and Undergrounds Tests 1965/66, N.D. (65), 1st Meeting, 22 January 1965.

45 TNA CAB 134/2241, Nuclear Requirements for Defence Committee, Item I Nuclear Policy and Undergrounds Tests 1965/66 Annex B to N.D. (65) 1 Nuclear Weapons Policy, Note by the Ministry of Defence, N.D. (65) 1st Meeting, 22 January 1965.

46 TNA PREM 13/1316, UK Nuclear Policy 1965–67. Healey to Zuckerman, 13 January 1967.

47 TNA CAB 134/2241, Nuclear Requirements for Defence Committee, Economy in Weapons Use of Fissile Material, Note by the Ministry of Defence, N.D. (64) 8, 22 June 1964.

48 TNA PREM 13/1316, Oliver Wright to A.P. Hockaday, MOD, 1 February 1965.

49 TNA AB 48/296, H.L. Lawrence-Wilson, AUS (R&D), MOD to A.E. Drake, UKAEA, 4 August 1965.

50 TNA DEFE 7/2364, Dr R. Press to Sir Solly Zuckerman, 16 February 1965.

clear to the NRDC. The Committee had subsequently recommended to Ministers that UK options on further tests in the US should be kept open. They had also recommended that the Polaris economy test should also proceed.[51] Ministers had not taken any decision on this by mid September 1965, the intention being to seek decision at the time of the coming defence review. However, this had been postponed and was re-scheduled for the end of September, but even this date was uncertain.[52] The scientific side of the MOD strongly supported the continuation of the research tests and intended to put its views to Ministers at the appropriate time. The scientists appreciated that possibly great damage would be done to collaboration if the UK decided to suspend tests. That said, in 1965 it had been UK policy to show that the UK's contribution to the West's strategic nuclear arsenal was both limited in size and dependent on the fulfilment of a cooperative programme with the US. Additionally, the UK wanted to take the initiative in steps towards general disarmament, including measures to prevent proliferation. In keeping with this policy the UK had deferred its programme of underground nuclear research testing.[53] However, this was only a partial reason; the main drivers were the defence cutbacks and the government's decision not to embark on a successor to Polaris or on any future tactical weapon.

Zuckerman made clear to the Americans at the annual 'stocktake' meeting in early October 1965 that the UK did indeed want to maintain its nuclear weapons technology at the highest level of competence through appropriate research programmes, conducting underground research tests where necessary. However, he also warned that no approval for any tests had been given, or was likely to be forthcoming pending the completion of the defence review.[54] It seems that Zuckerman may have actually told the Americans personally, with the full blessing

---

51 TNA AB 48/2965, Self-Initiated R&D programme AEA, A.E. Drake, UKAEA to H.L. Lawrence Wilson, MOD, 12 August 1965.

52 TNA AB 48/296, H.C. Hudspith to Mr Drake, Underground Tests, 13 September 1965.

53 TNA DEFE 5/164, COS 210/65 Chiefs of Staff Committee, Future Nuclear Defence Policy, Note by the Secretary, 10 December 1965; see also TNA DEFE 6/98, DP. 95/65 (Final), Chiefs of Staff Committee, Defence Planning Staff, United Kingdom Nuclear Policy Report by the Defence Planning Staff, 9 December 1965. TNA CAB 148/45, O.P.D. (O) (65) 81, United Kingdom Nuclear Policy, Note by the Ministry of Defence, 10 December 1965. See also TNA AB 48/296, A.E. Drake to Chairman, Defence Estimates target 1966/67, 1 July 1965 noted that 'the question of tests was left in abeyance last January and no move has been made since'.

54 TNA PREM 13/1316, Zuckerman to Prime Minister, U.S./U.K. Exchanges under the Nuclear Defence Agreement, 13 October 1965; Zuckerman had also been told in February 1965 by the USAEC that if the UK was not to be seen to be actively engaged in development and testing, the passage of nuclear information to the UK would slowly slide into the greyer area of dissemination. TNA PREM 13/652, Zuckerman, Chief Scientific Adviser to the Prime Minister, Pointers to Current US Moves in the Filed of Disarmament, 5 February 1965.

of the Prime Minister, that the UK had no interest in forward looking research into a post-Polaris nuclear deterrent.[55] Zuckerman was told by his US hosts that the trend of opinion in the JCAE was moving against further exchanges with the UK. The US technical community thus sought urgent UK assistance to help them show the JCAE that it could justify continued collaboration with the UK. Otherwise, they foresaw a fairly rapid withering away of the 1958 Agreement for Cooperation on the Uses of Atomic Energy for Mutual Defence Purposes. Planning in the UK, however, still appears to have been based on the assumption that the UK would continue to maintain a nuclear weapon capability and would continue R&D to maintain progress in warhead technology.[56] This was not enough for the US and the flow of information under the 1958 Agreement went into rapid decline.

Conclusions: The Resumption of UK Testing and Future Plans 1963–1965

The 1958 Agreement for Cooperation on the Uses of Atomic Energy for Mutual Defence Purposes had enabled AWRE to keep itself informed of new developments and policies. It had allowed Aldermaston to exert some influence on the design of weapons that might be adopted by British services, ensuring that they accorded as closely a possible with British ideas on reliability and safety. However, economic pressures resulting from the defence review by the 1964 Labour government and its decision that there would be no new strategic system after Polaris meant that further nuclear testing ceased for the time being. The UKAEA was not short of ideas for testing in this period and certainly saw a modest programme of testing as fundamental to continued access to US data. However, a resumption of testing would not be on the cards again until well into the later Polaris Improvement Programme in the early 1970s even though AWRE continued to seek permission for research tests throughout the remainder of the 1960s. The clear message from this review was that the UKAEA and AWRE would have wished to continue with a modest programme of nuclear tests. That these did not take place was down to economic and political considerations unrelated to any pressures for comprehensive nuclear test-ban treaty. In the years to come there were some promising developments in seismology that might suggest that one of the key obstacles to a CTBT could be overcome. We now turn to this issue and the role played by the government's Chief Scientific Adviser.

---

55 Personal correspondence.

56 TNA CAB 134/2241, Nuclear Requirements for Defence Committee, The Future of the Maralinga Range, Note by the Ministry of Aviation, N.D. (65) 8, 16 November 1965.

Chapter 10
UK CTBT Policy 1965–1973

Introduction

We saw in Chapter 6 how the solid foundations of UK seismic research were laid in the early 1960s. The 1964 Labour government of Harold Wilson was certainly committed to arms control, although we still need to see whether this commitment was of the same intensity and sustainability as we saw during the Macmillan years from 1958 onwards. To what extent did the Wilson governments of the period pressure the Americans to move on test-ban issues and what role did on-site inspection play in UK thinking? Was it still needed for an effective test-ban treaty? Had seismic research progressed enough to dispense with it all together? Whilst there was a *de facto* self-imposed national nuclear testing moratorium in this period, this was not caused by arms control or health imperatives as the 1958 to 1961 moratorium had been. We saw the rationale for this in Chapter 9, but we do need to see how UK test-ban treaty policy was pursued in these years as it provides the arms control backcloth to the issues that will be discussed in greater detail in the subsequent chapter. In this period the main activity on the weapons front was the deployment of the WE177B, decisions taken on procurement of WE177As for the Royal Navy and eventually the RAF and the deployment of Polaris as the UK's strategic deterrent in succession to the RAF's Blue Steel equipped V-Bombers.[1] Towards the end of the period, the UK decided to acquire a new version of the WE177, the Type C, but this could be done without the need for any underground tests, as it was a modification of an existing design.[2] A smaller secondary component was used to produce a lower yield weapon of about 200 kilotons, the primary being the same as for the WE177 As and Bs. Test-ban pressures, such as they were, had no impact on these developments whatsoever.

---

[1] The first WE177Bs entered service in 1966; orders for 66 WE177As for the navy were also placed in 1966 – see TNA AIR 2/18209, Background Note, Annex B to ACAS (Pol)/A890 dated 7 January 1969. The planned rate of delivery was to begin with two per month from January 1969. The order was reduced to 44 – see TNA AIR 2/18209, Background Note, Annex B to ACAS (Pol)/A890 dated 7 January 1969.

[2] TNA AIR 2/18210, J.G. Matthews DD Ops (S) (RAF) to DD Air Plans, RAF Germany Buccaneer Strike Capability, 15 October 1970; and TNA AIR 2/18210, J.W. Allan DOR3 (RAF) to DS9, 27 August 1970. Consideration was given to ordering WE177B weapons with a modified SIMON warhead (ASR 1195).

UK Seismic Research and CTBT Work in Geneva: 1965–1973

Western proposals for a CTBT formally rested with the Joint UK and US Memorandum tabled in the ENDC on 1 April 1963. This stated that, subject to certain conditions relating to the conduct of inspections, an annual quota of seven on-site inspections could be accepted on the territory of each nuclear weapons state in order to provide some deterrent against clandestine cheating. In September 1965, the UK tabled a working paper that outlined a scheme for monitoring underground explosions by means of a network of between 20 and 25 seismographic arrays.[3] The UK had been able following the UKAEA's own research programme to make detailed specifications of the array design and to site and engineering requirements, for a given detection and discrimination threshold.

Three years later the UK outlined a more political proposal in the shape of a novel quota initiative to phase out testing gradually; these ideas were outlined in a Working Paper for the ENDC presented on 20 August 1968.[4] This notion appears to have originated in some personal proposals from Frank Panton, MOD to the Foreign Office in early November 1967.[5] Panton's idea was that States Parties would agree progressively to reduce the yearly number of tests to an agreed minimum by 1973, possibly between three and six. However, the important thing for MOD at that time was that the UK should not propose anything that would create any antagonism in every US agency concerned with nuclear defence issues. MOD contacts with the US in the nuclear field at that time were such that the UK could not afford to incur any extra strain on UK-US relations, unless where the gains would be real, substantial and obvious.[6] After the successful conclusion of the Nuclear Non-Proliferation Treaty (NPT) in the summer of 1968, Ministers were anxious to pursue new UK disarmament projects, and a CTBT was the principal initiative identified by officials, along with something on biological weapons disarmament and PNEs. Ministers endorsed this.[7] The 'something on biological disarmament' was to lead to the negotiations that led to the Biological and Toxin Weapons Convention in 1972. Nothing came of the PNE ideas.

Fred Mulley, Minister of State, came to Geneva to outline a new approach to the thorny problem of on-site inspection; there might be a special committee to consider complaints of infringements and to assess the evidence produced in support of each complaint. Such a committee might be composed of three nuclear

---

3 United Kingdom, *Notes on UK research on techniques for the distinguishing between earthquakes and underground explosions*, ENDC/155, 9 September 1965.

4 United Kingdom, ENDC/232.

5 TNA FCO 10/136, Dr F.H. Panton, MOD to R. Hope-Jones, Foreign Office, 5 November 1967; R.C. Hope-Jones to Lord Hood, CTBT, 31 July 1968.

6 TNA FCO 10/136, Dr F.H. Panton, MOD to A. Stuart, Foreign Office, 29 November 1967.

7 TNA CAB 148/35, Item 2 Further Measures of Disarmament, OPD (68) 13th meeting, 12 July 1968.

weapons states party to the treaty, three non-aligned states and a nominee of the UN Secretary General or IAEA Director General. There would be a right of on-site inspection if the Committee decided by a majority of five to two that a *prima facie* case had been made. The test ban might be made a phased operation by having an agreed annual quotas of tests on a descending scale over a period of four or five years after which tests would be banned absolutely. The quotas might not be written into the treaty but would be fixed annually, possibly by a committee of the kind already suggested. Any CTBT would also need to be flexible enough to permit PNEs. Given the major loophole posed by PNEs, the diminishing quota should also extend to such explosions, but in the absence of any satisfactory progress in other disarmament fields, the quota should reduce to a minimum number only, not to zero.[8] In addition, the UK concept could also include experiments with verification methods during the tests conducted in the phasing out period.

It seems that this paper was tabled without full consideration by MOD. In the time available for consultations, MOD officials felt compelled specifically to reserve their position on the phrase 'strong evidence', which would be used to justify a request for an on-site inspection.[9] On the explanations given by the Foreign Office of the meaning of the UK proposal, and particularly of these words, the MOD was not satisfied that it made the best sense or was supported by the technical facts. The Russians did not like this proposal because it included such an express right of on-site inspection, something they continued to oppose.[10] It therefore went no further.

It might appear that the phasing out of tests via a quota system may have been linked to UK needs to keep open the option of any further underground nuclear tests that might conceivably be required if any Polaris improvement programme began. As we will see later in Chapter 11 work on what eventually became Chevaline started in 1967. There is no evidence to suggest that this particular idea was in any way connected to Polaris modernisation. However, as we will see any UK disarmament initiative, especially on CTBT, would not be pursued if it prevented the maintenance of an effective deterrent. MOD's view at the end of 1968 was that there was in fact little prospect of a test ban until such times as there was progress in strategic arms limitation between the superpowers. From 1968 until 1971, the main focus of multilateral action in Geneva was on a possible separate ban on biological weapons; this and the Seabed Treaty took up most to the available time in the Conference of the Committee on Disarmament (CCD), which had succeed the ENDC. Neither the US nor the USSR was prepared to change their positions on site inspections, which was of course the main reason why there was no concrete

---

8 TNA FCO 6/84, A paper on CTBT under cover of letter from Dr F.H. Panton, MOD to Mr R.C. Hope-Jones, Foreign Office, 24 October 1968.

9 TNA FCO 6/84, A paper on CTBT under cover of letter from Dr F.H. Panton, MOD to Mr R.C. Hope-Jones, Foreign Office, 24 October 1968.

10 Cmnd. 4141, *Further Documents on Disarmament: The Disarmament Negotiations July–December, 1968* (London, 1969), p. 7.

progress towards a CTBT in these years. However, technical work at the UKAEA continued on improving seismic verification capabilities. In fact this was the main thrust of UK policy at this time. Given US and Soviet reluctance to pursue a test ban in view of the uncertainties over whether there would be deployments, control or prohibition of ABMs in the late 1960s, the best hope was to carry on with UK technical research. Meanwhile, the UK could continue to impress upon both superpowers the need to make progress and the belief that the best way to achieve this was through technical talks.[11]

The UK tabled another technical working paper in Geneva in July 1970 aimed at determining what detection and identification capability could be achieved given the then state of the art in seismology and evasion. This outlined the capabilities of an extended array costed at about £15 million and which would be installed over a five year period.[12] The paper's sole purpose was to illustrate how recent SIPRI seismic experts' predictions of seismic detection and identification capabilities might in reality be achieved and at what cost and timescale.[13] Crucially the paper concluded that there was little point in defining a more elaborate and costly system because discrimination criteria for low magnitude events (4 mb and 4.5 mb) were not then proven. Moreover, in the larger countries the external portion of a seismic network might not detect signals from explosions of about ten kilotons and less fired in dry alluvium. The UK continued to work on more detailed studies on the siting, communications and system problems of a future global seismic array network.

Zuckerman, meanwhile, believed that a new seismic technique known as surface wave could improve underground event identification criteria and thus make on-site inspection unnecessary. Seismic waves from a shallow earthquake were broadly of similar magnitude, whether they travelled on the surface of the earth or through its centre. Whether the magnitude of a shallow earthquake was measured from the waves that travelled on the surface of the earth or through it, the magnitude of the earthquake would be the same. By contrast, the magnitude of an underground explosion as measured from surface waves was measurably smaller than that measured from the waves travelling through the earth. This difference permitted discrimination between the two kinds of seismic event. AWRE scientists had known of this capability, but had only recently developed the necessary instrumentation to observe the events in practice. However, by 1971 UK research was clearly pointing in the opposite direction. In June, Hal Thirlaway made clear to the CCD that discrimination criteria for low magnitude events were still not proven. Seismic signals from explosions of about ten kilotons and less fired in

11 TNA FCO 10/136, PMV (M) 3, 11 January 1968 PM's Visit to Moscow, January 1968, Disarmament (c) A CTBT Brief by the Foreign Office, 11 January 1968.

12 Cmnd. 4725, *Further Documents on Disarmament: The Disarmament Negotiations 1970* (London, 1971), p. 14; CCD/296 Working Paper by the United Kingdom on Verification of a Comprehensive Test-ban Treaty, 28 July 1970.

13 TNA DEFE 19/169, F.H. Panton, ACSA (N) to DCA (PN), Verification of the CTB, 27 July 1970.

dry alluvium might not be identified outside the country in which the explosion took place even by an improved array network, such as that posited by the UK.[14] Such conclusions cast grave doubts on the adequacy of purely national means of verification, the option still insisted upon by the USSR. For the UK this meant that there was still important scientific research and study required in the area of low-magnitude events in order to see to what extent it would be possible to sharpen further the present remote seismic detection and identification capabilities.

For the next few years the UKAEA continued to work away at the detection and identification problem.[15] Two more technical working papers for the CCD soon followed and were tabled in Geneva in 1972. The first of these concerned the estimation of yields of underground explosions from amplitudes of seismic signals, whilst the second described how new data processing equipment, a Seismic Array Station processor for use by individual seismic stations, could be used in monitoring underground explosions. This had been one of the principal efforts of the UK seismic research programme for some time. Such equipment could put states participating in the network proposed by the UK in a better position to assess seismographic evidence for themselves and could result in appreciable reductions in cost.[16] Such a system would not overload the data transmission capabilities which would mostly be taken up with background noise and earthquake signals.

A series of informal meetings attended by Western and non-aligned seismic experts was held in Geneva in July 1973 to examine the technical problems of CTBT verification. In preparation for these meetings, the UK submitted two working papers: one consisting of a review of the UK seismological R&D programme whilst the other was a study of the method of estimation of depth of seismic events.[17] Therefore, it is quite clear by the end of our period that the UKAEA was still investing considerable time and effort in overcoming the problems of seismic verification for a CTBT. The basic problem that had bedevilled the Conference on the Discontinuance of Nuclear Weapon Tests in the 1950s and early 1960s remained: relatively low-yield tests could not be identified with certainty.

---

14 CCD/327, 23 September 1971, Working Paper by the United Kingdom containing comments on the Canadian Study of the Seismological detection and identification of underground nuclear explosions (CCD/327) and on its implications for the expanded seismic system outlined in the United Kingdom Working Paper CCD/296.

15 Henry Hainworth, UK Ambassador to the CCD, CCD/PV. 541, 23 September 1971.

16 Cmnd. 5344, *Further Documents on Disarmament: The Disarmament Negotiations 1972* (London, 1973), p. 19; Working Paper by the United Kingdom on estimating yields of underground explosions from the amplitudes of seismic signals 25 April CCD 363/Rev. 1; and Working Paper by the United Kingdom CCD/386 22 August 1972, Working Paper on seismic data handling and analysis for a comprehensive test ban.

17 CCD/401, Working Paper by the United Kingdom on a review of Seismological Research and Development Programme, 28 June 1973; CCD/402 28 June, Working Paper on the Estimation of Depth of Seismic Events.

In the absence of any immediate treaty negotiations and the obdurate attitude of the USSR to on-site inspections and lack of US interest, there was really no alternative but to beaver away at the technical problems so that that when time became more propitious, solutions might be available. However, in view of the UK's own new testing requirements in connection with the Polaris Improvement Programme agreed in May 1972 as discussed in the following chapter, the UK was not going to be rushing out with immediate proposals for a test ban. We had thus come full circle. UK weapons programmatic requirements, as in the late 1950s, had to come first and only thereafter could a test ban become a top priority. In any case, the UK could only move if the USSR and US were also willing to work towards a test-ban treaty, and at this time they were not. Wilson, unlike Macmillan was not prepared to make a nuisance of himself in Washington over a test ban.

Pursuit of a Test Ban 1965–1973: Ministerial Engagement

Zuckerman told Michael Stewart, the Foreign Secretary and Lord Chalfont, Minister of State for Disarmament, in November 1965 that in his opinion there was no scientific reason for not concluding a comprehensive test-ban treaty without provision for on-site inspection. Chalfont himself was of a similar view even as early as January 1965.[18] However, Zuckerman amplified his statement by stating that present scientific methods could detect without inspection all underground events of military significance.[19] He had no fresh technical evidence to support this contention; he was simply expressing the view that there was only a small military risk that was well worth taking for political benefits.[20] This was eventually to lead to a change in UK policy, but MOD advice at that stage was that there was in fact no change in the basic technical position, namely that there would still be a significant number of unidentified events that, short of on-site inspection, could not be conclusively identified as either an earthquake or explosion. There would always be an element of risk in a test ban, but it was clear to Foreign Office officials in late 1965 that a CTBT would play an important part in preventing nuclear proliferation. In such circumstances, it might be worth accepting the risk that the Russians might cheat. The Americans were not too impressed with this argument. Congress was still a major obstacle, as it would insist on on-site inspections. It had also extracted the concession during the PTBT ratification hearings that the US would remain ready to resume atmospheric testing immediately if the Russians breached the treaty.

18 TNA FO 371/181368, IAD 1027/5, Chalfont to Secretary of State, 1 January 1964.

19 TNA FO 371/181453, R. McAndrew to Sir B. Burrows, 16 December 1965.

20 TNA FO 371/187543, J.E.D. Street to Lord Hood, 13 January 1966.

UK-US Ministerial Exchanges: Summer 1966

President Johnson had made it clear in the spring and summer of 1966, both in public and in private that he was interested in any possibilities of progress on disarmament. The UK objective was to show the urgent need to advance in the related fields of nuclear non-proliferation, nuclear sharing, nuclear guarantees and the test ban.[21] A fissile material cut-off treaty however was another matter; shortages of plutonium and HEU made the UK a trifle leery about any immediate treaty.[22] However, Johnson's room for manoeuvre on a CTBT was much less than it was on non-proliferation. Differences of principle between the US and USSR over on-site inspection was still an issue. Congress had insisted in 1963 that underground testing should continue and London knew that there was a considerable weight of US official opinion behind the assertion that the option to test underground had to be maintained in order to leave open the possibility of further development of ABMs. However, Dean Rusk (US Secretary of State) had told Michael Stewart in June 1966 that US hesitation about a CTBT was very firmly down to fear of Russian cheating, not the need to maintain a programme of tests for US defence purposes.[23] It was not possible for the US to sign a CTBT that left room for cheating. Congress would never ratify such a treaty.

Michael Stewart responded that his advice had been that cheating would be most unlikely because the Russians would realise that it would be hard to escape detection of explosions of a size of military significance. He recognised that some very small tests might take place without detection, but his advice was that the military significance of such tests would be so small that the political risk involved might be accepted.[24] Rusk disputed that small yield tests below 4.75 mb were of no military significance. Tests in this range were necessary for the development of tactical nuclear weapons, and it was precisely in this field that the US lead over the Russians was very marked.[25] In the early 1960s the UK view had been that the five to 20 kiloton yield range was the important one for weapon development, there was no need to test megaton weapons at their full yield to have confidence in their performance, unless it was for a radical new design.[26] Sweden had proposed a compromise on verification in the ENDC, which entailed a graduated process

[21] TNA CAB 168/15, Prime Minister's Visit to Washington, July 1966, Disarmament Brief by the Foreign Office, PMV (W) (66) 3, 22 July 1966.

[22] CAB 134/2241 Nuclear Requirements for Defence Committee N.D. (66) 1st Meeting 22 April 1966 Item 3 Cessation of Fissile Material Production for Weapons Purposes.

[23] TNA CAB 168/15, Prime Minister's Visit to Moscow, July 1966, Disarmament Brief by the Foreign Office, PMV (M) (2) (66) 4, 22 July 1966.

[24] TNA PREM 13/2440, Foreign Secretary and Mr Rusk, Canberra, 28 June 1966.

[25] TNA PREM 13/2440, Canberra telegram no. 967 to Commonwealth Relations Office, 29 June 1966.

[26] TNA DEFE 19/1, R.W. Snelling, UK Nuclear Test Delegation, Geneva to Dr R. Press, MOD, 11 May 1961. paragraph 1 and manuscript comment; The Scientific Aspects

of challenge that could lead to an on-site inspection. The US was contemplating linking this to a temporary suspension of tests. London's view was that if this were undertaken on a genuinely experimental basis, it would strongly support it as promoting a possible compromise on a CTBT. In general, the UK line at Geneva was that all ideas on moving the debate forward deserved further study and that London intended to examine them sympathetically.[27]

The Americans were also considering another 'middle ground' option; a threshold treaty without inspection banning only those tests giving a seismic signal above 4.75 mb. Although there were acknowledged problems in verifying such a threshold ban, there being no direct correlation between the magnitude of the signal and size of an explosion, the Foreign Office at least felt that if the US were to adopt this approach, the UK should give it its full support as being a positive step toward a CTBT. Zuckerman was not convinced however.[28] He told Wilson that messing around with thresholds would generate a good deal of suspicion in future, without advancing the cause of the CTBT. Another option for consideration was a trial system of inspection by invitation. Under this scheme, if a country thought that an event on the territory of another was suspicious, it would be open to the other to invite the former to come and see for itself that its suspicions were unfounded; a failure to invite could be taken as *prima facie* evidence that there had been a violation. This idea did not go very far.

UK-Soviet Exchanges

Lord Chalfont suggested that the Prime Minister sound out the Soviet leadership's attitude to this idea during his visit to Moscow in July 1966 and if they were open minded, then the UK could also pursue it with the US. It could also be combined with a scheme to press forward with the development of detection techniques.[29] The US, however, was not keen that Wilson raise CTBT issues in Moscow at that time, and the US Ambassador wrote to No. 10 to that effect on instructions.[30] Despite this, the Prime Minister was not prepared to ignore this issue. He certainly thought that the Russians would want to raise the problem, and if they did, he would need to respond and see whether there was any sense of substantive change in the Soviet position. He could reassure Washington that he would not be entering

of Nuclear Weapon Detection, AWRE Note, 1 May 1961 Roger Makins to Sir Hugh Stephenson, Foreign Office.

27 TNA PREM 13/2440, The Disarmament Negotiations, Foreign Office INTEL No. 40, 21 September 1966.

28 TNA CAB 168/15, Zuckerman to Prime Minister, 15 July 1966.

29 TNA PREM 13/2440, Chalfont to Harold Wilson, 9 July 1966.

30 TNA PREM 13/2440, David Bruce to A.M. Palliser PS/PM, 15 July 1966.

into any negotiations or changing current positions.[31] So, like Macmillan before him, Wilson was still prepared to push the boat out – although perhaps not quite as much – on the test ban even though the US might prefer that it remain firmly tethered to the quay.

Zuckerman Continues the Charge: Dropping the On-site Inspection Requirement

The Foreign Secretary met Zuckerman on 26 November 1965 during which the Chief Scientific Adviser told him that all military significant tests could be detected and that the UK should take the risk of signing a CTBT without provision for on-site inspection. AWRE however had recently reported that it would be possible to identify by seismic means alone all but five to ten per cent of earthquakes, but that the residue of unidentified incidents might be earthquakes, nuclear or even chemical explosions. London delayed the consideration of this issue following reports that the Americans were considering the possibility of putting forward the aforementioned proposal for a threshold treaty banning tests producing signals above a certain magnitude without inspection.[32] Following a meeting on 13 May 1966 with his officials George Brown, the new Foreign Secretary, agreed that a paper should be prepared for Ministers to authorise a change in UK policy to support a CTBT without any provisions for on-site inspection.[33] And if the Cabinet agreed, then the UK should try to persuade the US to adopt the same view.

Deadlock in Geneva appears to have been one of the catalysts for change. Swedish ideas that linked rejection of a challenge inspection to withdrawal from a CTBT were also influential.[34] The paper itself was based on a letter that Zuckerman wrote to Sir Bernard Burrows, Permanent Under Secretary of State at the Foreign Office, on 13 January setting out some arguments for a CTBT without on-site inspection.[35] The original intention was to put this paper to the Cabinet's Overseas Policy and Defence Committee in September 1966. Zuckerman welcomed the draft Foreign Office memorandum. It is instructive to note that Zuckerman made clear that he disagreed with the implication in parts of the draft paper that there was something achievable from the military point of view in continued underground

---

31 TNA PREM 13/2440, Prime Minister's Personal Telegram Serial No. T.246/66, Foreign Office telegram no. 6947 to Washington, 15 July 1966.

32 TNA FO 371/187544, J.E.D. Street to Sir B. Burrows, A Comprehensive Test-ban Treaty Without On-site Inspection, 9 August 1966.

33 TNA FO 371/187543, J.E.D. Street to Sir B. Burrows, A Comprehensive Test-ban Treaty, 19 May 1966.

34 TNA FO 371/187543, Prospects for a Ban on Underground Nuclear Tests, Despatch No. 1, UKDis Geneva, Harold Beeley, 7 January 1966.

35 TNA FO 371/187543, J.E.D. Street to G . Owen, PS/Sir Solly Zuckerman, Cabinet Office, 9 August 1966.

nuclear testing. He felt that there were always likely to be people who wanted to carry on developing weapons just for the sake of it. Although the Foreign Office draft paper suggested that dispersal of weapon design teams would be an issue, Zuckerman did not think it would be too difficult to reassemble them.[36] A view contested by MOD, but more of that in Chapter 11.

A revised draft that took account of most of the views from Zuckerman and Dr Press (ACSA) N, MOD and an accompanying submission to the Foreign Secretary was prepared on 8 September.[37] In considering the recommendation to support a CTBT without any provision for inspections, Brown was asked to take into account the fact that Ministers were about to be asked whether to go ahead with further R&D of nuclear weapons. There was not necessarily a contradiction between pursuing a CTBT on the one hand and, on the other, keeping UK nuclear weapons up-to-date as long as a treaty was unobtainable. However, development of UK nuclear weapons would require underground testing; and once a CTBT was signed, no further development of UK weapons would be possible beyond the Polaris programme and the WE177s that were already planned. The Foreign Office recommended that Ministers might want to consider the CTBT paper at the same time as the one about future nuclear policy.[38] (This latter paper is discussed in detail in the next chapter.) The Foreign Office saw several compelling arguments in favour of a CTBT. By preventing testing, a test ban would place a strict limit on the degree to which further warhead developments would be possible. It might well prevent both the US and USSR from deploying ABMs. And this of course was the main reason why the US no longer supported a CTBT actively in the mid 1960s, as new strategic weapons were needed to counter the ABM threat.[39]

Rusk told George Brown at the end of July 1968 that the US had a real problem with a CTBT at that time; the Americans needed to carry out further nuclear tests for Minuteman III, Poseidon and ABMs in order to negotiate effectively with the Russians about offensive and defensive missiles.[40] This was a change in the previous US position where a fear of cheating was the driver. George Bunn, then legal adviser to the US Geneva delegation, told the Foreign Office that the US had recently completed a survey of previous Russian tests and had decided that a number of large explosions were directly related to ABM development. The US had decided therefore that there was some risk of being left behind in the ABM race and that it was possible that a series of large test explosions would be needed

36 TNA CAB 168/15, Zuckerman to J.E.D. Street, 25 August 1966.

37 TNA CAB 168/15, J.E.D. Street to Lord Hood, A Comprehensive Test-ban Treaty Without On-Site Inspection, 8 September 1966.

38 TNA FO 371/187544, J.E.D. Street to Lord Hood, A Comprehensive Test-ban Treaty Without On-site Inspection, 8 September 1966.

39 Seaborg, *Stemming the Tide*, pp. 235–7.

40 TNA CAB 168/16, Washington telegram no. 2282 to Foreign Office, 25 July 1968.

in order to catch up.[41] This put paid to tentative US thinking that was moving in favour of a threshold treaty, at least for the time being as such a Treaty was eventually signed between the US and USSR in 1974.

Test Ban On-Site Inspection Paper and Tripartite Seismic Experts' Talks: Continuing Problems

There was obviously some problem over this paper. Brown held a meeting on 15 November 1966 at which it was decided to defer circulation of the memorandum as proposed. This bothered Zuckerman who was anxious that the official paper on Future Nuclear Policy should be considered at the same time.[42] Failure to do so could lead to less than coherent approach to all nuclear matters. Moreover, Zuckerman firmly believed that HMG could no longer publicly maintain that the detection and identification of underground nuclear explosions was not possible to at least 4.5 mb, which made it difficult to support any contention that a CTBT was necessarily dependent on on-site inspection. It is not clear from Cabinet Office papers precisely what happened, but on 1 November 1966, the Foreign Office sent a revised version to Dr Bob Press in the MOD. This was done in light 'of recent developments, and particularly of Sir Solly Zuckerman's discussions in Moscow and Washington'.[43]

Zuckerman had received an invitation to Moscow to discuss the technical problems of an underground nuclear test ban by the President of the Academy of Sciences. This emerged during a meeting of an 'International Assembly on Nuclear Weapons' held in Canada at the end of June 1966. Zuckerman and Chalfont had come away from this with the sense that the Russians might be willing to move on inspection issues, or in this case 'inspection by invitation'.[44] He wanted to impress upon the Russians the UK belief that a new scientific situation now existed, as result of recent advances in seismological techniques (surface wave) and that it had to be in the interests of all concerned to have them thoroughly discussed.[45] The Russians appear to have accepted this and agreed to a complete exchange of seismic information on the basis of existing and new seismological techniques.

Zuckerman undertook to organise tripartite talks in London to take the matter forward and he talked up the prospects of progress to the Prime Minister. In his

41 TNA FO 371/187543, A.C. Stuart to Mr Street, Threshold Treaty, 28 February 1966 and J.G. Tahourdin, UKDis Geneva to J.E.D. Street, Atomic Energy and Disarmament Department, 24 February 1966.

42 TNA CAB 168/15, Zuckerman to Secretary of State for Foreign Affairs, 16 September 1966.

43 TNA CAB 168/15, J.E.D. Street to Dr R. Press, MOD, 1 November 1966.

44 TNA CAB 164/631, Zuckerman to Prime Minister, 26 July 1966.

45 TNA CAB 168/174, Zuckerman's Visit to Moscow Tuesday, 20th September–Sunday, 25 September 1966.

view, the Russian changes in attitude to technical talks were likely to give Wilson considerable leverage in the political sphere.[46] This was a significant development as there had been no UK/US/USSR technical discussions since the Conference on the Discontinuance of Nuclear Weapons Tests in 1960. As George Brown was scheduled to go to Washington to discuss test-ban issues with Rusk, he suggested to the Prime Minister that the right way to tackle the proposal for technical talks would be for him to raise the matter with Dean Rusk first; approaches to US scientists could of course follow on afterwards.

Brown's key objective was to find out from Rusk whether the fear of cheating was the real reason for the US attitude, or whether it was a cover for a wish to carry on testing. If the later were true, then US sensitivities over the technical talks would be greater.[47] Officials briefed Brown to stress to Rusk that the UK believed that a CTBT was still urgently needed. It was of course important to weigh the needs of Western security and the risks of cheating against the potential gains, but the Foreign Office felt that the slow progress on the NPT and the dangers of a new round in the arms race if underground testing continued, weighed heavily on the side of seeking an early agreement.[48] Rusk agreed that the Russian change in position was significant, though surprising. He thus agreed that tripartite technical talks could take place and that US experts could make arrangements to participate.[49] However, he worried about the terms of reference for any US delegation attending such meeting as well as its purpose. Further inter-agency consultations were necessary and until these issues were resolved the US asked the British not to issue any invitation to the Russians.[50] One of the US concerns here was that work on surface waves was still classified and the data had been gathered from stations in third countries that were not aware of their true purpose. Indeed having been briefed by ACDA, Rusk felt that he had to clear the experts' meeting with the President.[51] Johnson eventually agreed in early November so exchanges could continue on the content of the draft meeting agenda.[52]

So all was going well, or so it seemed. Zuckerman and Brown were in for a shock when the Foreign Secretary spoke to Soviet Foreign Minister Andrei Gromyko in Moscow on 23 November 1966. It became painfully clear that the Soviet authorities were in fact not in favour of the tripartite meeting proceeding; Gromyko, speaking in English in response to pressure from Brown to agree the

---

46 TNA CAB 164/631, Zuckerman to Prime Minister, 26 September 1966.

47 TNA CAB 168/174, George Brown to Prime Minister, A Comprehensive Test-ban Treaty, Sir S. Zuckerman's Visit to Moscow, 27 September 1966

48 TNA CAB 168/174, Secretary of State's Visit to the USA, October 1966.

49 TNA CAB 168/174, Washington telegram no. 2822 to Foreign Office, 15 October 1966.

50 TNA CAB 168/174, Meeting in Dr Hornig's Office, 14 October 1966.

51 TNA CAB 168/174, R. Press, ACSA (N) to Sir Solly Zuckerman, Cabinet Office, 21 October 1966.

52 TNA CAB 168/174, Zuckerman to Lord Chalfont, 7 November 1966.

meeting even if it were unofficial, said that 'we definitely disfavour it'.[53] Gromyko remained impervious to all the counter arguments that the talks would help them reach the political decision, which he alleged was all that was now necessary in order to agree a CTBT. The Foreign Secretary concluded that this position probably meant that the Soviet military wanted to continue underground tests and had carried the day against the seismic scientists.[54] The Russians later claimed to Chalfont that once they had seen the draft agenda it suggested to them that the Americans were interested only in reopening an endless scientific argument and that they were not disposed to make early political decisions about concluding a test-ban treaty. The Soviet government had therefore decided that the talks would serve no useful purpose.[55]

Meanwhile Zuckerman became animated at the latest news from Washington when it was reported that the US had used one of the Ploughshare tests to see whether a nuclear explosion could indeed be effectively muffled in large cavity and thereby fox any detection system. The test was very small – about 350 tons of TNT equivalent. In his view, decoupling would in practice be almost impossible with any militarily significant nuclear yield. What had not been demonstrated was the practicality and cost of making cavities big and deep enough to achieve decoupling to mask a 100 kiloton test. He felt that the US had made the announcement to stir up anti-CTBT opinion. The laboratories, the USAEC and the military would always want to test in Zuckerman's view. A CTBT had became, however, even more urgent as the US and USSR moved closer to a more intensive phase in the arms race.[56] The Foreign Secretary agreed and told Wilson that that the UK would have to keep a wary eye on further developments and on US intentions as well as the implications for the test-ban negotiations as a whole.[57] So, the old arguments from the late 1950s were still there; de-coupling and the 'large hole theory' were very much alive and that some in the US were determined to keep them that way.[58]

---

53 TNA CAB 168/174, Visit of the Foreign Office Secretary to the USSR 22–25 November, 1966 Document No. 5 paragraph 6 p. 12; see Zuckerman, *Monkeys, Men and Missiles*, pp. 332–9 for Zuckerman's account of this episode.

54 TNA CAB 168/174, Moscow telegram no. 2147 to Foreign Office, 24 November 1966.

55 TNA CAB 168/174, P.W.J. Buxton to Atomic Energy and Disarmament Department, 21 December 1966.

56 TNA PREM 13/2440, Zuckerman to Prime Minister, 9 January 1967 and 7 February 1967.

57 TNA PREM 13/2440, George Brown to Prime Minister, Decoupling Underground Nuclear Explosions, 31 January 1967.

58 Bates, Gaskell and Rice, *Geophysics in the Affairs of Man*, p. 203. Even in 1962 with the UK effort and Vela Uniform programme making progress, the anti-test-ban forces were anxious that the Latter 'big hole' theory should be tested using a nuclear explosion. AWRE had set up an experimental array at Laramie Wyoming in order to make continuous records of the background seismic activity, specifically during earthquakes in the US.

The Test-Ban Policy Paper for the Overseas and Defence Policy Committee

The re-drafted paper on on-site inspections tried to incorporate the development of the surface wave technique as an integral part of the argument for dispensing with inspections. Recommendations to Ministers for the deletion of on-site inspections from test-ban treaty verification rested firmly on the technical arguments. Therefore the Foreign Office draft memorandum argued that if surface wave analysis could be shown capable of doing the job instead of a quota of inspections, then agreement on a CTBT without inspection ought to be in sight.

Support for a CTBT would mean the end of any further development of UK weapons, but the Foreign Secretary was prepared to argue that this consideration could not be allowed to affect UK policy if the possibility of agreement on a treaty became a real one. This paper was discussed at a Committee meeting chaired by Brown on 7 November 1966. A version of the paper was eventually finalised and circulated on 12 December to the Ministerial Committee on Nuclear Policy.[59] The Foreign Secretary's own view was that the new surface wave techniques offered a detection and identification capability that would serve as an adequate deterrent against cheating in a CTBT without on-site inspection. Soviet attitudes suggested that they were using US insistence on inspection as a way of avoiding a CTBT altogether. However, Brown did not think it would be wise to push the US to drop its own on-site inspection requirement. In light of the Soviet change of mind on Zuckerman's proposed technical talks, they would no doubt react strongly against such an attempt. Instead, the better course would be to press the US to intensify joint collaboration on the new techniques to the point at which it would be possible to identify most underground tests down to 4 mb. In these circumstances, therefore, it might then be possible to convince Congress that there was no longer any need for on-site inspection and that the Western powers could announce that they would support a CTBT without inspection provisions. If the Russians refused, that would put them in a bad light.

The Paper goes to Ministers

Brown's paper invited his colleagues to agree three points; first, that the UK should be prepared, when the opportunity offered, to accept and to propose a CTBT without on-site inspections; second, that the UK should say nothing to anyone for the time being to indicate this change in policy; and third, the UK should instead intensify its seismic research and urge the US to do the same, with the objective of identifying

AWRE intended to use the array to record the signals from Project Gnome, a five thousand ton nuclear explosion in a salt dome in New Mexico. TNA AB 16/3447, Lt Col L.G. Lohan, MOD to R.W.B. Truscott, UKAEA, London, 29 November 1961.

[59] TNA CAB 134/3120, A Comprehensive Test-ban Treaty, Memorandum by the Secretary of State for Foreign Affairs, PN (66) 5, 12 December 1966.

nuclear tests down to 4 mb. This represented a radical departure from traditional UK policy from the Macmillan era. The UK was changing only because it was now felt that adequate verification could be achieved by seismic means only.

Ministers agreed that the UK should be prepared to accept a CTBT without inspection.[60] The reason for this decision was that it was not only possible to detect, locate and diagnose underground disturbances of 4.5 mb and above, but also to detect any underground explosion of a device with a yield of one to ten kilotons, depending on geological environment. Below 4.0 mb it was not worth worrying about for the simple reason that explosions that would register so low were insignificant in a strategic military sense. A second argument that weighed with Ministers was that an isolated test was of little consequence; a third factor was that seismologists were improving their detection and identification techniques all the time, so that the gap between detecting and identifying events at 4.5 mb and 4 mb was bound to be narrowed.[61]

Ministerial Deliberations: Inspections and Treaties 1968–1969

Although the Cabinet Committee adopted the paper – nothing was done to execute its recommendations, which leaves one wondering why this decision was taken at all. There were no detailed discussions on a CTBT between the UK and US between the summer of 1968 and February 1969. In fact the MOD now faced serious difficulties in considering measures such as the CTBT as the UK did not yet have a comprehensive nuclear policy as decisions over Polaris modernisation were still unclear.[62] For the MOD it was important to be able to maintain the effectiveness of the Western deterrent in which the West's strength lay. Limitations on nuclear weapons should be linked with limitations in the conventional field.[63] It was clear then that little positive progress would be made because of the change in the US Administration (Nixon had just been elected) and because of the US attitude to the cessation of tests was very dependent on the strategic arms limitation talks that had yet to begin.[64] However, irrespective of this, the UK recognised in early 1969 that realistically an agreement on a CTBT was most unlikely until the US and USSR

---

60 TNA CAB 168/15, Zuckerman to Prime Minister, Item 1 Future Measures of Disarmament, OPD (68) 50, (PN (67) 1st Meeting, 11 July 1968.

61 TNA PREM 13/2442, Burke Trend to Prime Minister, Further Measures of Disarmament, (OPD (68) 50), 11 July 1968.

62 See Chapter 11 for a detailed discussion of the issues at stake.

63 TNA CAB 148/71, Defence and Overseas Policy (O) Committee Sub-committee on Disarmament OPDO (D) (68) 2nd meeting, 4 November 1968.

64 TNA CAB 168/15, A.D. Brighty, Foreign Office letter to A.M. Palliser, 10 Downing Street, 12 February 1969.

had reached an understanding on strategic arms imitation.[65] The MOD certainly did not think that any pressure from the UK would carry any real likelihood of progress to a CTBT.[66] Nor indeed, given the UK's dependence on the US in key areas, would advocacy of a position markedly different from the US serve UK practical interests at that point of the game. Not even the most advanced seismic systems available by the end of the 1960s would positively identify all explosions in the range of one to ten kilotons. In Zuckerman's view, it was because of this band of uncertainty that the argument about on-site inspection continued. The US still insisted that on-site inspections were essential, the Russians maintaining that national means would be sufficient.[67] Neither the US nor the USSR wanted to put an end to the arguments, and would probably not do so until the outcome of the strategic arms limitation talks had become apparent.

The MOD's formal technical view of the limitations of seismic detection without effective on-site inspection meant that a CTBT violator might well be able to undertake tests of about 50 kilotons, with little danger of having the offence effectively brought home to him by even the best of the monitoring capabilities. The result of tests with yields of this size could without technical difficulty be scaled-up to apply with reasonable confidence to improving weapons of megaton yield. In practice an evasion threshold at this level would place little assured inhibition on the development of new weapons systems and thus on the prosecution of the strategic arms race and on the possible disturbance of the East/West military balance.[68]

Conclusions: UK CTBT Policy and Activities 1965–1973

At the end of 1969 the UK's formal position was that a CTBT was still a major aim of HMG's policy, but it was recognised that in practice the effective conclusion of such a treaty depended on progress by both the USSR and the US towards a Strategic Arms Limitation agreement (SALT). There was not much the UK could do, despite Zuckerman's efforts and even a willingness to abandon an insistence on site inspection would not have altered the price of fish. Moreover, the UK could not afford to antagonise the US at this time on test-ban policy, as we shall see in the next chapter. A successful outcome in SALT would have a considerable influence on

---

[65] TNA CAB 148/71, Defence and Overseas Policy (O) Committee Sub-committee on Disarmament, OPDO (D) (69) 1st meeting, Item 3, 17 February 1969.

[66] TNA CAB 148/71, Defence and Overseas Policy (O) Committee Sub-committee on Disarmament OPDO (D) (69) 3 13 February 1969 Verification of the CTBT Note by the MOD, 12 February 1969.

[67] TNA CAB 168/15, Zuckerman to Prime Minister, 20 February 1969.

[68] TNA FCO 6/84, Comprehensive Test-ban Treaty UK Policy, OPDO (D) (69) 3, Cabinet Defence and Overseas Policy (Official) Committee Sub-Committee on Disarmament, Verification of and the Comprehensive Test-ban Treaty Note by the Ministry of Defence, 13 February 1969.

the extent to which both superpowers thought that that they would have a continuing need to test nuclear weapons; and on whether a reasonable compromise on the on-site inspection problem would be possible.[69] When officials in an interdepartmental Whitehall committee discussed this proposition, they noted that if it were to be suggested that the UK should abandon the CTBT as an objective of HMG's arms-control policies, Ministers would have to be consulted.[70] For much of this period the main arms-control focus was on the UK's new initiative on biological warfare that eventually saw the tabling of a draft treaty in July 1969. A CTBT slipped to the back of the queue, quite a contrast from UK policy in the late 1950s.

Blacknest's view at the end of 1969 was that there were high confidence limits of discrimination criteria above the threshold of three to six kilotons (4.5 mb), but as yet it was still not possible to have the same level of confidence for smaller yields. The level at which a violator could test in dry alluvium with little or no chance of being detected was thought to be about ten kilotons. All other evasion techniques, and especially those that would be necessary if tests of several tens of kilotons were found necessary, would be subject to increasing constraints of costs and logistics.[71] Despite a decade of research effort there was still no technical solution to the problem of detection and identification of low-yield tests. Nor was there any agreement on what would constitute military significant cheating. For the US, and the USSR especially, development of ABM warheads was a key objective. The prospects of an effective ABM system in the Soviet Union had potentially significant adverse consequences for the effectiveness of the UK deterrent with knock-on implications for UK testing policy. It is to this issue that we now turn.

69 TNA FCO 6/85, I.O.C. (69) 147, Steering Committee on International Organisations Twenty Fourth Session of the General Assembly Item 31: Urgent Need for the Suspension of Nuclear and Thermonuclear Tests, Report of the Conference of the Committee on Disarmament Note by the Foreign Office, 11 November 1969.

70 TNA CAB 148/71, Defence and Overseas Policy (O) Committee Sub-committee on Disarmament, OPDO (D) (69) 1st meeting, Item 3, 17 February 1969.

71 TNA ES 4/1167, AWRE Report No. 0 72/69, K.D. Bomford and H.I.S. Thirlaway, October 1969.

Chapter 11
A Static UK Nuclear Weapon Development Programme 1966–1973

Introduction

We cannot begin to understand exactly why the UK refrained from underground nuclear testing from 1965 to 1973 without following the lengthy debates that took place in Whitehall in this period over the future of Aldermaston and more particularly the future effectiveness of the UK deterrent in the face of potential Soviet ABM deployments. These two topics were closely related. This chapter will look at the nature and chronology of these debates in order to see when and why testing requirements were identified. Until such times as a clear decision was taken on what the UK intended to do on Polaris improvements, could there be any more testing? As we saw in Chapter 9, the combined effects of economic stringency and the absence of a decision on a future strategic system after Polaris meant that the UKAEA had been unable to secure approval for even basic research tests in 1965–66. This of course had direct implications on US readiness to exchange warhead information and we need to see the extent to which collaboration started to pick up again once work began in the UK on addressing ABM vulnerabilities and what might be done to mitigate these. We also need to see how requirements for testing affected UK test-ban policy.

Future Weapons Programme and AWRE's Role

Sir William Cook, now PC (N) in the MOD, minuted the Chief of the Defence Staff in early August 1966 on the future nuclear programme.[1] He noted that a UK programme of R&D aimed at a less vulnerable warhead for Polaris was the least that would provide a sufficiently substantial reason for the US to continue collaboration under the 1958 Agreement. Recent AWRE studies had suggested that comparative invulnerability of current warheads could be radically changed in the future through ABM developments able to cause damage at great distance by radiation. In Cook's view the UK would need:

---

1 TNA DEFE 19/197, D.C.S.A. (P) to CDS, Future Nuclear Programme, 2 August 1966.

1. an underground nuclear test to measure the effects of radiation on materials and warheads;
2. development and use of simulators to further the scientific understanding of the effects observed in nuclear tests; and,
3. development of techniques for reducing warhead vulnerability and the final demonstration of exposure of the improved warhead to radiation in an underground test.

The cost of such a programme would be about £15 million and would automatically cater for the R&D necessary for new weapons.[2] If the US were to collaborate on such a programme, then this would be sufficiently attractive to maintain the key scientific and technical skills at AWRE without which no future capability – thermonuclear, fission or rectification of warheads in service – was feasible. It might prove necessary in order to secure US collaboration to declare the positive intention to replace the Polaris warheads, subject to the outcome of the programme of work. If a CTBT entered into force, the proposed UK programme with the US could not be carried out. However, a ban would not affect the development of the small kiloton weapons envisaged by the UK at that time – a warhead for the stand off air to ground missile to replace the WE177 and for a field artillery rocket system – and would have no effect on the then current generation of weapons.[3]

MOD informed the NRDC in mid September 1966 that a decision was required on AWRE's future. If the UK were to maintain a nuclear capability without unacceptable expenditure, it would be essential to continue collaboration with the US. The minimum programme would need to include a strong programme of research, supported as necessary by underground nuclear tests, which might be aimed particularly at decreasing Polaris's vulnerability to counter measures. This was perhaps the most obvious area for new directed research. The most promising lead-in towards collaborative effort with the US, which was so indispensable on this project emerged from AWRE studies. These had suggested that the comparative invulnerability of current warheads could well be radically changed through the development of defensive nuclear warheads that were able to cause great damage at great distance by radiation,[4] echoes of the R-I problem that had bedevilled designers in the late 1950s.

Without a substantive research programme, it would be impossible to retain the number and quality of staff required at AWRE and the UK would be unable to solve

---

2 TNA DEFE 25/123, Chiefs of Staff Committee, COS 43rd Meeting, Future Nuclear Policy, 17 August 1966.

3 TNA DEFE 25/123, Copy of a minute from DCSA (P) to CDS, Future Nuclear Programme, 26 July 1966; TNA DEFE 25/123, Chiefs of Staff Committee, COS, Future Nuclear Policy, 43rd Meeting, 17 August 1966.

4 TNA CAB 134/2241, Nuclear Requirements for Defence Committee, Nuclear Weapon Development Policy, Memorandum by the Ministry of Defence, N.D. (66) 4, 7 September 1966.

even the technical problems that might arise from weapons already in service.[5] This programme would also be just about adequate in US eyes and persuade them to continue the 1958 Mutual Defence Agreement. In addition, it seems that the services had a number of possible new requirements for nuclear warheads in the kiloton range. When the NRDC discussed an MOD memorandum on 16 September 1966, Sir Bernard Burrows said that the Foreign Secretary wished to reserve his position. As for the Foreign Office, it agreed that a further programme would be required if the UK were to avoid a run down of AWRE, retain the ability to deal with problems arising from weapons already in service and maintain collaboration with the US. The programme should, however, be restricted to the minimum required and should not include any commitment to develop new weapons and weapons systems. UK policy in this field was to keep in line with disarmament and arms control policy.

The Foreign Secretary had it in mind to submit a paper to his colleagues on a CTBT and there might be advantage if the two policy papers were taken together – see previous chapter. It was important that any programme for R&D on nuclear weapons did not prevent the UK from adhering to measures that might be agreed in the disarmament field, such as a CTBT. Committee members felt that in the event of a CTBT the US would probably find alternative ways of continuing work on nuclear weapons development. The UK too could continue a programme of its own without testing and it might be in such circumstances that the US would value more highly any UK contribution.

But what would future UK weapon requirements consist of? As part of the MOD policy paper on options, the Chiefs of Staff recommended that UK design effort should continue at a level that would permit the development, if the need arose, of British warheads. The Chiefs also considered that effort should be directed towards warheads for possible UK weapons in, for example, the following fields:

1. A low level air-launched supersonic missile with a stand-off capability and variable yield nuclear warhead;
2. A retarded weapon for delivery from very low level with fractional kiloton yield;
3. An advanced nuclear depth bomb to match the increased threat from faster and deeper diving submarines;
4. Low-yield warheads for land surface to surface tactical weapons;
5. High neutron flux devices.

---

5 TNA CAB 134/2241, Nuclear Requirements for Defence Committee, Item 1, Nuclear Weapons Development Policy, N.D. (66) 2nd Meeting, 16 September 1966.

With the government's major defence budget review and compelling economic need for savings, these aspirations could be no more than a wish list; there was absolutely no prospect for such a programme ever being agreed.[6] And it was not.

Future UK Programme

Towards the end of 1966, Denis Healey circulated drafts of papers that he had prepared himself on the future nuclear weapons programme and sent these to the Prime Minister's office; Burke Trend as Cabinet Secretary was the only other recipient. Healey thought that given the political sensitivities Wilson would want to see and comment on the drafts before he circulated them to other Ministers.[7] Healey argued that there were sound military arguments for retaining AWRE. At the lowest level, he felt that the UK had to continue to meet the minimum need of being able to deal with problems that could weaken the operational efficiency of the current nuclear weapons stockpile. At a higher level, he argued that it would be unwise to discard the capability for developing nuclear weapons. AWRE's current programme focused on three activities: completing the current development programme, post design services on weapons in service and a forward research programme.[8]

However, as the first two activities wound down, a further programme would be needed. If this were to be as useful and economical as possible, it would need to be related to specified objectives of potential military value and which would also be calculated to attract continued cooperation with the US. Healey pointed out that until the UK was able to tell the US that it had taken a decision in principle to continue an R&D programme at AWRE, then the UK would not be in a position to discuss what programme would be best suited to meet common needs. Tellingly Healey reminded the Prime Minister that in the last two years uncertainty about the future AWRE programme had led to a progressive decline in US readiness to give information of value.

The MOD had considered whether it would be feasible to restrict a future programme to the development of fission weapons only. If no work were undertaken on thermonuclear weapons, then the UK would receive virtually no worthwhile information from the US even if Aldermaston did continue with general research. Without advanced work, it would become increasingly difficult to retain staff of the necessary calibre and Aldermaston would rapidly cease to be viable. This would mean that the UK would not have the capability to solve problems that might arise with weapons already in service. For these reasons, therefore, the best

[6] TNA CAB 134/2241, Nuclear Requirements for Defence Committee, Revised paragraph 18 to ND (66) 4 Annex to ND (66) 2nd Meeting, 16 September 1966.

[7] TNA PREM 13/1316, Healey to Prime Minister, Nuclear Weapons and Propulsion Programme, 7 November 1966.

[8] TNA PREM 13/1316, Nuclear Weapons Development Policy, Memorandum by the Secretary of State for Defence, 7 November 1966.

option would be to begin research on a replacement for the Polaris warhead. There was no guarantee that such a programme would tempt the US, but it was the best estimate of the minimum that would be required. Such a programme would need occasional underground tests, possibly every 18 months to two years. And as noted above, if there were to be a CTBT, the MOD felt sure that the US would continue work on nuclear weapons and might in such circumstances value more highly a UK contribution. This then was the minimum requirement if the UK were to keep AWRE viable, maintain US cooperation and to ensure that the UK could sustain its current warheads and keep open the option of future warheads. It was this minimum requirement that Healey invited his colleagues on the new Ministerial Nuclear Committee on Nuclear Policy to endorse.

Wilson, Trend and Healey discussed this minute on 14 November 1966 and it seems that Healey agreed to share the draft with the Foreign Secretary, preferably before he went to Moscow and afterwards to circulate to the new Committee before its first meeting, the date for which was dependent on the availability of a Foreign Office memorandum on non-dissemination (the initial term for non-proliferation), which itself could not be finalised until after the Foreign Secretary had returned from Moscow.[9] Zuckerman, once he had digested the Defence Secretary's memorandum, was highly critical. He did not think that there was any justification for the proposed programme.[10] It would only make sense if there were firm intentions to develop new and advanced nuclear weapons and there was no such understanding. He did not think the Polaris warhead was vulnerable or that missile defence would ever be so effective as to nullify the current Polaris force's capabilities. Nor would Healey's minimum programme, which entailed an underground nuclear test every 18 months to two years, be compelling enough to entice the US. Finally, the cost of maintaining a very expensive cadre of scientists at AWRE was not worth it and the resources – in the order of £15 million per year for the overall programme – would be better spent elsewhere. Given the defence cuts and overall economic situation, Zuckerman felt that the UK could not afford the costs of exploiting any new knowledge that might be gained from endorsement of Healey's minute.

Ministerial Committee on Nuclear Policy

The first meeting of the Ministerial Committee on Nuclear Policy took place in January 1967. It considered two papers central to our story: a paper by the Foreign Secretary on the CTBT and Healey's paper on nuclear weapons development policy. Trend, in his brief for Wilson, agreed that as a basic decision of foreign policy, it was clearly right to maintain UK efforts to promote a CTBT regardless of

[9] TNA PREM 13/1316, William Reid, Cabinet Office to R.J. Dawe, No. 10 Downing Street, 23 November 1966.

[10] TNA PREM 13/1316, Zuckerman to Prime Minister, 8 December 1966.

the effect on the UK weapons programme. Indeed, there was no alternative; the UK still had to be in favour of such a treaty.[11] The Foreign Office memorandum also suggested that the UK should be prepared to support a test ban without any on-site inspection provision, but that for the present the UK would keep this position to itself (see Chapter 10). This did not convince Trend. The Soviet refusal to take up Zuckerman's offer of talks on seismic verification, US insistence on the need for on-site inspection given Soviet ABM developments, suggested that the UK would gain little from such a change in position especially since it could not be used as a bargaining counter.

As to Healey's paper, the balance of advantage lay with accepting the recommendations. Trend thought that the balance of the conflicting arguments – Healey and Zuckerman – was a matter of specialised judgement. It was very difficult, Trend pointed out, to form a firm view without trying to ascertain more clearly whether a purely research effort on the scale envisaged would, or would not, suffice not only to retain AWRE's skill base, but even more importantly sustain US cooperation on a worthwhile scale. If the US indicated sufficiently in the affirmative, then Trend's instinct was to maintain the UK's existing capability in working order for at least some time. Given the uncertainties over the future shape of Europe, the fate of nuclear disarmament and the long-term direction of US policy in relation to the growing Chinese nuclear threat, caution was the wisest counsel. However, accepting Healey's recommendation would need to be subject to the essential proviso that it should embody no commitments either to design or to produce a new generation of weapons to say nothing of a far more costly new generation of delivery systems.

The Ministerial Committee appears to have endorsed most of Healey's recommendations when it met on 9 January 1967. In particular, Ministers agreed that the UK should retain a capability for R&D in the nuclear weapons field; that after consultation with the US a programme should be devised which would allow work at AWRE to continue on a collaborative basis beyond 1970–71, and that the programme should be reviewed in light of US reaction to it after six months.[12] In the main this meant maintenance of a viable AWRE and of the effectiveness of Polaris. Zuckerman commented to the Defence Secretary that if the decision taken proved to be a step towards a new generation of weapons, or the devising of new weapons, it would mean the first step towards the advancement of the UK nuclear effort in the development of nuclear weapons since 1963.[13] The Committee also proposed consultations with the Americans and further examination of research proposals.

[11] TNA PREM 13/1316, Trend to Prime Minister Ministerial Committee on Nuclear Policy, 6 January 1967.

[12] TNA DEFE 13/544, William Cook to Secretary of State, Nuclear Weapons Development Policy, 16 March 1967.

[13] TNA PREM 13/1316, Zuckerman to Healey, 20 January 1967.

UK-US Consultations January 1967

Penney and Cook visited the USAEC and DOD on 20 January to begin negotiations for an extension of the weapons collaboration programme and to prepare the way for negotiations on UK fissile material requirements.[14] Dr Gerry Tape, the USAEC Commissioner responsible for US/UK collaboration, gave Cook and Penney the impression that he was pleased that the UK had taken a positive decision to continue a programme of weapons work at AWRE, thus unstitching the indefinite position currently clouding collaboration and that he would try to go some distance to meet UK wishes. It is clear from these exchanges that radiation vulnerability of nuclear warheads was still a key problem; the US had a major programme to address this. Although there would be some barred areas, the US would assist the UK where it could; any nuclear test would have to be a joint one, but this suited the UK anyway. It was also evident that electronic and electrical components were perhaps even more vulnerable than the nuclear warhead itself. It was not clear to the British however just how far the US was prepared to help if the UK programme was limited to research work on hardening the Polaris warhead with no commitment to further weapon development and production.[15]

The Americans asked what the UK attitude was to penetration aids and to Poseidon – the US second generation SLBM. Cook and Penney said that the UK would like information on penetration aids for Polaris to consider alongside the possibilities to hardening against radiation. As for future arrangements, Tape proposed that a directors' meeting should be held to look at the details of future joint research work. Such meetings between the Laboratory directors had been a feature of UK-US nuclear collaboration, but had been in abeyance for about two years. Zuckerman, who had been excluded from this trip, made his own visit to Washington ten days later and no doubt arranged for the UK Ambassador to make clear to his US counterparts that he was responsible for coordinating nuclear matters in the civil and military fields to the Ministerial Committee on Nuclear Policy.[16] The UK was aware that penetration of ABM defences would depend on the full exploitation of the knowledge arising from the study of the vulnerability of warheads to nuclear radiation. This was a separate and more fundamental problem than penetration aids, and was one that the UK was most anxious would achieve effective collaboration with the US in order to give British warheads the greatest degree of 'hardness' in the most economical way.[17]

---

[14] TNA PREM 13/1316, Healey to Prime Minister, Nuclear Weapons and Materials Policy, Visit to USA, 20 January 1967 by Sir William Penney and Sir William Cook, 25 January 1967.

[15] TNA CAB 164/931, Note of a Meeting in Sir Burke Trend's Room, 2 March 1967.

[16] TNA PREM 13/1316, Summary Report of My Discussion in Washington during the period 30 January to 2 February 1967.

[17] TNA DEFE 13/700, A.U.S. (NS) to PS/Secretary of State, Implications of Ballistic Missile Defence System, paragraph 10.

US officials told Zuckerman that if the UK decided to modify its Polaris warheads they ought to test the resultant modification.[18] The first proposals for what was to become Super Antelope envisaged incorporating that system within the existing Polaris warhead leaving it essentially unmodified and therefore without the need for further nuclear testing.[19] Wilson later remarked to Healey that it had been a mistake to exclude Zuckerman from the Penney/Cook visits as they had brought back a differing account of US Administration views.[20] Dr Tape told Zuckerman that the earlier difficulty over collaboration stemmed directly from a lack of declared UK intent to apply new technological knowledge to develop new nuclear weapons. He added that the most recent attempt to see what exchanges might now be possible, in the light of information reported by Penney and Cook, had demonstrated that the same basic difficulty remained. In this respect, Dr Tape had received little encouragement from other branches of US government. Indeed his enquiries had shown that the issue of technical exchanges on nuclear weapons had now acquired a political quality, which made it increasingly difficult for the USAEC to play as significant a role as they had hitherto done in arranging warhead information exchange. The question of invulnerability came up again, in this case simulated experiments in vulnerability studies. Although Dr Tape thought that this would have some value, especially as an insurance against the possibility of a CTBT, he noted that it remained very difficult to obtain the energy intensities available in full-scale nuclear experiments and to apply them in experimental circumstances that would lead to reliable interpretation of the experimental results. However, this was certainly an area for further UK-US collaboration. Unless a research programme was specially directed to the design of new weapons, Dr Tape felt that the gap between AWRE and the US labs would become wider than it already was. This added to the difficulties of restoring effective collaboration. The key question was whether the UK proposed to proceed to a new weapons system for which new and advanced information would become applicable.

Future UK-US Collaboration on the Edge

Zuckerman told Wilson that unless the UK showed that it intended to proceed beyond the present Polaris programme, any further exchanges of information would be strictly limited to the modification of the A-3 missile, and to a critique of whatever proposals AWRE might make about British re-entry warheads for these missiles.[21] If a UK-US laboratory director's meeting were to take place it would

[18] TNA DEFE 13/544, S. Zuckerman, US/UK Information Exchange in Nuclear Weapon Technology, Summary report of my Discussion in Washington during period 30 January to 2 February 1967.

[19] Personal Correspondence.

[20] TNA PREM 13/1316, Letter to P.D. Nairne, MOD, 14 March 1967.

[21] TNA PREM 13/1316, Zuckerman to Prime Minister, 21 February 1967.

be possible to discover more precisely what information the US authorities would be prepared to provide about the hardened A-3 missile. If the UK wanted further collaboration, then there would need to be a high-level political request. So that the US could then decide what would be appropriate to meet British needs, John Foster, Director Research and Engineering, DOD subsequently telephoned to tell Zuckerman that the UK could obtain at extra cost, the A-3 missile as modified for US use. He also confirmed that it would be for the UK to put forward its own technical proposals for modifying the British re-entry bodies and that the US would be prepared to offer a critique of UK suggestions. He added that the Americans might be able to give additional and valuable assistance derived from their own work in this field – a field Zuckerman thought alluded to penetration aids. However, it would be up to the UK to ask for this.[22] If the UK were to accept what was being offered, Zuckerman thought that it should only do so in the realisation that the steps taken would almost certainly be the last in the context of the special relationship that had operated since 1958.

Healey, therefore, told Wilson that what was now required was a top-level exchange between himself and McNamara and then between the Prime Minister and the President.[23] In light of the apparent muddle in understanding exactly what the US views were, Wilson decided that there would need to be another meeting of the Ministerial Committee before Healey went to Washington to see McNamara in an attempt to obtain a clear and definitive view on US intentions.

Healey meets McNamara

Healey recommended to Wilson that he should make the following points to McNamara privately.[24] First, he would make clear that HMG had decided to maintain AWRE to ensure the efficiency of current and planned nuclear weapons; there being no decision on any new generation of weapons. Second, AWRE was intending to embark on work relating to the hardening of the Polaris warhead – a matter of the greatest importance for the continuing effectiveness and credibility of the UK deterrent. Third, the UK would recognise the need to confine information and other help to Polaris, but would still hope that the arrangement would not be interpreted too narrowly. Exchanges on threat analysis would be essential to AWRE's task in defining its response to the ABM problem. Since the UK had already decided to review the future AWRE programme in mid 1967, the nature of the US response to these points would be an important factor in that review. It also seems to have been recognised by the UK that underground nuclear tests would

22 TNA PREM 13/1316, Zuckerman to Prime Minister, 21 February 1967.

23 TNA PREM 13/1316, Letter to P.D.N. Nairne, MOD, 14 March 1967.

24 TNA PREM 13/1317, Healey to Prime Minister, Nuclear Weapons Policy, 20 March 1967.

probably be necessary to give effect to this.[25] However, at that time no decision had been taken to commit the UK on the full engineering development of a new or improved Polaris warhead, still less to embark on a manufacturing programme.[26] These points were circulated to all members of the Ministerial Committee prior to a meeting in April. Zuckerman remained unconvinced by much of Healey's arguments and he set out at length the counter arguments to any UK involvement with an improved/hardened Polaris missile/warhead in a minute to the Prime Minister on 23 March 1967.[27]

Nuclear Policy Committee: April 1967

Trend briefed the Prime Minister before the April Ministerial Committee meeting on the arguments made by both the Chief Scientist and the Defence Secretary.[28] Trend suggested that if the full details of Zuckerman's counterarguments became clear, the Committee would probably feel that it would be premature to take any decisive step about the hardening of Polaris at that or indeed at any point. HMG would need to be a good deal clearer about UK approaches to Europe, where the chances of success could well be affected by the extent to which UK depended on the US as regards nuclear weapons and which could be taken to contradict British efforts to promote an acceptable NPT. Issues of deployment of the Polaris force east of Suez were also still in the melting pot – how to support the fleet for example through the role of support ships. It was essential now, irrespective of the character of Healey's meeting with McNamara, that the Committee suspend its earlier provisional decision to maintain a certain level of effort at AWRE at annual cost of £17 million. It should also defer any further attempt to close the issues in question until the UK had clarified its own thinking on the aforementioned issues and Wilson had met the President.

The Committee agreed the terms of Healey's speaking note for McNamara. Healey had his meeting with the US Secretary of Defence on 6 April and the reaction was 'helpful'. At McNamara's request, Healey undertook to set out in writing the points that he had made.[29] McNamara expressed the view that a new Presidential Determination under the 1958 Agreement would probably not be needed in order to enable the exchange of information to take place in the Polaris field. McNamara also agreed that Zuckerman, Cook and Penney should come

---

25 TNA DEFE 13/544, C.A. (P) William Cook to Secretary of State, 16 March 1967.

26 TNA DEFE 13/544, Healey to George Brown, 30 March 1967.

27 TNA PREM 13/1316, Zuckerman to Prime Minister, Nuclear Weapons Policy, 23 March 1967.

28 TNA PREM 13/1316, Trend to Prime Minister, Nuclear Weapons Policy, 31 March 1967.

29 TNA PREM 13/1316, Washington telegram no. 1085 to Foreign Office, 6 April 1967.

to Washington to settle the arrangements. Healey sent a draft letter back from Washington for urgent consideration in London and a text was agreed after careful deliberation by the Foreign Office, Trend, Dunnett, Penney and Zuckerman and approved by the Prime Minister.[30] The final text was sent to McNamara on 10 April 1967 and the key message was:

> HMG have no plans for the adoption of Poseidon or for a generation of weapons beyond Polaris. They are therefore concerned with the problem of maintaining the viability of the Polaris A3 missile; and I mentioned that we already know a fair amount about the scope and character of the ABM problem. The immediate requirement, as we see it, is to obtain the technical information we need to enable us to make the best possible assessment of the cost and advantages of the various steps of hardening Polaris.
>
> We are anxious, therefore, to have an exchange of information with you in three particular areas:
>
> (a) Threat analysis – an assessment of the nature of possible Soviet ABMs as related to the United Kingdom Polaris A3 system;
>
> (b) Overall evaluation of the vulnerability of the UK Polaris A3 system;
>
> (c) The technical problems and design to provide penetration capable of overcoming the above threats. This would include the various concepts and information concerning the hardening of sub-systems and components for the UK Polaris A3, including the warhead.[31]

Technical Exchanges: UK Requirements

The letter also proposed that Cook, Penney and Zuckerman travel to Washington to work out the detail and before doing so they would send out a list of detailed questions that the UK would like to discuss. Zuckerman sent these to Foster on 17 April.[32] He hoped that some of the broader questions could be covered in a way that would enable the UK to obtain a better idea of the range of the problem of hardening the A-3 missile and re-entry body. Guidance on likely additional production costs would also be welcome. In Zuckerman's view, configuring of the

---

[30] TNA PREM 13/1316, Foreign Office telegram no. 3306 to Washington, 7 April 1967.

[31] TNA PREM 13/1316, Healey to McNamara, 10 April 1967.

[32] TNA PREM 13/1316, Zuckerman to Dr John S. Foster, Director of Defense Research and Engineering, Evaluation of the Vulnerability of the UK Polaris A-3 System, 17 April 1967.

costs of the hardened missile and the possible acquisition of penetration aids would be most useful. One of the suggested areas for follow up discussion on the technical problems and designs to provide penetration capable of overcoming the ABM threat included methods of assessing the various ABM effects using laboratory simulators and underground nuclear tests. Hardening of UK warheads, however, would almost certainly require an underground nuclear test or tests.[33] US research on hardening re-entry bodies had been checked by massive experimental work, including 12 underground tests over at least two years. However, it was thus quite clear that any future UK nuclear tests were off the agenda for the immediate future and that the UKAEA's programme of research tests were abandoned by this stage.

Zuckerman informed the Prime Minister on his return, that what emerged very strongly was that the US would not provide the UK with detailed information, or allow detailed discussion, which related to the vulnerability of warheads unless and until the UK decided to go ahead with the application of such knowledge to British warheads.[34] Measures to reduce the A3's vulnerability to some ten to 20 kilometres without serious weight penalty seemed to be possible however, but the UK had no data from nuclear tests of either the primary effects, or the value of possible countermeasures to x-ray damage from ABM warheads.[35] AWRE knew that the US had been studying this problem and had probably carried out several underground nuclear tests to obtain experimental data. Much of this work had been done since 1965, but the US had not passed any of this information to the UK under the 1958 Agreement.

The UK team did not receive detailed information on warhead vulnerability distances related to x-ray threats; nor would the UK be granted detailed discussion of this problem until London indicated the extent to which HMG planned to harden the Royal Navy's Polaris weapons in overcoming the total threat. In no circumstances would the US tell the UK what they were building into Poseidon and Minuteman in order to counter possible Soviet defensive measures. This left the UK with three decisions. Did the UK want to buy the hardened A-3 missile plus US penetration aids, while hardening the UK warhead? Did the UK want the hardened A-3 missile but no penetration aids while hardening the UK warhead? Or did the UK want to buy the hardened A-3 missile and do nothing to harden the UK warhead? Wilson's Private Secretary commented on Zuckerman's note that it might indicate that the UK had come to the end of the road or nearly the end of the road, in cooperation with the US; and if it did so, what should the UK do about it. Did HMG give up, or begin to think about the French? These difficult

33 TNA PREM 13/1316, Possible Hardening of Polaris A3 Missile System, Summary Record of US/UK Discussion in US Department of Defense on Friday, 28 April 1967.

34 TNA PREM 13/1316, Zuckerman to Prime Minister, Possible Hardening of Polaris A-3 Weapons System, 5 May 1967.

35 TNA AIR 20/11782, Nuclear Warhead vulnerability and US/UK collaboration, Annex to CA (P)/118/67, 5 April 1967.

questions were to be central to debates over the future of the deterrent over the coming years.

The issue of what to do next was remitted to the Ministerial Committee and Healey prepared another paper setting out the options and costings. A draft was prepared on 15 May 1967 – the tenth anniversary of the first Grapple test.[36] This pointed out that if the UK wanted to retain US cooperation and not to lose the momentum created by UK approaches, the Committee would need to reach a decision on hardening earlier than it had originally envisaged – autumn 1967 was the timeframe in mind. There were two reasons for this. First, in light of the progress made at official level, it was necessary for the Prime Minister to have a clear position on this issue before he met President Johnson at the beginning of June. Second, if the UK were to decide that it had to obtain US cooperation it would be best to engage in detailed discussions as soon as possible since there was some risk that the Americans might retreat from their present attitude if HMG ran into difficulties with them on wider defence and foreign policy issues later in the year – probably an allusion to withdrawal from East of Suez.

The Defence Secretary's minute made three programmatic recommendations. The first of these was that the UK should decide to buy hardened missiles at an estimated additional production cost of $22 million; that the UK should purchase hardened re-entry bodies at an estimated additional production cost of $5 million and that AWRE should be instructed to undertake the necessary work of hardening the warheads. Furthermore, the UK should also seek detailed discussions with the Americans to enable AWRE to continue work on warheads and decoys. This was needed primarily in order to have fuller information about costs and about the possibility of doing the maximum amount of work in this field in the UK. There was no need for any underground tests in this immediate work programme. In fact, such was the importance of decoys, Healey also suggested that the Prime Minister should be invited to emphasise this point to Johnson in his June meeting. However, this draft was not as ambitious on decoys as the Defence Secretary had originally proposed; on reflection the more extensive programme involving penetration aids, which he had outlined earlier to Wilson and estimated to cost about $85 million, was dropped.[37] Meanwhile at the working level, the US agreed that a small AWRE team could visit Lockheeds as soon as possible in order to gain more information on the cost of making US penetration aids in the UK. The experts' brief was to decide how much of the penetration aid packages it would be sensible for the UK to make and to prepare a rough estimate of the costs including those for necessary proving trials.[38]

---

[36] TNA PREM 13/1316, PN (67) 3, Polaris: possible Hardening Programme Memorandum by the Secretary of State for Defence.

[37] TNA PREM 13/1317, P.D.N. Nairne, MOD to A.M. Palliser, 10 Downing Street, 15 May 1957.

[38] TNA DEFE 11/437, V.H.B. Macklen, ACSA (N) to DUS (Pol) et al., Polaris Hardening, Note for the Record, 25 May 1967.

Once again, Zuckerman was not overly impressed by the Defence Secretary's arguments; in an initial reaction he felt that the suggestion that the UK should go for simple hardening of warheads without adding penetration aids was the thin end of the wedge, the thick end of which was not even argued. There was no point in hardening if there were no decoys since there was the problem of the vulnerable terminal part of their flight. Zuckerman told the Prime Minister that the steps suggested by the MOD would merely open the door to very much greater expenditure for no strategic or political reason that could be logically argued.[39] The Minister of Technology, Anthony Wedgewood Benn, remarked that, although the Defence Secretary's paper talked about the possibility of producing the US decoy system more cheaply, RAE had just completed its own study. This demonstrated the feasibility of a British penetration aid system that did not require the removal of one of the three warheads and was likely to be cheaper that any American system.[40]

After further reflection, Zuckerman and Trend jointly minuted the Prime Minister to set out their opposition to Healey's proposal.[41] They were firmly of the view that the MOD proposals represented a critical step whose significance was contrary to HMG's political interests in Europe and the world at large. It would make the UK even more dependent on the US and open the door to an even greater political and technological subservience without any compensating advantage. Domestically and internationally it would be seen – for it would be impossible that the outlines of any action would not become known – to be a step that reinforced the Polaris Sales Agreement rather than re-negotiated it. Moreover, the proposals were devoid of strategic significance, technically inadequate and open-ended financially. Such a step would in fact be an elaboration of the UK's nuclear weapons that would necessitate the resumption of underground tests, something that would inevitably reflect on the UK's good faith in the disarmament field and the stand taken in the NPT negotiations. And for these reasons they urged Wilson to reject Healey's proposals.

The Nuclear Policy Committee discussed Polaris hardening on 18 May 1967. Wilson summed up the mood of the meeting when he recorded that colleagues generally accepted that there was a strong case for deciding at that point to purchase hardened missiles and re-entry bodies for Polaris and that it had now become urgent to choose between hardened and unhardened missiles. Wilson told Johnson in June that the UK was not going to buy Poseidon SLBMs.[42] However, in light of the Prime Minister's meeting with President Johnson and de Gaulle and the outcome of the Defence and Overseas Policy Committee meeting on the

---

39 TNA PREM 13/1317, Zuckerman to Prime Minister, 15 May 1957.

40 TNA PREM 13/1317, Anthony Wedgewood Benn to Prime Minister, 17 May 1957.

41 TNA PREM 13/1317, Trend/Zuckerman to Prime Minister, 17 May 1967.

42 TNA PREM 13/1317, Record of a Conversation between the PM and the President, 2 June 1967; and Extract of a conversation between the PM and the President of France at Versailles, 19 June 1967.

defence review on 26 June, Healey felt that the best course was to obtain his colleagues' agreement for the missile-hardening proposal. They could postpone consideration of other hardening measures – warhead and decoys until Ministers saw the general review of the Polaris programme then envisaged for October 1967.[43] He submitted a draft minute to this effect for Wilson to consider. After consulting Trend and Zuckerman Wilson accepted the proposal provided other Committee members agreed. He accepted too that the UK should buy hardened missiles, but solely because these were cheaper than the unhardened type and on the assumption that the costs relationship between hardened and unhardened missiles had not changed. This agreement did not extend to hardened re-entry bodies; something that Downing Street knew would certainly distress Healey.[44] This matter could be reviewed in the autumn when the overall Polaris programme was to be considered.[45] All Committee members agreed to Healey's proposals, and therefore the Prime Minister endorsed the recommendation to proceed with the order for hardened missiles.[46]

Review of Nuclear Policy: Summer 1967

The review of Polaris turned into a comprehensive re-examination of all outstanding issues of nuclear defence policy and requirements.[47] Nuclear testing requirements could not be settled until these larger questions were settled. Wilson minuted the Foreign Secretary to commission the work and to pull together the divergent strands that had already been reviewed.[48] The review was to be processed through the DOP (O) C. The starting point was the decision that the UK should not develop a new generation of nuclear weapons to succeed Polaris or purchase Poseidon.

The review, notwithstanding the announcement that Polaris would be retained into the 1970s, set out the case for and against adherence to this decision covering the political, strategic and non-proliferation aspects as well as European considerations. Given that Polaris would continue for 15 to 20 years, the review would also look at the case for further hardening beyond the decision already taken

---

43 TNA PREM 13/1317, Polaris: Possible Hardening Programme, P.D.N. Nairne, MOD to A.M. Palliser, 28 June 1967.

44 TNA PREM 13/1317, Manuscript note from Palliser to Prime Minister, 6 July 1967.

45 TNA PREM 13/1317, Palliser to Nairne, Polaris Possible Hardening Programme, 7 July 1967.

46 TNA PREM 13/1317, Prime Minister's Personal Minute M.88/67 to Secretary of State for Defence, 17 July 1967.

47 TNA PREM 13/1317, Trend to Prime Minister, Nuclear Weapons Policy, 20 July 1967.

48 TNA PREM 13/1317, Prime Minister's Personal Minute No M 94/67 to Foreign Secretary, Nuclear Weapons Policy.

to buy hardened missiles. Whatever was decided would in large part determine the level of effort required at AWRE and therefore its future. This latter issue would also be a central theme in the review. Any further underground testing would of course be bound up in the detail of the review, especially if it were decided to proceed with warhead and re-entry body hardening for which some testing would almost certainly be necessary, the actual number of tests depending on the hardening options selected. If re-entry body hardening was pursued to reduce its liability to damage by x-rays and to shield the warhead and electronics from x-rays one or two tests might suffice.[49] If advanced hardening was adopted – a radical re-design of the re-entry body and warhead to a standard higher than that developed by the US, then one or two tests for several years would be needed.[50]

Further testing in the Foreign Office view would prejudice the UK's position on disarmament, especially in the NPT negotiations then still underway in Geneva, but equally a waning capability could well detract from the strength of the UK's negotiating position. This illustrates the extent to which the pros and cons were finely balanced. Testing in any case would cost up to £10 million over ten years. The UKAEA were adamant that if a hardening programme were not undertaken it would be very difficult to retain key skilled staff at AWRE who would still be needed if the establishment were to be viable and able to maintain current weapons in use.[51] Moreover, since scientific collaboration with the US was needed for nuclear weapons, if the UK were to reduce significantly its work on strategic weapons, then the US would cut off nuclear information.

Nuclear Policy Review: Review of Outcome

The Ministerial Committee met on 5 December 1967 to consider the results of the officials' extensive autumn labours.[52] This was a massive tome, the main body

[49] TNA CAB 164/394, Extracts from statement to Congress by Dr J. Foster, DDRE, DOD noted that in 1968 four of the US effects tests were designed to establish the x-ray structural vulnerability of re-entry systems for Polaris, Poseidon and Minuteman, 13 June 1969. Further x-ray effects tests were planned over the next 18 months. (Memorandum prepared by V.H.B. Macklen, 2 July 1969.)

[50] TNA PREM 11/3162, Nuclear Weapon Policy, R.N. Heard for Secretary of COS to CDS, CNS, CGS, CAS, Annex A to COS 1702/11/8/67 Polaris Improvement Programme – Contribution B to Paper for DOP (O) Committee, 11 August 1967.

[51] TNA CAB 148/55 Defence and Overseas Policy (O) Committee Defence Review Working Party 1967, OPDO (DR) (67) 37th Meeting Nuclear Policy, 8 September 1967.

[52] TNA CAB 134/3120, PN (67) 4th Meeting Cabinet Ministerial Committee on Nuclear Policy, 5 December 1967 British Nuclear Weapons Policy; for details the Defence and Overseas Policy (Official) Committee's Defence Review Working Party meetings see TNA CAB 148/55, Defence and Overseas Policy (O) Committee Defence Review Working Party, 1967 37th, 41st, 42nd, 44th, 45th, 46th, 47th, 48th, 49th, 52nd, 54th and 55th meetings held between 8 September and 27 November 1967; the draft memoranda are in TNA CAB

of which was over 30 pages in length in addition to more than 20 pages in a note from the study's secretaries.[53] As far as Polaris improvement was concerned, there were three options: the UK could develop and produce a decoy system that would be allied to an advanced warhead hardening programme; buy US decoys and hardening measures used for the US programme or harden the weapon system against x-ray damage but without any decoys. Some underground nuclear tests were required regardless of the courses of action pursued, but the actual number depended upon the degree of US assistance.[54] There was no agreement between Whitehall departments on the main question whether HMG should maintain the UK deterrent or abandon it as quickly as possible. Nor was there agreement on whether, if the balance of advantage were to be seen by Ministers to lie in retaining Polaris, it would be worthwhile to go to the expense of improving it. However, the Committee agreed that there was no advantage in retaining the rest of the UK's nuclear capability while giving up Polaris.

On the main questions, whether the UK should retain a weapons programme and if so whether it should be improved, the Committee was divided. The Treasury and the Department of Economic Affairs wanted to abandon Polaris at once, or if it were to be retained, then it should not be improved. It seems that the Treasury's underlying motive in this was whether it could close Aldermaston; the more that could be made of this argument the better in the view of Treasury officials. For them the real issue was one of closure not continuing review.[55] The Board of Trade and the MOD's Chief Scientific Adviser thought that Polaris should be retained, but unimproved. The Foreign Office, Commonwealth Relations Office and MOD believed that the capability should be retained and improved; the Ministry of Technology agreed to retention and pointed out that without improvements AWRE's ability to sustain the current capability would be in jeopardy. Consequently, there were no agreed recommendations as consensus had not been possible.

Wilson summed up the lengthy and somewhat inconclusive Committee discussions. Further studies would need to be undertaken to clarify the requirements and costs of alternative programmes for hardening the Polaris warhead and of penetration aids for the system. The nuclear programme review paper produced

---

148/60, Defence and Overseas Policy (O) Committee Defence Review Working Party 1967 Serials 57, 58, 62, 63, 69 and 70. These drafts were then considered by the Defence and Overseas Policy (Official) Committee see TNA CAB 148/80, OPDO (67) 7th Meeting 13 October 1967; OPDO (67) 8th Meeting ,8 November 1967; OPDO (67) 9th Meeting, 30 November 1967 and OPDO (67) 10th Meeting, 13 December 1967.

[53] TNA CAB 134/3120, PN (67) 6, Cabinet Ministerial Committee on Nuclear Policy British Nuclear Weapons Policy, Note by the Secretaries, 1 December 1967.

[54] TNA CAB 148/55, Defence and Overseas Policy (O) Committee Defence Review Working Party 1967, British Nuclear Weapons Policy, Note by the Secretaries, OPD (DR) (67) 58 (Final), 10 October 1967.

[55] TNA T225/2923, J.A. Patterson to Mr Nicholls, Mr Bancroft, Mr Hawtin, British Nuclear Weapons Policy, OPDO (67) 17, 2 November 1967.

by the Cabinet Office had noted that even if the UK opted for the cheapest and indigenous route to hardening, the UK would still need to use the US test facilities at Nevada. In his forthcoming discussion with McNamara about the extent to which further United States information might be forthcoming about penetration aids for Polaris, the Defence Secretary could take the line that there had been no change in the planned Polaris programme. However, the UK's economic situation was such that further reductions in planned defence expenditure might be needed and in such circumstances, no particular element of the programme could be regarded as sacrosanct. At the same time consideration should be given to possible terms of reference and membership for an inquiry into the minimum scale of effort that would be necessary at AWRE and related establishments on the alternative hypotheses that HMG decided to retain the Polaris programme as already planned, or any steps that were to be undertaken in addition such as warhead hardening. The UKAEA had warned that Aldermaston might cease to be viable in the absence of scientifically challenging work, such as the hardening of Polaris, with the consequence that the UK's nuclear capability as a whole could be put at risk.

The Committee endorsed the course of action suggested by the Prime Minister. It also agreed to resume its discussion on the options before the end of the year. There was a tacit agreement that for the present at least that the UK should retain Polaris. However, hardening was a different matter and this was left for a decision in mid spring 1968 in light of the outcome of studies to clarify the various alternatives.[56] It appears that this study of improved re-entry systems examined eleven different options assessed against the then existing system comprising of three soft re-entry bodies each fitted with the UK ET 317 warhead.[57] MOD looked at options ranging from relatively simple hardening of the existing warheads through the adoption of the US Antelope Penetration Aid scheme, to the UK development of an essentially Polaris Re-Entry System containing multiple small hard warheads with or without penetration devices. This was the favoured option and would require harder and more widely dispersed re-entry bodies, necessitating re-design of the ET 317 warhead. Underground testing would certainly feature in these options at some stage.[58] Furthermore, it does seem that the UKAEA had submitted proposals in the 1968 Defence Nuclear Costings estimate for two tests per year, later changed to three tests every other year.[59] These proposals were not accepted.

---

[56] TNA CAB 165/601, Trend to Prime Minister, Nuclear Policy, (PN (67) 8, 4 January 1968.

[57] TNA DEFE 19/191, D.C. Fakley, Head of D Sc 6 to DCA (PN), 14 May 1975; although the warhead was redesigned, the secondaries from ET 317 were recovered and refurbished for the KH 793 programme – the Chevaline warhead. Most secondaries were simply re-clad. Warhead Safety Co-ordinating Committee 130th Meeting, 4 June 1974.

[58] TNA DEFE 13/548, William Cook, CA (PR) to Secretary of State, Polaris Improvements, 9 April 1968

[59] TNA T 225/2924, J.A. Patterson to Mr McDonnell, Mr Nicholls and Mr Bancroft, 19 May 1968.

Polaris Hardening

Healey had to report to Wilson in June 1968 that the AWRE/RAE study on Polaris hardening and decoys that he had thought would be ready in six months – he had mentioned this at the Ministerial Committee meetings in December 1967 – were delayed as further engineering information was needed from the US. The US had refused to discuss engineering feasibility with the UK when related to the US's own missile system designs. There had been two month's delay in getting Dr John Foster, Defence Advanced Research Projects Agency (DARPA), Department of Defense (DOD) to agree to detailed discussions and these had only taken place in May 1968.[60] The new US Defense Secretary Clark Clifford had emphasised to the UK that it would not be possible to obtain a clearer idea of what the US would be able to provide until after the Presidential elections in November 1968. This was no great tragedy as Healey expected the final technical report from AWRE/RAE to be ready in the autumn, which would then give Ministers time to consider its findings before approaching the US again in the spring of 1969.[61] However, Wilson took the view that more immediate decisions were required on Polaris and AWRE's future plans.[62] There were compelling financial reasons and the possibility of continuing uncertainty over future loading for the Royal Ordnance Factories (ROFs) at Burghfield and Cardiff in connection with the nuclear programme.[63] In fact, it had always been Ministers' intention to settle both the Polaris improvement and the future of AWRE questions when the government dealt with the defence programme generally, the then current financial situation (devaluation of the pound) made it more than ever urgent that this be done.[64] So what about AWRE's longterm future?

Kings Norton, the Future of AWRE and Polaris Improvement

The Lord Kings Norton (Chairman, Metal Box Co.) review of the future of AWRE was presented on 31 July 1968 to the Minister of Technology and the UKAEA Chairman. Norton's committee had been asked to examine and report on the minimum scale of cost and effort required for defence purposes at AWRE,

---

[60] TNA DEFE 13/548, William Cook CA (PR) to Secretary of State, Polaris Improvements – Study Report, 15 May 1968.

[61] TNA CAB 165/601, Healey to Prime Minister, Nuclear Issues in the Defence Field, 5 June 1968.

[62] TNA CAB 165/601, Prime Minister's Personal Minute No. M 39/68 to Secretary of State for Defence, Nuclear Issues in the Defence Field, 26 June 1968.

[63] TNA CAB 165/601, Anthony Wedgewood Benn to Prime Minister, 17 June 1968.

[64] TNA CAB 165/600, Burke Trend to Prime Minister, Nuclear Issues in the Defence Field, 24 June 1968.

the ROFs at Cardiff and Burghfield.[65] The committee examined three alternative hypotheses that:

1. it would be decided to limit the weapons programme to what was essential for the maintenance in an operational and safe condition of current and presently planned warheads and to other work such as test-ban monitoring;
2. in addition to (a) it was decided to harden the Polaris warhead by one or more of the methods that were being investigated; and
3. in addition to (a) and (b) it was decided to incorporate penetration aids in the Polaris system or to modify the warhead.

The committee concluded that on any of these hypotheses it would necessary to retain the existing range and quality of staff and facilities. However, after 1970 it would be possible through normal management action to reduce AWRE's Weapons Group support staff. None of the Polaris improvements projects would involve an increase in the planned minimum size of the Weapons Group. There would still need to be additional expenditure for hardware for the developmental programmes and on tests in the US. It was also telling that the Kings Norton report concluded that if the UK options for the third hypothesis were pursued in the UK combined with diversification work, this would be sufficient to retain staff interest and employ all the resources to the full.

Two alternative programmes were looked at for Polaris Improvement; the US Antelope system and the UK Super Antelope. The US option entailed fitting a hardened 'sock' over the re-entry body with the electrics packaged in a screened container. One of the existing warheads would be replaced by a decoy system. The kit for this option would be bought from the US, with AWRE responsible for the development of internal shields for warhead electronics and for conducting, in collaboration with the USAEC, at least one underground nuclear test to improve the efficiencies of the shielding.

In contrast, the UK Super Antelope scheme would harden the existing re-entry body to a level substantially greater than the US option. This entailed development of a new re-entry body skin into which the existing warheads would be fitted. The shape and ballistic characteristics of the new re-entry body would be similar to the US Antelope scheme so that the minimum of flight trials would be needed. Development of Super Antelope would require a substantial effort and vulnerability tests in the US, which would probably be undertaken on the basis of the UK sharing research objectives in US tests. If it were decided that hardening alone were insufficient, it would be possible to associate the US Antelope decoy system with

[65] TNA CAB 134/3121, Report to the Ministry of Technology and the Chairman of the Atomic Energy Authority by the Working Party on Atomic Weapons Establishments, July 1968 Volume 1, Chairman Lord Kings Norton, July 1968.

the UK Super Antelope re-entry body, in which case a decision would be needed to start US production in 1970.

Alternatively, it would be possible to develop a UK decoy system as part of a joint AWRE/RAE undertaking. If the UK took this approach, there were two important advantages for the UK. First, it would save foreign currency exchange costs and lead to tax revenue from value added tax on materials and equipment supplies in the UK; and, second, where a new field of technology, such as hardening was concerned, AWRE would not gain real technological understanding if a US system were adopted instead. As part of the third hypothesis, one programme considered by Kings Norton was for development of a small re-entry body and development of a complete new warhead of compact design. As for test requirements between 1968 and 1978, there would be none under the first hypothesis; two under the second and adoption of the US Antelope system; three under the third if the UK Super Antelope were adopted; and four if a new small super hard warhead and maintenance were required.

Dissent: A Minority Report

There was, however, a minority report. Its existence supported the suspicion of some in Whitehall that the Committee had swallowed the AWRE argument, based entirely on statements presented by Aldermaston scientists. We should note in this regard that the MOD's Assistant Chief Scientific Adviser's (Nuclear) (ACSA (N)) objective in 1967 had been to retain a real nuclear weapon R&D capability at AWRE in cooperation with the US, so this is not surprising.[66] In the view of one contemporary, Zuckerman had initiated this enquiry with the aim of emasculating AWRE.[67] Zuckerman minuted Wilson to tell him that the draft reports that he had seen reflected AWRE views and did not really constitute a critical and independent analysis of the issues of expenditure and optimum deployment of scientific personnel.[68] He later wrote that the majority report was a whitewash.[69] Lord Rothschild, one of the Committee members and author of the minority report, felt that he could not put his name to the final report.[70] He could not accept the open-ended financial commitment that underpinned the second and third

66 TNA AIR 20/11782, V.H.B. Macklen, ACSA (N) to DCAS, Future Nuclear Weapons, 18 April 1967.

67 Personal correspondence.

68 TNA PREM 13/2493, Zuckerman to Prime Minister, AWRE Enquiry, 25 July 1968.

69 TNA PREM 13/2493, Zuckerman to Prime Minister, 12 September 1968.

70 TNA CAB 134/3121, Report to the Ministry of Technology and the Chairman of the Atomic Energy Authority by the Working Party on Atomic Weapons Establishments, July 1968 Volume 2, Chairman Lord Kings Norton, July 1968, Letter from Lord Rothschild to the Chairman of the Working Party.

hypotheses because it was not possible to predict with certainty the outcome of the necessary R&D; they would also include increased expenditure in the form of trials and development work elsewhere. Nor was Rothschild very impressed with the Director of AWRE's assertions that the current scientific cadre was required for the minimum maintenance programme. Perhaps more importantly, he was far from convinced that there were any strategic arguments in favour of hardening and decoys given the size of the US stockpile and that it was inconceivable that the UK would ever launch Polaris in circumstances other than a US nuclear attack on the USSR. A modified ejector system to disperse the warhead over a wider area would be an expensive and open-ended commitment.

This looks like the sorts of arguments against Polaris improvement that Zuckerman was making,[71] a suspicion evidently shared by officials at the time.[72] However, Rothschild conceded that without a thorough investigation that would have taken many months, the Committee had to accept that putting AWRE on a care and maintenance basis appeared to achieve negligible economies. Although Lord Kings Norton presented his report on 31 July, Wilson wanted circulation of the report delayed until the beginning of September. He was keen that there should be an opportunity for Ministers to consider it before they responded to either Kings Norton or Rothschild. Those who had already received copies in the Ministry of Technology were instructed to keep its content to themselves.[73] Further delays ensued and consideration of the report was further postponed until December; this was all down to the financial doubts about the case for Polaris improvement.

Wilson's Concerns

Wilson himself was dissatisfied with the outcome. In his view it was not possible for those Ministers who, despite the majority report, remained sceptical of the conclusions reached to argue this case effectively simply on the basis of Rothschild's minority report. He had consistently taken the view that this subject required more critical analysis by the Treasury. In his view, it seemed incapable of subjecting the demands of organizations such as AWRE to the same exhaustive scrutiny it devoted to other government departments.[74] Wilson wanted to see closer Treasury review of the figures and to prepare a critical analysis that had to be available

[71] TNA CAB 134/3121, Atomic Weapons Establishments, Note by Officials of Ministry of Technology, Ministry of Defence and the UKAEA.

[72] TNA DEFE 19/98, V.H.B. Macklen to DUS (P), Independent Enquiry in to AWRE, 27 June 1968. It turned out that the Rothschild memo was typed on a typewriter in Solly Zuckerman's office. TNA ES 13/1, Victor Rothschild to Robert Armstrong, Cabinet Office, 1968.

[73] TNA PREM 13/2493, Note for the Record, D. Gruffydd Jones, 8 August 1968.

[74] TNA PREM 13/2493 UK Nuclear Policy in defence field; work of AWRE Aldermaston in connection with Polaris Programme: part 3 1967 8 November to 19

before Ministers discussed the report; and to be sure, Zuckerman was to brief Treasury officials. This effort and the fact that the Chancellor had seen the report first were to be kept secret from other departments and the fact that the Treasury report would be based on Zuckerman's briefings.[75]

Zuckerman kept up his remorseless attack on the Kings Norton report and its arguments. He emphasised that the UK could not proceed to hardening and penetration aids without additional flight tests and a programme of underground nuclear tests. Such events could not be kept secret and would have damaging consequences both in terms of domestic political reaction and from the NPT perspective.[76] If the Committee decided to proceed with work on hardening, then it would start on an open-ended commitment. In view of the lack of nuclear testing AWRE had been doing some work in the mid and late 1960s on methods of simulation that would reduce the need for expensive underground testing. Such ideas were of interest to the US, who over the last year had been helpful to UK reviews on the hardening question.[77] It seemed that subject to the view of the new US Administration, DOD and the USAEC seemed prepared to assist the UK with information, equipment trial facilities and access to US underground nuclear trials, once the UK took a decision to harden its Polaris A3 warheads.[78]

Nuclear Policy Committee Reviews Kings Norton

On 20 December 1968 the Committee took note of the Kings Norton Working Party Report on AWRE, which indicated certain possibilities for rationalising associated work in various supporting, as well as other defence research establishments, once the future nuclear weapons programme was decided. Progress on these MOD studies were set out in PN (69) 7. Three major and undecided issues separately and collectively affected the proper rationalisation of AWRE and its associated establishments. These were the extent to which the UK was likely to engage in further R&D on production of new or improved nuclear weapons, with particular reference to hardening; the extent to which work could be transferred away from

February 1970, Note for the Record, PM meeting with Zuckerman 13 September 1968, 16 September 1968.

75 TNA PREM 13/2493, Note for the Record, 25 September 1968.

76 TNA PREM 13/2493, Zuckerman to Prime Minister Atomic Weapons Establishments, PN (68) 9, 19 December 1968.

77 TNA PREM 13/2493, M.J.V. Bell, MOD to A.M. Palliser, No. 10 Downing Street, Brief on Technical Discussion with the Americans on Polaris Hardening, 19 February 1969.

78 TNA DEFE 13/548, M.J.V. Bell, MOD to A.M. Palliser, 19 February 1969.

Aldermaston and whether it was to remain a public corporation or become a government establishment.[79]

Despite the minority report, officials in MOD, the Ministry of Technology and the UKAEA recommended that Ministers accept the report's first four conclusions. In a note from the Defence Secretary and Minister of Technology to the Ministerial Committee on Nuclear Policy, Healey and Benn recalled that Ministers had agreed in December 1967 that further studies were needed to quantify the requirements and costs of alternative programmes for hardening the Polaris warhead and of penetration aids.[80] A detailed technical study started, but although not available by early December 1968, it was expected to be ready by the end of the year. At the 20 December 1968 meeting, the Committee was also told that the decision on whether to improve Polaris would need to be deferred until the UK had information about the nuclear policy of the new US Administration and about the likelihood of the US and USSR reaching agreement to limit or stop deployment of ABM systems.[81] Therefore, the picture was incomplete, which meant that a final decision on Polaris and on when and how it should be improved would have to be deferred again. It seems however, that there was agreement to proceed with a Super Antelope project definition phase, which the US agreed to support until its completion at approximately the end of 1972.[82] No final decisions were taken therefore on whether to leave Polaris unimproved and to accept a decline in its effectiveness, or to proceed to full development of Super Antelope or seek US agreement to the conversion of the UK's Resolution class submarines to carry Poseidon.

Until such times as a final decision was taken, there would be no further underground nuclear testing, not even it seems general research tests as had originally been proposed by the UKAEA back in early 1963 and again in 1965. AWRE and MOD noted at that time that if the UK wanted to stay in the nuclear weapons business it was most probable that further underground tests from 1969 would be required. Moreover, apart from any specific UK requirements, if the UK was to continue to enjoy US cooperation in R&D, then the UK would be expected to participate in future test programmes and bear some of the cost. If any UK test

[79] TNA CAB 165/601, Zuckerman to Prime Minister, Future of Atomic Weapons Establishment (PN) (69) 7), 8 July 1969.

[80] TNA CAB 134/3121, Ministerial Committee on Nuclear Policy Atomic Weapons Establishments, Note by the Secretary of State for Defence and Minister of Technology, PN (68) 9, 5 December 1968.

[81] TNA CAB 165/600, Chairman's Briefs, Zuckerman to Prime Minister, Future of the Atomic Weapons Establishment, (PN (69) 7, 8 July 1969; TNA EG 7/73 Report to the Minister of Technology and the Chairman of the Atomic Energy Authority by the Working Party on Atomic Weapons Establishments (Chaired by Lord Kings Norton) dated July 1968: Volume 1 (majority view by members of the working party) together with Volume 2 (dissenting view by Lord Rothschild) 1966–71 PN (68) 4th meeting, Ministerial Committee on Nuclear Policy, Minutes of a Meeting held on 20 December 1968.

[82] TNA CAB 130/533, GEN 69 (71), 1st meeting, 13 December 1971.

programmes were thought useful by the US, then it would be ready to renew the 1958 Agreement for continued use of the NTS.

Polaris Improvement Programme

Sir William Cook reviewed the report of the technical study group on the Improvement of Polaris, established by the PN committee at the end of 1967. He recommended that the UK could not leave Polaris unimproved in the face of Soviet ABM developments, since the initial SALT talks held out no prospect of the complete abandonment of ABM systems.[83] Healey asked MOD officials to prepare a draft minute to go to the Prime Minister, Chancellor and Minister of Technology which would seek agreement to keep the option of improvement open until after the General Election (June 1970). This would be done by means of a feasibility study, which would make the best use meanwhile of spare capacity at AWRE, on the understanding that a decision whether to go forward with such a programme would be put to the Nuclear Policy Committee at about the end of the year.[84]

Research and exploratory work on the need to improve penetration capability of UK Polaris had shown that the US Antelope modification had serious shortcomings. However, AWRE believed that it would be possible to improve design of the decoy systems, allied with some further hardening of the warhead, so that satisfactory penetration capability could be achieved without resulting in either a new generation weapons system or intolerable costs. The US had given the UK much assistance, information, and agreed with UK conclusions.[85] However, once Healey and Benn discussed the issue they agreed that on the assumption of a summer election, it was not going to be feasible to make an approach to the Prime Minister or Chancellor for approval to undertake a project definition study until after the election was over. Healey still felt that it was extremely important to avoid any major delay in proceeding with the feasibility study, there was no great harm in marking time for one or two months, provided that AWRE could be getting on with essential preliminary feasibility work in-house. This required the cooperation of a small number of the staff in the Ministry of Technology, the Royal Radar Establishment and the RAE. Benn agreed and issued the necessary instructions to the relevant personnel.[86]

---

83 TNA DEFE 13/700, A.R.M. Jaffray (PS) to Secretary of State, Polaris, Improvement, Meeting with CDS, CA (PR), PUS, CNS and DUS (P), 4 May 1970.

84 TNA DEFE 13/700, Report of the Technical Study Group on the Improvement of Polaris A3, A.R.M. Jaffray, 5 May 1970.

85 TNA DEFE 13/700, V.H.B. Macklen DCA (PN), Polaris Improvements draft minute from Secretary of State to Prime Minister, 6 May 1970.

86 TNA DEFE 13/700, Healey to Minister of Technology, Polaris Improvement, 18 May 1970; I.T. Manley PS, Minister of Technology to A.R.M. Jaffray, 2 June 1970.

A Change in Management: The Tories Take Over

Shortly after the Conservative government of Edward Heath took office in June 1970, Lord Carrington, now Minister of Defence, sought Prime Ministerial authority for a project definition study, estimated to cost £4 million, which would keep open the option of making essential improvements to Polaris.[87] It was now clear to the MOD that if the Russians continued to improve their ABM system, the threat posed by the UK deterrent would be virtually eliminated by the mid 1970s unless the warheads were hardened and decoys included. However, effective measures could be taken in time provided work began immediately on the project definition stage of a UK system of warhead hardening and penetration aids (Super Antelope). No decision about full development and deployment need be taken until the end of the project definition stage, which was expected to take between one and two years. The outcome of the SALT and ABM talks would then be clearer. Failure to take this decision would render Polaris's deterrent credibility to zero and would mean that the flow of nuclear information from the US would dry up; the UK would also lose as a result a valuable card for discussion with the French on European defence.

The virtual shutdown of advanced UK nuclear weapons work after production of the existing Polaris warhead was followed by an almost complete shutdown by the US on the transmission of advanced R&D information. This shutdown was complete in the important field of warhead vulnerability. However, the build up since 1967 of a small but technically competent research effort at AWRE on vulnerability studies had improved the situation considerably.[88] The costs of the project were already in the Defence Long Term Costings, which means that the Labour government must have agreed to take this approach too. Super Antelope had four advantages over the US alternative. First, it could be expected to provide a better exchange ratio (the number of ABM warheads required to destroy incoming warheads) because the warheads would be harder than in the case of Antelope. Second, the UK system would counter effectively Soviet defence tactics because the decoys would be hard and less sensitive to radar discrimination. Third, the system would be based on UK developments thus giving the UK a basic technology in these areas; and finally, it would cost much less than in dollars.

Zuckerman however was still against any modernisation on strategic and economic grounds: the UK deterrent was puny in comparison to the Soviet stockpile and he could never imagine a credible scenario in which the UK would start a nuclear war with the Russians. As for the economics, the costs would

---

87 TNA PREM 15/1359, Carrington to Prime Minister, Improvement of the Polaris System, 16 July 1970.

88 TNA PREM 15/1359, Annex A to Carrington Minute, The Current State of the Polaris Force. The US Navy decoy readiness programme for Polaris led to studies of UK options, which culminated in an RAE Project HR 169, which had reached project definition stage by 1966. Personal Correspondence.

soon multiply way beyond their current estimates, which of course proved to be the case.[89] He urged Heath to seek further detailed studies and not to rush into a decision on the MOD request.[90] Now Heath appears to have acceded to this since the Ministerial Committee on Nuclear Policy met on 20 July and invited Carrington to put forward further proposals on the issue, taking into account a firmer assessment of the SALT/ABM Treaty process and some differences of opinion on technical points that had been expressed in discussion. Carrington reverted on 21 October 1970 emphasising the urgency of a decision to proceed with a project definition study.[91] On SALT and the ABM Treaty, it seemed to Carrington that the latest reports indicated that there was no chance of the Americans and Russians agreeing to no ABMs, at least in the near future. As for the technical differences, Zuckerman and Cook as well as Carrington had discussed these. In the end the Chief Scientific Adviser did not dissent from the need to take action to keep UK options open.

Thus far there had been little reference to nuclear testing, but is seems that this was the reason behind Carrington's sense of urgency. An essential part of the study – accounting for about £1 million of the £4 million gross cost – was the exposure of weapons materials to the effects of nuclear explosions. In 1970 the UK had participated in the US programme of underground experiments, which were elaborate and expensive. The UK could meet its own requirements within them at relatively small cost; to mount national experiments, even with US help, would be ten times more expensive. It was therefore essential for the UK to bid for its requirements within the US planning schedule. Firm commitments were required in November 1970 for participation in the series of US experimental underground nuclear tests planned for 1971 and 1972, each of which was designed to cover a different aspect of the hardening problem. Some of that programme had already been approved in the US, provisional arrangements had to be made for discussion between the UK and US relevant authorities in a matter of weeks.

Carrington was convinced that the UK should not miss this advantageous offer to participate in the US test series. An essential part of the UK plan to keep options open was to maintain cooperation with the US. Washington was aware of UK delays and was wondering what its ultimate intentions were. Carrington wanted to be able to tell Melvin Laird (US Defense Secretary) that HMG had now decided to proceed with the project definition study; he asked for approval to this course of action. Heath agreed on 27 October 1970 that an immediate start could be made with the Super Antelope project under the strictest security and at the

---

89 House of Commons, *Ninth Report from the Committee of Public Accounts Session 1981–82 Ministry of Defence Chevaline Improvements to the Polaris Missile System* (London, March 1982).

90 TNA PREM 13/2440, Zuckerman to Prime Minister, 17 July 1970.

91 TNA PREM 15/1359, Carrington to Prime Minister, Improvements of the Polaris System, 21 October 1970.

estimated cost described; and that Laird could be told too.[92] The two-year study was to end on 31 October 1972, with the cost now at £7.5 million instead of the originally estimated £4 million.[93] Treasury approval only existed for £6.1 million. The decision to buy space in certain underground US nuclear tests gave added impetus to reviving UK-US nuclear collaboration.[94]

February 1972: Definitional Studies Continue

Carrington provided an update for the Prime Minister in February 1972.[95] At that point, the project had proceeded satisfactorily, in terms of both substance and timing. The UK had received considerable US help in light of the formal Presidential Determination to assist under the 1958 Agreement. UK studies had confirmed that it would be feasible to develop Super Antelope in Britain, with US assistance particularly over the provision of underground trials facilities. One of the reasons for the costs increase had been because the UK had to pay an earlier advance for the trials facilities than had been expected. Some initial flight trials were also now envisaged at the end of the two-year study. The programme of work had been intensified in order to reach earlier decisions of certain design options, with a view to keeping long-term costs down.

Everything was on track for the government to take a decision by 1 November 1972 about proceeding with Super Antelope and the long-term future of the deterrent in general. The UK also had reassurance that there were would be no adverse implications of any 'non-transfer' provisions that the US might agree with the Russians in the next round of SALT negotiations; in other words, provisions which would impose constraints on what assistance the US could provide to the UK in terms of missiles and nuclear warhead-related data.[96] Other factors to be considered would be whether Poseidon might be an alternative, albeit a much more expensive one. Moreover, it would be important to look at the Poseidon successor – the C4. Future discussions with the French over a possible successor system would also need to be considered.

---

92 TNA PREM 15/1359, R.T. Armstrong, 10 Downing Street to A.R.M. Jaffray, MOD, 27 October 1970.

93 TNA PREM 15/1359, Super Antelope Financial Statement Annex B to Minute from Carrington to Prime Minister, 9 February 1972.

94 TNA DEFE 13/752, V.H.B. Macklen DCA (PN) to Private Secretary to Secretary of State et al., 18 November 1970.

95 TNA PREM 15/1359, Carrington to Prime Minister, Improvement to Polaris System, 9 February 1972.

96 TNA FCO 46/1002, FCO telegram no. 998 to Washington, 1 May 1972 and Washington telegram nos. 1458 and 1459 to FCO, 3 May 1972.

In order to keep work going at Aldermaston Carrington sought approval for a further £9 million to keep the Super Antelope project open. The US programme of underground nuclear tests planned for 1973 and 1974 offered a unique opportunity to participate in trials essential to the project. By 1969 the flow of information under the 1958 agreement had slowed to a trickle. This meant that US data on nuclear weapons effects, Soviet understandings and capabilities in this area and warhead vulnerability to such effects and capabilities did not become available until AWRE started its own work to look into this problem and see what could be done about it.[97] The US payments schedule required the UK to commit itself in April 1972 to the tune of £4.7 million. Unless this was taken up, delays of at least one year would ensue and the UK might have to pay the entire costs of a later series of tests.[98] Once again, Heath agreed to Carrington's proposal following advice from the Cabinet Secretary. The Foreign Secretary and Chancellor also agreed, but the later was worried about the long-term implications for the defence budget.[99] Trend argued that unless the UK was prepared to allow the deterrent's credibility to erode gradually until it ceased to enjoy any independent capability in this field – HMG would have to decide in October 1972 to adopt either Super Antelope or Poseidon. Furthermore, whichever of these was preferred, the UK had to accept the proposed expenditure of £9 million, mainly on participation in the US programme of underground nuclear effects tests. These were indispensable if the UK was to keep the Super Antelope decision option open, but it would also be relevant to the development of Poseidon if that option were adopted instead.[100]

Admiral Levering Smith had mentioned to the Chief of the Defence Staff (Admiral Hill Norton) that he thought the UK had not appreciated that help with Super Antelope might in fact be more difficult for the US than making Poseidon available.[101] His main reason was that Super Antelope could be held (and was so held by some people in the US) to be in aid of a UK national deterrent whereas Poseidon could plainly be represented as part of the Allied deterrent. He had added as collateral that he had express instruction that no such assistance on

---

97 Personal correspondence.

98 TNA PREM 15/1359, Carrington to Prime Minister, Future of Strategic Nuclear Deterrent Appendix 1 to Annex B Deterrent Super Antelope, 1 April 1972. The work had also led to a decision to concentrate on one hardening method for the re-entry bodies. This used three dimensional matrix of quartz fibres bonded together with phenolic resin (3D quartz phenolic or 3DQP) TNA PREM 15/1359, Financial Statement Annex B to Minute from Carrington to Prime Minister, 9 February 1972.

99 TNA PREM 15/1359, Anthony Barber to Prime Minister, Future of the Strategic Nuclear Deterrent, 5 May 1972.

100 TNA PREM 15/1359, Burke Trend to Prime Minister, Future of Strategic Nuclear Deterrent, 17 April 1972; R.T. Armstrong, 10 Downing Street to Robert Andrew, MOD, 15 April 1972.

101 Director Strategic Systems Projects US Department of the Navy; US end of the PSA channel for the UK/US information exchange on Polaris and related matters.

Super Antelope, including even the arrangement of flight tests, was to be given without approval from the White House itself.[102] Ministers and officials were also worried about the prospect of any 'non-transfer' provisions in the SALT talks on US abilities to provide assistance on Super Antelope. Levering Smith had said that he was quite sure that it would not only be inappropriate but impossible to deal with any UK request on Poseidon until the current US-USSR summit meeting had finished and the small print of any SALT agreement carefully studied.[103]

The Need for UK Underground Nuclear Tests

AWRE always reckoned in the Super Antelope context that it might be necessary to change the Polaris warhead and that if it was so decided, then the UK would need two or three underground tests.[104] These would be in addition to the nuclear effects tests in the US in which the UK was already participating.[105] A joint UK-US Working Group (JOWOG 26) started work in late 1969 under the rubric of the 1958 Mutual Defence Agreement to cover UK participation in US effects tests.[106] A requirement for underground effects tests with materials and components to be used in Super Antelope had been recognised earlier and an extensive cooperation programme had taken place with the US between 1970 and 1973.[107] The first US/

[102] TNA DEFE 13/1050, CDS to CNS CSA and DUS (P), Polaris Improvement, 25 May 1972.

[103] TNA DEFE 13/1050, CDS to CNS CSA and DUS (P), Polaris Improvement, 25 May 1972.

[104] The issue on the warhead for KH 793 was whether the UK went for a low impulse heat shield route, using materials such as beryllium or carbon or a high strength high shock impulse route. It was found that the former could not be made as hard as desired from a systems point of view, but the latter would require significant changes to the warhead design. Personal Correspondence. In the event the underground effects' tests confirmed the use of 3DQP as the material to be used in the REB shell. TNA DEFE 19/207, The Project Definition Study of Super Antelope (KH 793), Report of the Project KH 793 Review Board, 3 October 1972.

[105] TNA PREM 15/1359, Carrington to Prime Minister, Nuclear Warhead Tests, 15 May 1972. There appear to have been at least ten US effects tests between 1970 and 1972 if one assumes that these took place in tunnels rather than shafts, but it is not clear how may of these involved the UK although one source notes that the UK bought space in at least seven flux tests. Cochran, Arkin, Norris, Hoenig, *Nuclear Weapons Databook Volume II*, pp. 168–70. Personal correspondence.

[106] TNA DEFE 19/191, D.C. Fakley, Head of D Sci 6 to DCA (PN), Chevaline – Historic Summary of Events, 10 April 1975.

[107] Personal correspondence. See open description, retained file title TNA ES 12/256, US/UK Agreement on Use of Atomic Energy for Mutual Defence Joint Working Groups JOWOG 26 (underground effects testing): results of UK experiments in Diana Mist experiments: interim reports, 1 January 1970–31 December 1970.

UK stocktake meeting since 1965 had taken place between 7 and 11 February 1972. One of the key points to come out of this was that the UK should supply the US with a tentative programme specifying likely UK requirements for underground effects tests post-1972 arising out of the latter phases of the Super Antelope programme. The objective of this was to allow the US to comment technically and in Anglo-American discussions, to agree without commitment a technical programme for political approval on both sides.[108] UK officials also made clear at this meeting that development plans also included the need for a new warhead test.

Whether such a test was in fact necessary as of late January 1972 was not clear as it might have been possible to complete the project without one; it would be foolish however not to cater for the possibility that it would turn out to be essential. In any case, there were other reasons that made a British test highly desirable.[109] These reasons remain classified, but the lack of testing and the knowledge and experience that came with it for some seven years must have been a key factor. As we saw the UKAEA had a major test research programme in mind until postponed by the first Wilson government. Cost estimates had always allowed for some testing. The need for a test arose from the fact that the warhead used in the present unhardened Polaris was probably susceptible to attack by a copious flow of neutrons from new types of ABM warhead that the Russians might develop.[110] This could be countered by changing the design of the warhead.[111] It seems therefore that the Polaris warhead re-design was driven primarily for nuclear hardening reasons.[112]

Technical discussions with the US had shown that the Americans thought it was plausible that the Russians might develop enhanced neutron warheads and consequently they strongly urged the UK to harden the Super Antelope warhead in anticipation. UK knowledge of French testing suggested that they used an unspecified material (possibly boron) in their devices, though their reason for

---

[108] TNA DEFE 13/1050, Chief Scientific Adviser to Secretary of State, US/UK Stocktake Meeting, 23 February 1972.

[109] TNA DEFE 13/1050, Chief Scientific Adviser to Secretary of State, Nuclear Warhead Work: US Cooperation, 21 January 1972.

[110] Neutrons would heat high Z materials (high atomic number materials such as tungsten and uranium) and the risk of damage would depend on the phase of the alloyed material. X-rays and other radiations would affect other aspects of the overall design and would need to take these into account. Personal Correspondence.

[111] The exact addition is retained in the file under Section 3 (4) of the Public Record Act 1967. The US knew in the mid 1960s that its ballistic missile warheads and guidance systems were vulnerable to x-ray damage in exo-atmospheric conditions from x rays from an exploding ABM nuclear warhead over thousands of miles. Rex Pay, New Effort Aimed at X-Ray Protection, *Technology Week*, 2 January 1967, pp. 10–12.

[112] TNA DEFE 19/240, UK Stockpile Reliability in the Absence of Nuclear Experiments, presentation by Dr P.G.E.F. Jones, Deputy Director/Chief Warhead Development, AWRE, 1 February 1976.

doing so was probably not necessarily to deal within neutron vulnerability. Unless the UK made the change, or had put itself in the position to do so, the UK alone of the Western nuclear powers would be particularly vulnerable to enhanced neutron ABM warheads. MOD did not think that the French knew of the UK problem. Therefore, tests were a highly desirable part of the UK nuclear programme, not only to round off the ability to deal with Russian developments, but also in view of possible future collaboration with the French. Current planning envisaged that the UK would need to conduct tests in late 1973 or early 1974. The US told the MOD that it would probably take 18 months to two years from the date of the request to the date on which the first test could take place at Nevada. For this reason an approach to the US authorities was urgent and Carrington sought approval to make the necessary arrangements for two to three tests in the Super Antelope context.

Heath agreed that the nuclear tests should proceed.[113] Sir Herman Bondi wrote to Dr John Foster on the same day to seek authority for US and UK project planners to discuss the necessary support, including facilities and assistance for underground effects tests, parts and contractors' services for underground tests.[114] Bondi told Foster that because of discussion with US experts, AWRE now thought it wise to change the UK's present warhead. UK ideas on the design of a warhead for Super Antelope had yet to crystallise, but continuing exchanges with the USAEC greatly helped AWRE's efforts. As a result he would be writing to Commissioner Larson on the Commission to advise him that he would like to request facilities for two or three underground device tests to take place early in 1974. Britain was already dovetailing its underground nuclear effects tests programme into the US equivalent.[115]

AWRE's expectation in September 1973 was that warhead drawings would be frozen by June 1975.[116] The KH 793 programme (the new code name for the improvement project) included two tests, although if the first were successful the second would not be required. It seems that a re-appraisal of the vulnerability of the ET 317 warhead to neutrons from an ABM explosion meant that it needed to be modified and then proved by an underground nuclear test.[117] There was also some thinking that the completed system would require a final test as a confirmatory

[113] TNA PREM 15/1359, R.T. Armstrong to Robert Andrew, MOD, 16 May 1972.

[114] TNA DEFE 13/1050, CDS to CNS CSA and DUS (P), Polaris Improvement, 25 May 1972; CSA Herman Bondi to Dr J.S. Foster, 11 May 1972.

[115] TNA DEFE 13/1050, CSA's Report on Nuclear Matters Discussed during visit to USA 9/10 March 1972, 24 March 1972.

[116] TNA DEFE 19/182, M.A. Mann, Head of AWD 5 to ACSA (N), 30 October 1973.

[117] TNA DEFE 19/207, The Project Definition Study of Super Antelope (KH 793), Report of the KH 793 Project Review Board, 3 October 1972. The warhead also had to be able to withstand the mechanical shock that was imparted to it when the REB was subjected to x-ray effects from an ABM burst. Laboratory simulation tests had indicated that the modified warhead would be able to survive the shock well enough, but underground effects tests were needed to confirm this.

demonstration of the success of the hardening programme.[118] Nixon endorsed the UK request for up to three tests on 15 August; the first of these, code name Fallon, took place at Nevada on 23 May 1974.[119] Officials had hoped at the start of December 1973 that if the first test worked sufficiently well then it would be possible to cancel the second one.[120] This could save about £1 million. The test appears to have been delayed because of slippage in the supply of 3D quartz phenolic (3DQP) heat shields provided by the US firm AVCO because of a large order from the US.[121] It also appears that this test was originally intended for February and had been postponed two and half months to permit inclusion of a design change.[122] It is also clear therefore that the resumption of testing and the Polaris modernisation programme were instrumental in reviving the 1958 Agreement.

---

[118] TNA DEFE 19/182, M.G. Davies WLR3 TPC Visit Report 3rd Vulnerability Working Group, 27 September 1973.

[119] Norris, Burrows, Fieldhouse, *Nuclear Weapons Databook Volume V*, pp. 402 and 404, footnote 68. There was some concern about this test in the Labour Cabinet and the Prime Minister noted that some of his colleagues felt that there should have been a discussion about the need to test in the full Cabinet. TNA CAB 128/54 CC (74) 21st Conclusions Minute 3, 27 June 1974. The decision, he reassured his colleagues, had been taken solely in order to keep open the option of retaining a credible deterrent and did not prejudge any policy decisions. Wilson added in a subsequent note to his colleagues that he, the Defence Secretary, Foreign Office Secretary and Chancellor had to decide when they met on 5 April whether the planned test should proceed; the previous government had set the timetable. He said that the UK knew from technical assessments of all the available information on the nature and rate of development of Soviet ABM developments that UK missiles would have to be given better penetration capability. This could be done without MIRVing, but it did involve some testing. Once the need had been accepted, the timing had been governed by the tight logistic programme for modification of the missile and the need to fit into the US test programme schedule. A decision not to test would have prejudiced the effective fulfilment, in terms of time and money, of one of the options that the UK might wish to take. Wilson went on to state that there would be no need for a further test before HMG had decided, in the context of the defence review, on the UK's future defence policy as a whole. TNA CAB 129/178, C (74) 85, Nuclear Testing, Note by the Prime Minister, 31 July 1974. Wilson, Callaghan (Foreign Secretary), Healey (Chancellor) and Roy Mason, however, had decided on 5 April to delay an announcement about this test. Mason had argued that this was desirable in light of the First NPT Review Conference. TNA PREM 16/1181, Record of a Meeting held at 10 Downing Street, 5 April 1974. The Americans also agreed to make no announcement about the test, John Hunt to Prime Minister, 3 May 1964.

[120] TNA DEFE 19/182, KH 793 Project Review Board Minutes of Meeting held in Polaris Management Centre, MOD, Chairman's summary – ACSA (N) Dr F. Panton, Monday 3 December 1973.

[121] TNA DEFE 19/182, KH 793 Project Review Board, Report on Technical Progress, Note by the Secretary DC Fakley, 28 November 1973.

[122] TNA DEFE 19/182, Record of a Special Super Antelope Review Meeting (Roy Dommett, RAE and Peter Jones, AWRE present) held on 6 September 1973.

The UK Resumes Testing: The Nine Year Self-imposed Moratorium Ends

This 'fairly small' test appears to have demonstrated the performance of the basic design yield of the new hardened warhead.[123] AWRE deemed this test successful and the scientists were overwhelmingly satisfied that the device tested was one-point safe, a view confirmed by the USAEC. Consideration then turned to reviewing requirements for further tests, either to improve the Polaris warhead by a large factor, or to prove the safety of the existing design.[124] The Nuclear Weapons Safety Committee appears to have accepted AWRE's safety advice and its chairman (Sir William Cook) planned to recommend that a safety test was not required. This meant that the UK could use the other two slots allocated for underground nuclear tests by the US for other research purposes; any further test for Polaris would have to be held before the end of 1975 when the warhead drawings were due to be frozen.[125] Warhead and Re-entry Body (REB) designs were 'chilled' by early September 1975, which meant that formal approval was required for any subsequent modifications.[126] Warhead production was due to begin by the summer 1976.[127] Roy Mason, Secretary of Defence, wrote to Wilson on 14 May 1975 to

[123] TNA DEFE 19/19, Draft paper to KH 793 Project Review Board, Chevaline 1974 Development Cost Plan, Note by DAWD dated 16 June 1975; TNA DEFE 13/1039, F.H. Panton, ACSA (N) to DCA (PN), Nuclear Test, Draft Note from Secretary of State to Prime Minister, 17 February 1976; TNA PREM 16/1181, Record of Conversation between the Secretary of the Cabinet and the American Secretary of State, 26 April 1974. 'Fairly small' was Kissinger's term when told of the planned yield; the actual yield remains classified.

[124] TNA PREM 16/1181, Background Note, UK Underground Nuclear Tests. Effects tests were also planned for a further three events before the end of 1976. It is also worth noting that although UK tests had been required primarily for improvement programmes, AWRE aimed to extract as much information from each test. None of the UK nuclear tests since 1964 has had a single or simple objective. The nuclear tests themselves were the culmination and proof of a long series of non-nuclear testing, which had enabled the UK to exercise the expertise of the design teams and provide reasonable proof of its continued validity. It takes more than a year to plan and execute a test. See TNA AIR 8/2785, B.M. Norbury, Head of DS 11 to APS/Secretary of State, CTB: Warhead Stockpile Safety and Reliability, 3 April 1978, attachment Comments of the Foreign Secretary's minute to the PM of 31 March.

[125] TNA DEFE 19/208, KH 793 Project Review Board, 4 October 1974.

[126] TNA DEFE 19/170, D.B. Janisch, DAWD to ACAS (N) Chevaline Status report to Secretary of State, 2 September 1975. The warhead also had a fuzing capability for airbursts at three different altitudes and for a surface burst. The underground nuclear effects test programme was nearing completion.

[127] TNA DEFE 24/895, Draft letter from PUS to Mr Pliatzky, Treasury, Chevaline, June 1976. In this regard it is odd to see that Callaghan, Prime Minister, told the Cabinet in 1976 that a UK underground test that year was to 'improve the effectiveness of our existing system'. TNA CAB 129/59, Cabinet Conclusions 3 August 1976, p. 2. TNA DEFE 13/1039, F.H. Panton, ACSA (N) to DCA (PN) Nuclear Test, 17 February 1976. A draft Note from

note that the 1974 test results had been very satisfactory. AWRE was sure that the UK could manufacture an intrinsically safe warhead to meet the specifications required. However, the Fallon test of a radically new design had not given the scientists sufficient information to be able to guarantee to keep it in a reliable and safe condition over the expected ten to 15 years of service life. To acquire this information a further one or two tests of the basic design were needed. As a bonus these tests might also enable the UK to economise on the manufacture of the warhead, especially in its use of plutonium. To fit the development programme the next test would have to be held in mid 1976.[128] The Prime Minister agreed to authorise a provisional booking of a slot in the US 1976 underground test programme.[129] A third test was planned for the autumn 1978, which was intended to further develop the warhead design to make a major reduction in its weight. If successful, this would allow the Chevaline system greater range and thus a greater amount of sea room for operational deployment.[130] The date of this test was eventually brought forward in order to avoid political complications in relation to the recently agreed US, USSR and UK test-ban treaty negotiations.[131] Echoes of 1958 are clear here. In the event the test took place on 11 April.

Continuing Collaboration with the US

Trend saw Kissinger in Washington on 28 July 1972 to discuss aspects of possible and continuing US assistance for Polaris improvement work.[132] Further UK-US discussions were arranged for 10 August. MOD wanted to explore two main questions with the US: its attitude to the provision of assistance on a Poseidon

Secretary of State to Prime Minister noted that the design was a new departure for the UK and that a further test was necessary to give greater depth of knowledge of the design that would enable AWRE confidently to cope with any in-service problems. In this context Callaghan's comment is understandable. Macklen told the Secretary of State that although an intrinsically safe warhead to meet the Chevaline specifications could be manufactured following the successful 1974 test, AWRE needed to conduct one or two more tests in order to be able to guarantee to keep the warhead reliable and safe during its expected ten to 15 years in-service life. V.H.B. Macklen, DCA (PN) to PS/S of S Chevaline Nuclear Warhead Tests, 27 February 1976. A further test might also enable the designers to economise in the use of plutonium. Draft Minute from Defence Secretary to Prime Minister, 3 March 1976.

[128] TNA PREM 16/1181, Roy Mason to Harold Wilson, Polaris Improvement Programme – Nuclear Testing, 14 May 1975.

[129] TNA PREM 16/1181, P.R.H. Wright, 10 Downing Street to J.F. Mayne, MOD, Polaris Improvement Programme – Nuclear Testing, 28 May 1975.

[130] TNA PREM 16/1181, UK Nuclear Weapons Programme, 30 May 1977.

[131] TNA PREM 16/1181, Fred Mulley to Prime Minister, Planned Date for the Possible Next British Nuclear Test, 2 June 1977.

[132] TNA DEFE 24/896, Strategic Nuclear Policy Options: Discussions in the White House, Brief for DUS (P), 8 August 1972.

conversion, should HMG decide to pursue that option; and whether the US would be prepared to give the UK sufficient access to the US Poseidon warhead and to sell the non-nuclear components of the warhead and the re-entry body shells. In addition, London wanted to know what US plans were for phasing out Polaris and Poseidon from their own strategic forces, in view of their possible repercussions of those plans on future UK intentions. As for Poseidon, the UK wanted to know whether it could assume that any package would include the sale of the missile complete with the 'bus' and any necessary supporting systems in the ship under the Polaris Sales Agreement such as the sale for the re-entry bodies under the 1958 Mutual Defence Agreement and detailed information on the design of the US warhead and the sale of non-nuclear parts, also under the 1958 Agreement.

A British team came away from a meeting in Washington in September 1972 with information on four modernisation options. These were:

> Option 1: unequivocal support for the Super Antelope Programme – the US had already cooperated in providing underground test facilities.[133] This implied straightforward assistance in accomplishment of considerably more hardening, penetration and re-entry vehicle dispersal. Nonetheless, total R&D costs, not previously included in Super Antelope cost estimates, could be in the range of $200 million.
>
> Option 2: modify the Polaris re-entry system by adapting it to accept five Poseidon re-entry vehicles. This would probably enhance penetration capability through achieving higher degree of dispersal. However, it would not include a Poseidon bus, or bus technology. The approximate cost of this option could range from $150 to $175 million. For the warhead the UK would require complete US design information as well as extra fissile material for the seven live warheads (plus three inert) per missile.[134]
>
> Option 3: sale of Poseidon missile with Poseidon RVs and without a MIRV bus.[135] This would enhance penetrability against ABM defences. The cost estimate for this was approximately $500 to 550 million.
>
> Option 4: Acquisition of ULMS-1 (Trident C-4) with Poseidon bus. This would provide flexibility and added range. The costs in this case were estimated to be approximately $550 to $600 million. This was unclear since Schlesinger had emphasised at the outset that Option 3 and 4 would not involve MIRV

[133] TNA DEFE 24/896, The Way Ahead for Polaris, Extract from OR (N) 2 Mtg/72, 21 September 1972. This must be referring to effects tests where equipment and materials are exposed to radiation resulting from underground tests of nuclear devices.

[134] TNA DEFE 24/896, DEPC (N) Meeting 27 September 1972 Brief for CSA, DUS (P) and C POL (PE), The Way Ahead for Polaris, 22 September 1972.

[135] MIRV – Multiple Independently Targetable Re-entry Vehicle.

technology, subsequent clarification revealed that the MIRV capability including bus was not incorporated in any of the options discussed.[136]

In Options 2, 3 and 4 the warhead design and non-nuclear components would be obtained externally. However, the UK could assume that the US would be ready to give access to design information. Warhead manufacture would be in the UK. Submarine modification designs and associated missile and navigation and fire control systems would also be obtained externally.[137]

Kissinger emphasised to the UK team that the US Presidential election was posing problems, but if Nixon were returned to the White House and HMG wanted to pursue the Poseidon option, then the Administration would want to do what it could to help.[138] These new options however would now require careful scrutiny and they could well be subject to further discussion with USAEC experts before MOD could reach any conclusions about their potential value to the UK. This would be best done as part of the then current programme of reviewing the strategic nuclear options that was to be submitted to Ministers in the autumn of 1972. As the options were evaluated it was clear to officials that Super Antelope would have to continue in any interim phase otherwise given time factors, there would be a risk that the UK would be left with no effective deterrent capability at all in any reasonable timescale. MOD officials believed it was essential that Ministers agree to this, but that the recommendations to them would state that what was now known as the Hybrid version was the preferred solution (namely the non MIRVed Poseidon with Super Antelope on top). This however was still only a concept; there were no details of exactly how this would function. An extensive programme of flight trials would be required. Indeed as it would later transpire, the costs for flight trials rather than those directly related to nuclear warheads expenditure, would be the main reason why Super Antelope/Chevaline overall costs increased.

November 1972: A Key Date

November 1972 was the turning point. Carrington finally submitted his paper on options to the Prime Minister, Foreign Secretary, Chancellor, Cabinet Secretary and Chief of the Defence Staff on the future of the deterrent and the possible options.[139] Carrington did not doubt that the UK should improve and eventually replace Polaris, but there were three overlapping questions to be addressed. What

136 TNA DEFE 24/896, AUS (Defence Staff) to ACNS (P) and ACSA (N), Strategic Nuclear Policy Options: White House Visit, 4 September 1972.

137 TNA DEFE 24/896, UK Nuclear Deterrent: Information provided by White House, 10 August 1972.

138 TNA DEFE 24/896, P.D. Nairne, MOD to Sir Burke Trend, 11 August 1972.

139 TNA PREM 15/1359, Carrington to Prime Minister, Strategic Nuclear Policy, 6 November 1972.

was the best option on military, technical and financial grounds for improving the current force? How should the UK approach the longer term question of replacing Polaris and how did the answers to this affect the choice in the first question? What were the implications of the answers to the first question and the second for future relations with the US and French? There were now three choices clearly set out with costings: Super Antelope, Hybrid and UK Poseidon.

Carrington thought that Hybrid was the best option for the UK – even though the costings and technical elements had been studied in least detail – provided the US was willing to sell the missiles under the Polaris Sales Agreement. A new agreement might fall foul of any 'no transfer' or 'non-circumvention' clauses in SALT II, which was why the UK had been so keen to ensure that the US did not agree to any such inhibiting arrangements.[140] He recommended that this should be pursued in two parallel stages. At official level, the UK should clarify further with the Americans the technical and financial implications of both Super Antelope and Hybrid options in order to secure a firmer basis for a final choice. At the political level the UK should explore the wider policy implications of both the shorter term and longer term options for future relations with the US and France to secure a clearer political basis on which to shape future strategic nuclear policy. The hope then would be to take a firm decision in early 1973 on Polaris improvement as a basis for negotiating an agreement with the US. The government would also want to be in a position to decide at the same time how it wished to carry forward further exploration of the policy and programme required for a successor system. Heath should also inform Nixon of the UK decision and he should sound out his French counterpart on possible avenues for Anglo-French collaboration.

Trend advised caution on Hybrid.[141] The uncertainties might suggest that it would be premature at that point to take a decision, even in principle. It was important to guard against a compromise that would give the UK the worst of both worlds. The UK needed to know more about the degree of assistance that might be forthcoming from the US in terms of development and production and about expenditure involved. In addition, the UK need to probe more deeply on US plans for in-service lives of Polaris and Poseidon, since it was on these that the validity of any assertions about additional expenditure and the period of effectiveness of Super Antelope ultimately depended. In Trend's view, these considerations meant that Heath should expose UK thinking frankly to Nixon and ascertain just how far the US would be prepared to support the UK on each of the options. Meanwhile, it was important that the Defence Secretary should not commit himself any further with the French in the nuclear context than was strictly necessary until such times as the UK had a better idea of Nixon's intentions. Work on Super Antelope should

140 TNA FCO 46/824, A.P. Hockaday, Cabinet Office to C.M. Rose, FCO, SALT – Draft Brief on No Transfer, 19 September 1972; TNA FCO 46/829, C.M. Rose to Permanent Under Secretary and Private Secretary, SALT: Anglo-US Bilateral Talks, 25 October 1972.

141 TNA PREM 15/1359, Trend to Prime Minister, Defence Policy/Strategic Nuclear Options, 10 November 1972.

continue uninterrupted as proposed by Carrington and agreed by the Chancellor. Trend suggested that the Prime Minister should confirm his own concurrence at the meeting scheduled for Tuesday 14 November 1972.

At the meeting, Carrington made clear that the Chiefs of Staff regarded the deterrent as the corner stone of UK security. If it was to remain credible, it had to be capable of penetrating the defences around Moscow. Intelligence estimates suggested that by 1974–75 the radar coverage would extend over the whole of the North Atlantic. Polaris would then have insufficient firepower to penetrate a 100 missile defence as permitted by the ABM Treaty.[142] Conclusion of the ABM Treaty had made UK calculations that much easier, but the problem was not completely solved as Macklen explained to Carrington in mid June 1972.[143] It was important for the UK that the ABM Treaty had included a separation distance between the two permitted ABM complexes of 1300 km, which meant that in the worst case the UK deterrent would only have to face the 100 launcher no reload capability ABM complex around Moscow. However, the permitted radar coverage was also sufficient to allow the Russians to use the ABM system around Moscow to cover large areas of the Western USSR. The 100 ABM limit would be sufficient to negate either one or two boats worth of unimproved Polaris missiles. The warheads were close enough to be rendered ineffective by one successful ABM explosion. Since the UK could expect on reliability grounds either 13 or 26 missiles to threaten the defended area of Western Russia, there were more than sufficient ABM launchers available after allowance for their reliability factor. Therefore, the ABM Treaty did not remove the need for the Polaris Improvement Programme. However, the Treaty did set a limit to the UK task over a longer period. Both Super Antelope and Poseidon (with 12 warheads) would fulfil the Joint Intelligence Committee's criteria for an effective deterrent threat. Poseidon was marginally superior solely on penetration arithmetic grounds. Thus the ABM Treaty enabled the UK to define more closely its improvement requirements against the then current Soviet systems.

As soon as possible after 1975, it would be essential to have a new system capable of penetrating such defences and if possible embodying some reserve capacity against other possible improvements. Penetrative capability was the important criterion; in other words the ability to defeat any defence that might be mounted to ensure that an incoming nuclear warhead reached its target. The arguments for the various options, both pro and con were rehearsed.

On balance, the MOD view was that Hybrid was the better option provided the US was prepared to supply Poseidon missiles (which would give an extra 400 miles of range) and that further clarification with the US of the technical and financial implications of the Super Antelope and Hybrid options reinforced the provisional conclusion. Although the Foreign Secretary agreed that they had to maintain Polaris as the ultimate deterrent, the Chancellor felt that the costs had to be included in any

142 TNA PREM 15/1359, Burke Trend to Prime Minister, 15 November 1972.

143 TNA DEFE 24/783, V.H.B. Macklen to PS/Secretary of State, SAL Agreement and Polaris Improvement, 14 June 1972.

consideration of the total defence expenditure; he was evidently not prepared to make a special case for the modernisation that had been the MOD preference.

Heath summed up the discussions, which had also looked at the prospects for collaboration with the French and the technical risks of Hybrid and the costs of establishing separate underground testing facilities. It was clearly necessary to improve the penetrability of Polaris and Hybrid – the code name changed to STAG in early 1973 – seemed to be the best bet subject to the aforementioned caveats. He would therefore raise the possibility with Nixon at their next meeting and see whether the US would supply Poseidon missiles and details on the technical and financial aspects of Super Antelope in order to provide a firm basis for the final choice. Heath took the view that it was very important that the UK should not appear as in the Skybolt case, to be presenting the French with a *fait accompli*, which the UK had arranged with the US without reference to French interests. It was important too for the French to understand that the final decision had not been taken and the prospect of UK collaboration on a successor system would not be foreclosed. Carrington could make this point to his French counterpart, Michel Debre. As for the financial aspects, it would be necessary to give this further thought and to the degree of reference to the Cabinet which would be appropriate.

Heath meets Nixon: February 1973

Heath met Nixon on 1 February 1973 in Washington to discuss the UK deterrent.[144] He explained the UK preference for STAG, but added that uncertainty on its feasibility mean that a final decision was not possible without further discussion with the US on its technical, logistic and financial implications and left an *aide memoire* setting out in detail the UK options and thinking.[145] Nixon asked whether British thinking had extended to possible collaboration with the French. London had decided to defer the question with the French until after their 1972 election, and although the Prime Minister Georges Pompidou had always seemed disinclined to discuss the issue, Debre was more open to discussion. Nixon was sympathetically disposed to continued collaboration with the UK, although this would involve him in some difficulties with Congress and probably the USSR during SALT II, which was scheduled to start soon. He added that if it were possible to find a way of dealing with the Poseidon question, then he would support making it available to the UK. That said, further technical discussions had to come first without prejudice to a definitive decision that still had to be taken. These would be organised through James Schlesinger, who was only just ceasing to be head of the USAEC.

---

[144] TNA PREM 15/1359, Record of a Discussion at the Old Executive Building Washington DC, British Nuclear Deterrent, 1 February 1973.

[145] STAG was a hybrid of the Poseidon missile with a UK Super Antelope front end. It had been suggested as an alternative to Super Antelope during 1972, but not pursued in the end. It was originally described simply as Hybrid.

Shortly thereafter Professor Bondi met Schlesinger to review the technical issues and their discussions were successful, so successful in fact that they resulted in an expansion of choices open to the UK from three to six. This entailed any one of the combinations of two forms of warhead and three forms of missile; the talks also allowed the UK to set aside some of the technical and logistical snags which the MOD thought might arise.[146] All of this was subject, as Nixon had made clear, to his over-riding decision on political feasibility with Congress. The next step for UK officials was to confirm the costs of the various options as a prelude to a Ministerial decision on Polaris Improvement.

The US view was the other options would be a better bet for the future effectiveness of the UK deterrent; these included the Mark III re-entry body carried on Poseidon whose small size and re-entry speed made it an extremely difficult target for the defence; the Poseidon C3 missile and the Trident C4. If the UK wanted either, then they could probably be made available, although the Poseidon MIRV capability could not be. For any of the options likely to be available, some modified missile guidance and warhead-dispensing facilities would also have to be developed. These and their implications for the weapon system as a whole would be the main areas of technical cost uncertainty.[147]

Schlesinger had stressed that great political difficulty could arise if there were to be any discussion whatever of the options other than with the chosen few US experts selected by Kissinger before the President had given a specific decision in principle. This decision would require a specific request from the UK to the President through Kissinger. Once Nixon decided, it would then be possible to start the necessary detailed decisions with the agencies involved. Such a decision would not be forthcoming until the UK had committed itself to the solution that it wished to study in detail. More detail was therefore required to follow up the recommendations on the Mark III re-entry body. Carrington suggested to Heath that a UK expert team should go to Washington to discuss the US strategic assessment of Soviet capabilities and check UK estimates of the costs of a possible Mark III re-entry body. Once this had been done, it would be possible to come to a decision either to go for STAG or the Mark III. Heath agreed.[148]

Schlesinger's proposal had put the UK in a bit of a spot. Officials could not abandon the results of two year's careful study, which had embraced but rejected the kind of solution that Schlesinger had proposed without a fuller understanding of the reasoning that had informed the recommendation, hence the need for urgent technical discussions with US officials.[149] Schlesinger thought that the STAG

---

[146] TNA PREM 13/1359, Trend to Prime Minister, Nuclear Deterrent, 16 February 1973.

[147] TNA PREM 15/1359, Carrington to Prime Minister, Improvement of the Strategic Nuclear Deterrent, 21 February 1973.

[148] TNA PREM 15/1359, Bridges to R.J. Andrew, MOD 23 February 1973.

[149] TNA PREM 15/1359, Foreign Office telegram no. 525 to Washington, 26 February 1973.

project was not the best solution, principally because of the inadequacy of the re-entry speed. There were two better options: either Polaris with Mark III or Poseidon with Mark III and it was the US estimate that Britain could build the Mark III re-entry body to a US design within three years. In any case, UK experts (Macklen, Jennings, MOD and Peter Jones, AWRE) would be welcome to come and discuss the details with Schlesinger's team.[150] These included penetration calculations, discrimination and warhead hardening, dispensing and production, costs, timescale and relevance of existing inter-governmental agreements.[151]

Trend was concerned that the UK seemed to be going round in circles on this issue – as indeed they were; the US had made suggestions, but the UK could not find out the necessary details to make a decision, and the President would not make a determination until the UK had indicated which option it wanted to pursue.[152] However, the meeting eventually took place on Friday 9 March 1973. The US team was extremely helpful and the talks clarified a number of important issues for the UK. MOD and AWRE thought that the US seemed motivated by a desire to give the UK the best advice as they saw it; they were not trying to push London into adopting any particular solution.[153]

Four key issues emerged. First, the US had no new intelligence about terminal defence that had not previously been made available to the UK. The only qualification to this was a reference by the Russians in the SALT talks to their ability to deploy a Sprint type of terminal defence system, which might be significant.[154] Second, the need for a prior UK commitment to Mark III was not driven by any US reluctance to let the UK have the information without some assurance that it would be used. Instead, the problem lay in the publicity that would be likely from the moment the process of obtaining Presidential Determination to release the necessary design information was set in train. Third, with the use of Poseidon missile, de-MIRVing was likely to remain the most difficult and crucial problem. The UK was unlikely to find a cheap solution. Given the political constraints that the US had indicated, there was likely to be little saving in the cost of the STAG solution compared with a Mark III re-entry body/warhead, system to set against the advantage of the latter's invulnerability to any foreseeable development of Soviet ABM defences. Fourth, the UK could not usefully launch the studies that Heath had asked for in the memorandum that he had left with Nixon in February until the UK decided whether to ask for Mark III re-entry body information; and, if this was done, until a Presidential Determination to release it had been given.

---

[150] TNA PREM 15/1359, Summary Record of a Conversation between HM Ambassador and Dr Kissinger, 1 March 1973.

[151] TNA PREM 15/1359, FCO telegram no. 608 to Washington, 6 March 1973.

[152] TNA PREM 15/1359, Trend to Prime Minister United Kingdom Deterrent, 5 March 1973.

[153] TNA PREM 15/1359, R.J. Andrew, MOD to Burke Trend, Cabinet Office, Polaris Improvement Programme, 14 March 1973.

[154] Sprint was the US ABM interceptor missile.

No timescale for this was possible. MOD now started to evaluate the pros and cons of the various options in light of the new data brought back before preparing recommendations for Heath and other Ministers.[155]

A Narrowing of the Options: May 1973

However, by May 1973 it seems that the choice was now down to two: Super Antelope or what officials now called Option M – de-MIRVed Poseidon C3 with the Mark III re-entry body.[156] MOD officials concluded that all three options for improvement, Super Antelope, STAG and Option M, would provide the penetration capability currently required. In military terms, Option M was the most attractive since like STAG, it offered operational and logistical advantages over Super Antelope, but it was also technically more advanced and carried a higher technical risk.[157] And it appears as if Option M was now the front runner both in terms of technical capacity (longer range and capacity to overcome both exo and endo atmospheric interception) and maintenance (availability of continued US technical support until the 1990s). US advice was that Super Antelope would not be able to cope with endo atmospheric interceptors should the Russians decide to deploy them. The strong advice from the UK Ambassador in Washington was that if the UK were to disregard this advice and not accept Option M, then the US would no longer regard the UK in the same league as the US and downgrade British significance accordingly. However, the only US evidence for Russian intent on such interceptors was that one chance remark made during the SALT talks. Even the maintenance argument was not without flaws. US future intentions were still unclear as the Underwater Long Range Missile System (ULMS) system was in the works – this would be Trident, and moreover the Americans had told the British that they had Polaris spare parts 'running out of their ears'.

Trend advised Heath that the case for Option M was not proven and it was at least arguable that for the next ten years Super Antelope would suffice against any form of ABM defence that the Soviet government was likely to be able to deploy in that period. Furthermore, the US advice was that Option M should not be 'anglicised' but bought outright. Nor would there be enough fissile material around to meet the Mark III re-entry body/warhead requirement. Trend thought that on balance Super Antelope was still the best all round bet for Britain, in terms of cost effectiveness, strategy and maintaining a degree of independence from the US. Ministers met on 7 May 1973 to review the options and decide on the next steps.

---

155 TNA PREM 15/1359, R.J. Andrew, MOD to A.A. Acland, Polaris Improvement Programme, 23 March 1973.

156 TNA PREM 15/1360, Trend to Prime Minister, 2 May 1973.

157 TNA DEFE 13/1038, Options for Improving the Polaris Force (Memorandum by the Ministry of Defence), 1 May 1973.

Carrington circulated a minute to his colleagues before this meeting and a paper setting out the arguments and issues.[158] If costs were not an option, then Carrington felt that Option M was the best choice. It was one moreover strongly supported by the Chiefs of Staff. Furthermore, if costs were a major factor, then Super Antelope now had the edge over STAG since it would provide the improved penetration capability required. More importantly, it would present smaller financial problems for the defence budget. Going for Option M would also increase UK dependence on the US and consequently the extent to which the UK would be at risk if developments in SALT placed new constraints on the ability of the US to assist the UK and there were clear signs that the Russians were going to press the non-transfer issue hard in SALT II. The UK faced a dilemma.[159] The US had made a very good offer and would find it hard to see how it could be rejected; acceptance would tie the UK closer to US apron strings and further complicate any future collaboration with the French.

So in the finest tradition of deferred decisions, Carrington admitted that he could not make any recommendations at that time on the options and suggested to Heath that yet more talks with the US would be needed to elicit views on strategic, technical and financial aspects; French views should also be sought. Once this had been done Ministers (himself, the Chancellor, Foreign Secretary and Prime Minister), could take stock again and possibly take a decision in principle. In the interim, it was essential that Super Antelope programme maintain its momentum and a further £17.5 million should be allocated for the next six months. However, a decision on Option M was still needed by 1 July 1973 and the planned meeting with Kissinger in London on 10 May 1973, Heath's meeting with Pompidou and the Cabinet discussion on public expenditure on 24 May, would help things move towards a decision.[160]

Kissinger met Trend and other UK officials on 10 May in London. Trend handed him a UK *aide memoire* that sought confirmation of a number of understandings and assumptions on the possible deployment of terminal defences around Moscow, the in-service lives of Polaris and Poseidon in the US Navy and the extension of the Polaris Sales Agreement. Kissinger undertook to provide considered answers to these points and also on the precise nature and possible timescale of the requirements for a Presidential Determination and the agreement of the JCCAE and on time factors related to the present Polaris production programme. He would try to elicit answers to these before the Heath saw Pompidou on 21 and 22 May 1973. Kissinger was also adamant that the US would not enter into

[158] TNA PREM 15/1360, Carrington to Prime Minister, 2 May 1973.

[159] TNA PREM 15/1365, Prime Minister to Secretary of State for Foreign and Commonwealth Affairs, Strategic Arms Limitation Talks, Prime Minister's Personal Minute No. M 57/72, 27 June 1972.

[160] TNA PREM 15/1360, Trend to Prime Minister, 4 May 1973.

a SALT agreement that would invalidate arrangements previously established for the provision of missile systems to the UK.[161]

Kissinger was as good as his word and responses were provided promptly. MOD officials were then able to prepare a memorandum for the Prime Minister containing an analysis of the new information.[162] This confirmed that there was indeed no new evidence of Soviet intentions to deploy an exo-atmospheric interceptor. Carrington remarked that it now looked as if the US might be willing to entertain the provision of Poseidon missiles with a MIRV capability, something that would make the cost look much better. It was now clear to the UK that, if the UK chose Option M, the US would be prepared to authorise the sale of Poseidon minus the warheads under Article IV of the Polaris Sales Agreement, thus avoiding the need for a new agreement. This accorded with MOD's strong preference.

For MOD, the US response did much to smooth even further the path towards the introduction of Option M if this was decided upon. The further information about the way in that the Soviet ABM capabilities were open to improvement lent support to the assessment already made in London of the lesser risks and greater military advantages offered by Option M compared to Super Antelope. Carrington urged Heath that now was the time to take a decision.[163] The Poseidon production line was planned to start shutting down in July or August 1973. The US indicated too that it had rejected Soviet demands for a ban on transfers of strategic offensive weapons systems and they did not expect that possible SALT limitations would affect any of the UK options. If the UK were to acquire Poseidon, the MOD's Chief Scientific Adviser noted that its acquisition might have unfortunate consequences on the arms control atmosphere.[164]

However, MOD officials did wonder whether the Americans would be able to maintain their resistance.[165] At official level, at least MOD thought that none of the developments since the last Ministerial discussion had a determining effect on the final choice. There were now fewer questions marks about the willingness of the US to provide Option M if the UK wanted it. The French had begun to nibble at the bait of cooperation with the UK. The element of military risk in going for Super Antelope remained a matter of subjective judgement, depending on what guesses one was prepared to make on Russian intentions on ABMs. Political arguments for

[161] TNA PREM 15/1360, Record of Meeting Sir Burke Trend/Dr Kissinger, The Strategic Nuclear Deterrent, 10 May 1973.

[162] TNA PREM 15/1360, Further Consultation with US Memorandum by Ministry of Defence Officials, 8 June 1973, TNA DEFE 13/1038, Polaris Improvement: further Consultations with US Memorandum by Ministry of Defence Officials, 8 June 1973.

[163] TNA PREM 15/1360, Carrington to Prime Minister, Polaris Improvements, 8 June 1973.

[164] TNA DEFE 13/1038, Future of Our Deterrent, *Aide Memoire* by CSA, 15 June 1973.

[165] TNA DEFE 13/1038, Polaris Improvements: further Consultation with US Memorandum by Ministry of Defence, 5 June 1973.

not tying the UK to US apron strings in this field for another decade were still valid and the desirability on financial grounds for going for the option with the least impact on costing still applied.[166] The Chief Scientific Adviser on the other hand believed that in pursuing either Option M or full Poseidon the US Administration would have to put such a request to the JCCAE with the consequent risk that this would delay work on Super Antelope. Unless the Administration and the UK could be very confident of a positive answer, there were very real risks.[167]

In advising the Prime Minister, the Cabinet Secretary now moved in the direction of endorsing Option M. Although Trend found it difficult to say exactly why, he suspected that it would be unwise to discount too much the latest US warning about the risk of Soviet terminal defences that would fall short of Sprint but would still outstrip Super Antelope.[168] His inclination was to tell the US at once that, although a final decision had to be reached, the US responses to UK questions had made London recognise Option M's advantages very clearly. The Chief of the Defence Staff had the same view: the military and technical advantages over Super Antelope were clear-cut: greater range and sea room, US support and low technical risk as the system was largely proven.[169] The Chief of the Naval Staff was of a similar mind.[170]

Ministers Meet: 12 June 1973

Ministers discussed the US response on 12 June. They concluded that before deciding between the options, it would be desirable to discover whether, in light of recent SALT developments (US opposition to a ban on SLBM MIRVs and refusal to accept any non-transfer clauses), there had in fact been any change in US attitudes towards the supply of fully MIRVed Poseidon to the UK. Lord Cromer, HMA Washington was tasked to find out from Kissinger.[171] He was also invited to offer his own assessment of Nixon's ability to secure Congressional agreement to any Presidential Determination that may be required, both for de-MIRVed and MIRVed Poseidon. Ministers wanted to decide on the options before the end of July; hence the urgent need for Kissinger's views.

---

[166] TNA DEFE 13/1038, W.F. Mumford to Secretary of State, 11 June 1973.

[167] TNA DEFE 13/1038, Future of Our Deterrent, *Aide Memoire* by CSA, 15 June 1973.

[168] TNA PREM 15/1360, Trend to Prime Minister, The Nuclear Deterrent: Polaris Improvements, 11 June 1973.

[169] TNA DEFE 13/1038, CDS (Lord Hill Norton) to Secretary of State, 22 June 1973.

[170] TNA DEFE 13/1038, Record of a Discussion in the Secretary of State's Room, Polaris Improvements, 23 June 1973.

[171] TNA PREM 15/1360, FCO telegrams nos 1276 and 1277 to Washington, Personal message for the Ambassador from Sir Burke Trend, Improvements of the Strategic Nuclear Deterrent, 15 June 1977.

The Washington Embassy's tentative assessment of the probability of obtaining Poseidon was that de-MIRVed option might well be possible, but instinct told Richard Sykes (Minister) that a fully MIRVed Poseidon would be much more problematic. An authoritative assessment would be needed from the White House. Kissinger's advice was much bleaker. Giving the UK MIRVed technology would cause major problems with Congress. His understanding of the legal position was that only a Presidential Determination and not specific Congressional authority was necessary to give the UK MIRV and the Mark III, but Congress would undoubtedly seize on the issue, and the Administration could not stop them from discussing it. In short, he felt that a Congressional uproar was certain and that the President would much prefer not to have to face this in his present post-Watergate weakened position. It would also cause problems with the Russians in SALT, but these were not insuperable.[172]

When pressed by the Ambassador, Kissinger thought that if Heath made a personal appeal to Nixon there might be a 50–50 chance, but that was only his personal opinion. Lord Cromer gained the very strong impression that Nixon would be mightily relieved if the UK did not press the MIRV question.[173] Schlesinger had taken the view that he could not see why the UK would want MIRVed Poseidon; the only advantage over the Mark III was that it gave greater accuracy. Kissinger said that if the UK wanted to send experts to discuss differences in yields and accuracies, then that could be arranged. Officials were worried that this raised doubts as to whether they had accurately represented to Ministers Option M's technical parameters. They had assumed that yields were the same, which would have implications for weight, range, penetrability and the number of warheads that could be carried. This had to be resolved urgently and Macklen was despatched to Washington once again to clarify the matter with US experts.[174]

In light of the news from Washington Carrington noted that Ministers had to assess together the operational, technical and logistic considerations that pointed in favour of Option M, the financial considerations that pointed in favour of Super Antelope and the political considerations, which were finely balanced. Important amongst these was the desirability of sustaining a demonstrable UK nuclear weapon technology capability, which pointed in favour of Super Antelope and the related question of Britain's position as an independent nuclear power in the eyes of Europe and the rest of the world. But these also included the interlocking factors of the credibility of alternative deterrent systems, the desirability of deploying the most militarily effective system, the future of European defence and UK relations

[172] TNA FCO 46/1002, Tom Bridges, 10 Downing Street to Michael Alexander, FCO, 30 April 1973. This describes UK reactions to latest Soviet moves on non-transfer issues.

[173] TNA PREM 15/1360, Washington telegram no. 2055 to FCO, Personal for Sir Burke Trend, Improvements to the Strategic Nuclear Deterrent, 1 July 1973.

[174] TNA PREM 15/1360, A.P. Hockaday, DUS (Policy and Programmes), MOD to Sir Burke Trend, Cabinet Office, 3 July 1973; FCO telegram no. 1391 to Washington, 5 July 1973; FCO telegram no. 1415 to Washington, 9 July 1973.

with the US and France. Trend still thought that the balance was in favour of Option M even though it was the more expensive of the two choices left on the table.[175] Across Whitehall in the MOD, Carrington had made clear to officials that in reviewing and submitting the options, they were not to make a recommendation in favour of one or other option.[176] He wanted to remain neutral in the presentation of the issue to his colleagues.[177]

MOD and AWRE Views: Future Capabilities and Testing

What of the potential impact on AWRE of all of this? Whichever of the options was selected by Ministers, there would clearly be significant effects on R&D at Aldermaston as well as on the UK's overall capability to sustain current and possibly future generations of nuclear warheads. This was also critically important in the final decision on Polaris Improvement. We can see the first signs that AWRE and its allies in the MOD clearly preferred Super Antelope; at least that is the impression from a key memorandum put to the Permanent Under Secretary by Vic Macklen now DCA (PN) on 15 June entitled 'Maintenance of British Nuclear Weapon capability at AWRE if the UK buys US Poseidon/Option M'.[178] This was also submitted to Carrington. AWRE would still be required to provide surveillance and maintenance for a Poseidon warhead plus the WE177s; 15 years of experience had shown that deterioration could and had occurred that affected both safety in storage and reliability in operation. Surveys had shown that it was necessary to keep a considerable amount of support R&D over a wide range of disciplines in order to cope with life and storage problems as they began to emerge. A large effort, covering both research and production, was required to fulfil these tasks. However, this type of work would not place the kind of demand upon warhead design teams that would encourage either the retention or recruitment of the necessary skilled personnel.

A new and challenging programme was required and although its precise details were unclear, it would be essential if the UK were to decide upon a new strategic system. Nevertheless, a number of its features were identifiable. These could include: vulnerability studies for Option M and Poseidon, investigations into novel ways of generating nuclear reactions for weapons purposes, novel penetration

---

[175] TNA PREM 15/1360, Trend to Prime Minister, The Nuclear Deterrent: Polaris Improvements, 13 July 1973.

[176] TNA DEFE 13/1038, A.P. Hockaday (DUS (P)) to PS/Secretary of State, Improvement of the Strategic Nuclear Deterrent, 6 July 1973.

[177] TNA DEFE 13/1038, W.F. Mumford PS/Secretary of State to DUS (P), Improvement of the Strategic Nuclear Deterrent, 9 July 1973.

[178] TNA DEFE 13/1038, V.H.B. Macklen to PUS, Maintenance of the Nuclear Capability in event of Option M, 15 June (memo attached). Macklen had returned from UKAEA Risley to the MOD with Sir William Cook.

techniques, the development and design of progressively harder warheads against improving defences, the development of new nuclear testing techniques against the possibility of a CTBT and as an essential supporting activity, the continued participation in underground nuclear flux and device tests under arrangements to be agreed with the US. Such a programme would be totally ineffective without the ability to have occasional nuclear tests of the concepts produced by theoretical work. Moreover, against the background of a possible CTBT or a limitation of underground tests it would be most unwise to postpone the research programme whatever the timescale for future successor systems. Whilst techniques could be foreseen which might allow nuclear warhead development without significant nuclear yields, this prospect was still some way off and the more data gathered from the existing kind of underground tests the better AWRE's knowledge and capability would be.

If the right calibre of staff was to be attracted to AWRE and to this kind of work, it was as vital to hold out some prospect to design teams that their work would come to fruition to meet a British requirement. If there were no real prospect of a successor system beyond Option M because of further recourse to US designs, this would ultimately destroy any future UK design and production capability. Moreover, if the UK bought either of the US options and decided that it must retain as full a nuclear weapon capability as possible, then it must also support a full and challenging research and testing programme at Aldermaston. Without such a programme partly because of political pressure on nuclear tests and partly because it would take years to recreate existing design potential, there would be a risk that UK capability would be lost forever. Indeed the loss of highly skilled staff would in the end diminish UK capabilities to maintain current warhead stockpiles.

By July 1973 the UK position on testing was such that London now had no wish to see a CTBT enter into force for at least five years or so. During that time there were plans to test UK devices and participate in US effects tests as we have seen. Policy was thus to maintain a low profile, leaving initiatives to other countries and keeping in step with US policy.[179] If a test ban were to enter into force, it would greatly affect the UK's ability to advance weapons design and development. What would otherwise be a feasible development might prove impossible; and although some development might be practicable, compromises of one sort or another would be have to be accepted.[180] Thus any UK CTBT initiative at that time, in the MOD view, would run counter to defence interests; would be unlikely to be successful; it would antagonise the US and would complicate the problem of maintaining an effective UK nuclear force. This view did not change until 1977 when the trilateral test-ban talks began between the UK, USSR and US and even then there

[179] TNA DEFE 19/169, Carrington to Edward Heath, 27 July 1973.

[180] TNA DEFE 11/791, P.M. Stanford, Commodore, Secretary COS Committee, Comprehensive Test-ban Treaty, 28 October 1975.

were testing pressures that made balancing competing foreign and defence policy objectives difficult.[181]

Ministers Meet Again: 26 July 1973

Ministers met on 26 July to discuss the options. This meeting led to the Defence Secretary, in consultation with the Chancellor, circulating a note that showed the implications of different levels of expenditure for the defence budget as a whole according to whether the decision was taken in favour of Option M or Super Antelope.[182] As if to complicate things further the US indicated to Chief of the Defence Staff (CDS) Sir Peter Hill Norton in late August that it might be possible to spread the costs of Poseidon acquisition over a longer period, and it would therefore be better suited to meeting the UK's financial problems.[183] Macklen once again trooped across the Atlantic to find out what this was all about, as it now seemed that the US was thinking in terms of MIRVed Poseidon for the UK. Such an approach would however lengthen the duration of the programme and increase the overall costs.[184] Trend was convinced that the UK still needed to consider fully the alternative of Option M and further cost details were still required. It would also be necessary to get a clear view from Kissinger as to whether MIRVed Poseidon was now indeed a possibility and his agreement that another UK team should visit Washington to talk about Option M and its finances.[185]

A Ministerial meeting to review the choices had been tentatively arranged for 12 September and before this Carrington wrote once again to Heath to summarise the current position on the options. Despite the indications that CDS had had about the possibility of the UK receiving MIRVed Poseidon, the advice from Washington was that the prospects were not that much better than they had been the last time Kissinger had been asked for a view.[186] The choice still lay between Option M and Super Antelope and the core arguments about effectiveness, cost and dependence on the US had not really altered since the last time that Ministers had reviewed

---

[181] TNA DEFE 19/181, Fred Mulley to Prime Minister, British Nuclear Test Programme, 26 October 1978 and David Owen to Prime Minister, British Nuclear Test Programme, 31 October 1978.

[182] TNA PREM 15/1360, Carrington to Prime Minister, Defence Expenditure, 17 August 1973.

[183] TNA PREM 15/1360, Message No 1 No 10 to Chequers Prime Minister from Christopher Roberts, 25 August 1973.

[184] TNA PREM 15/1360, Notes on a meeting held at Crystal City, Alexandria VA, 28 August 1973.

[185] TNA PREM 15/1360, Trend to Prime Minister, United Kingdom Deterrent, 31 August 1973.

[186] TNA PREM 15/2038, Carrington to Prime Minister, Improvement of Strategic Nuclear Deterrent, pp. 3 and 4 missing, 1973.

the problem.[187] The Cabinet Secretary advised Heath that the arguments were still slightly in favour of Option M, but it would be necessary to fund work on Super Antelope until such times as the US was able to confirm that it could supply the de-MIRVed missiles.

12 September: Another Inconclusive Meeting; Decisive Meeting on 31 October

Another inconclusive Ministerial meeting duly took place on 12 September even though Carrington had argued that it had become urgent for Ministers to take a decision.[188] At that point the operational, technical and logistic considerations continued to point in favour of Option M, but the political considerations, which included the arguments pointing in favour of Super Antelope, of the future of a UK nuclear weapon technological capability remained finely balanced.[189] Financial considerations suggested adoption of Super Antelope. The core problem was finance and the need to reduce the defence budget, specifically the proportion of GNP devoted to defence. The Foreign Secretary was not willing to sacrifice conventional forces given the damaging consequences for NATO defence postures. Heath summed up the meeting by stating that maintenance of a credible nuclear deterrent had to be regarded as of very great importance in the wider political context. However, before choosing between the options, further consideration was required on the implications of possible changes in the level of defence expenditure as a whole. In the meantime, Super Antelope work would have to continue.[190]

Come October 1973 the pendulum was swinging decisively in favour of Super Antelope. It was clear to the new Cabinet Secretary (Sir John Hunt) that Option M would be in trouble in view of developments (Watergate) in Washington since September. These had raised very grave doubts about whether the UK could still rely on the Administration's ability to honour the offer of Option M.[191] Super Antelope though still less satisfactory than Option M in operational, technical and logistic terms, still constituted an adequate deterrent. It would also involve lower expenditure in the period up to and including 1977–78 as well as lower dollar

---

187 TNA PREM 15/2038, Trend to Prime Minister, The Nuclear Deterrent, 11 September 1973.

188 TNA DEFE 24/895, Carrington to Prime Minister, Improvement of Strategic Nuclear Deterrent, 10 September 1973.

189 TNA DEFE 24/895, A.P. Hockaday, DUS (P) to PS/Secretary of State, Improvement of Strategic Nuclear Deterrent, 10 September 1973.

190 TNA PREM 15/2038, The United Kingdom Deterrent Minutes of a Meeting, M/17/2, 12 September 1973.

191 TNA PREM 15/2038, John Hunt to Prime Minister, 29 October 1973.

expenditure.[192] London was also most anxious to avoid having to make any cuts in conventional forces committed to NATO.[193] So, when the small group of Ministers met again on 31 October it was now little wonder that the Prime Minister summed up the discussion to say that there was general agreement that that right course was to develop Super Antelope. By that stage, MOD had extensive plans for Super Antelope testing; this entailed ground testing of sub-systems in the UK and US and flight testing of sub-systems at Woomera using Skylark and Falstaff missiles, together with Super Antelope pad firings at Canaveral, Florida and submarine firings in the Eastern Test Range.[194] MOD had already received assurances of support from DOD for the programme as a whole, from the USAEC for warhead development and nuclear device tests, and from the Defence Nuclear Agency for nuclear vulnerability and nuclear effects tests.[195] These tests exposed selected materials in re-entry bodies and warheads to x-ray damage, something that helped start the flow of much needed information to AWRE.

Thus after six years of detailed discussions and analyses and seemingly endless technical exchanges with the US, a definitive decision on the future of the strategic nuclear weapons programme had been taken at long last. The decision to resume testing had of course been taken before Super Antelope was confirmed and the first UK test since 1965 took place in May 1974 as noted above. By then the US and USSR were on the point of agreeing a threshold nuclear test ban treaty. When Kissinger first mentioned the possibility to British officials, he gave a firm assurance that British interests in regards to nuclear weapons testing were being taken fully into account.[196] This assurance was further discussed between officials in Washington and London was satisfied that any limits that the Americans might be prepared to accept on their own testing would not be so restrictive as to cause difficulties for the UK. That said, the UK did not think that a threshold treaty would be seen internationally as a valid step towards a CTBT. They were right.

The Test Ban and the Polaris Improvement Programme 1971–1972

MOD unease on test-ban issues appears to have surfaced in early 1971. Although there appeared to be no requirement to test the Polaris warhead at that point, the UK did need in the course of the improvement programme, to expose materials,

---

[192] TNA PREM 15/2038, Future of strategic nuclear deterrent, M/17/2, 1 November 1973, Cabinet Defence Expenditure, Minutes of a Meeting on 30 October 1973.

[193] TNA PREM 15/2038, Washington telegram no. 190 to FCO, 17 January 1974.

[194] TNA DEFE 25/334, Controller of the Navy to CDS, Polaris Improvement Programme, 12 October 1973.

[195] Papers on PM's letter to Nixon with request for assistance and his reply not on file; appear to be retained pending review. Date 6 April 1974.

[196] TNA PREM 16/93, A.A. Acland, FCO to The Lord Bridges 10 Downing Street, 19 June 1974.

parts of the re-entry system and re-entry bodies to radiation environments in underground tests and for this the UK was totally dependent on the US, since if the UK went elsewhere it would take a minimum of three years to learn how to do all of this.[197] It is clear that provision of test facilities was particularly important for the timetable of the work.[198] FCO Disarmament Department officials discussed this issue with ACSA (N) in April 1971 and learned that there were factors that inhibited the UK from playing a more positive role in the search for a CTBT. This was reflected in instructions sent to UK's disarmament delegation in Geneva (UKDis) on 21 April, which noted in particular that there was no support in London for any UK working papers on the threshold idea (a limit of the number or yield of tests) or on a complaints procedure for dealing with suspect tests.[199] The delegation of course would still have preferred to be seen to be taking a constructive part in general CTBT discussions.[200] Despite UKDis Geneva's wishes there were as seen from London important British defence interests, actual and potential, that inhibited the UK from playing too active a role on CTBT at that time. As Frank Panton ACSA (N) commented at the end of 1971, in the case of a CTBT it could be argued that it was better to travel hopefully than to arrive. Certainly, with the UK's current and future need for, and dependence on, US underground facilities, the UK had no scope for new initiatives in this area.[201] As for the UK's defence interest, all would depend on the timing of the defence programme in relation toward progress towards a CTBT.[202] Limitations on tests in the form of annual quotas, however, would be compatible with UK defence interests whereas a complete ban or moratorium would not be.[203] In fact, it was quite clear by the following year – early 1972 – that from the UK's point of view, whatever option chosen for the future of the deterrent, the UK could not possibly contemplate a test ban at any time during the next five years.[204] This proved prescient as the UK joined the US

[197] TNA DEFE 13/752, UK Strategic Nuclear Force – Short Term Working Party Report, 3 June 1971.

[198] TNA DEFE 13/752, Defence Secretary to Prime Minister, Improvements to Polaris System, 9 February 1972.

[199] TNA FCO 66/366, D.M. Summerhayes to W.N. Hillier-Fry, UKDis Geneva, 21 April 1971.

[200] TNA FCO 66/366, W.N. Hillier-Fry to D.M. Summerhayes, Disarmament Department, 17 May 1971.

[201] TNA FCO 66/366, Dr F.H. Panton ACSA (N) to Mr D.M. Summerhayes, Disarmament Department, 9 December 1971.

[202] TNA FCO 66/367, M.J. Llewellyn Smith to Mr Marshall and Mr Summerhayes, Comprehensive Test Ban, 11 February 1972 .

[203] TNA FCO 66/367, D.M. Summerhayes to Mr Rose, Comprehensive Test Ban, 11 February 1972.

[204] TNA FCO 66/367, C.M. Rose to Mr Summerhayes, Test-ban Treaty, 15 February 1972.

and USSR in Trilateral Test-ban negotiations in 1977.[205] If the Americans were to agree to a test ban or moratorium, with the implication that the UK could not use the NTS, the UK would be unable to maintain the effectiveness of its deterrent. It was therefore vitally important that the UK should do nothing to encourage or support a ban and moratorium.

Officials need not have worried. Hal Sonnenfeldt (National Security Council) and Phil Farley (ACDA) confirmed to Clive Rose (Assistant Under Secretary, FCO) that the US had no current intention of taking any CTBT or moratorium initiative.[206] However, that would only apply for the current Administration, they could not commit on what might happen after the election, so there was still some residual anxieties in London.[207] It was possible to conceive of a future Democratic Administration under, say, Edward Kennedy, cutting the ground from under UK feet.[208] What was now needed from the MOD was an indication of the possible time-scale and extent of the UK testing programme. Such information would enable the FCO to reassess its CTBT policy. However, it seems that the level of uncertainty over Polaris modernisation meant that there would be further decisions to come before the UK could make any confident assessment of what would be needed in the way of a test programme. On certain assumptions it was possible that the UK might even need to be able to test throughout the 1980s.[209] This meant that the UK should do nothing for the next few years on a test ban. However, given the UK's previous level of activity on this issue, it could not disengage altogether, since to do so would seriously undermine British credibility in Geneva in the Conference of the Committee on Disarmament. As Rose noted, it might be that the national interest would be best served by adopting a policy of tapering off, so that, at the end of the next few years, no one would expect the UK to resume its former position as a leading advocate of a CTBT. Over the road in the MOD, Frank Panton, ACSA (N)'s personal view was the UK defence needs were such that it would not be in British interests at that time to initiate or promote progress towards a CTBT.[210]

205 TNA FCO 46/1829, Negotiations for a Comprehensive Ban on Nuclear Tests: Part I Narrative, Percy Craddock to David Owen, 21 April 1978.

206 TNA FCO 66/367, Washington telegram no. 680 to Foreign Office, 24 February 1972.

207 TNA FCO 66/367, C.M. Rose to Mr Summerhayes, Comprehensive Test Ban, 28 February 1972.

208 TNA FCO 66/367, M.J. Llewellyn Smith to Mr Summerhayes, 7 March 1972.

209 TNA FCO 66/367, C.M. Rose to Mr Summerhayes, Comprehensive Test Ban, 16 March 1972.

210 TNA FCO 66/369, Dr F.H. Panton, MOD to C. Rose, FCO, 2 May 1972.

Non-nuclear Tests

As in the late 1950s and early 1960s, AWRE had to have recourse to explosive testing of implosion systems from time to time. These were normally performed in the explosive test chambers in the southeast corner of Aldermaston following closure of the Maralinga range in 1966. By mid 1972 and in light of Polaris modernisation requirements, AWRE were considering a proposal to develop a UK version of the US XW 58 warhead, but this revived the need for firings with natural uranium in which the explosive charge weight was greater than could be accommodated safely in the test chambers.[211] Full scale firings with uranium were needed to reduce the development risks associated with the current weapon development programme namely Super Antelope.

The obvious place for such test was Foulness where the necessary diagnostic instrumentation was available. Such tests were urgent and the MOD did not want to trouble the Americans by asking to conduct such trials at the NTS.[212] The FCO saw no grounds to object on either an arms control or legal basis.[213] There was no prospect of such event being caught up in CTBT negotiations. The departments agreed that the proposal should be put to the Official Committee on Nuclear Policy so that a detailed submission might be put to Ministers. However, FCO officials still worried about the presentational aspects of the AWRE's plans as they felt there was always a risk that criticism of UK policy on the test ban would escalate, especially in Parliament.[214] Public opposition to French atmospheric testing and HMG's reluctance to condemn it amplified these concerns. In the end AWRE were able to run the test firings in the explosive test chambers at Aldermaston. It seem that given the risks of embarrassment involved in open-air firings these tests were not to be held except in response to an actual and demonstrated defence need.[215]

Conclusions: A Static UK Nuclear Weapon Development Programme 1966–1973

The uncertainties surrounding decisions on maintaining the effectiveness of the deterrent – political, economic and technical – were such that no UK nuclear tests

---

[211] TNA FCO 66/371, E.F. Newley, Director AWRE to V.H.B. Macklen, DCA (PN) MOD, 5 June 1972.

[212] TNA FCO 66/371, C.M. Rose to Mr Summerhayes, Firing at Foulness of Assemblies containing natural uranium or depleted uranium, 20 June 1972.

[213] TNA FCO 66/371, N. Marshall to C. Rose, Firing at Foulness of uranium or depleted uranium, 27 June 1972.

[214] TNA FCO 66/371, D.M. Summerhayes to Mr Fenn, Science and Technology Department, 12 December 1972.

[215] FCO 66/371, draft letter to Dr F. Panton, MOD attached to submission from J. Taylor, STD to Mr Rose, 20 December 1972.

were conducted until clearer decisions were taken on the need for a new warhead design for an improved front-end of Polaris – Super Antelope/Chevaline. This opened the door again to meaningful exchanges and collaboration with the US. There were, however, no requirements for new tactical warheads at that time as fulfilment of the planned deployment of WE177As for the RAF and Royal Navy was still underway, and would not be complete for some years to come. And as we have seen the requirement for the WE177C did not result in any testing either. Fortunately for the UK there were not the same domestic and international pressures against testing that had existed in the 1950s, so it was easy to resume in 1974. That is not to say that there were no political problems as the Labour Cabinet voiced complaints about the UK resumption.[216] However, there were no CTBT negotiations in the offing or underway, and it was easy too to disengage quietly from active support for a test-ban treaty in Geneva without too much political cost. Nuclear defence requirements, as they had done between 1954 and 1958, now took precedence over a test ban.

[216] TNA CAB 128/54 CC (74) 21st Conclusions Minute 3, 27 June 1974; and TNA CAB 129/178, C (74) 85, Nuclear Testing, Note by the Prime Minister, 31 July 1974.

Chapter 12
Conclusions

We can see very clearly from the key years 1954 to 1958 that nuclear test-ban pressures were not only a major concern for Ministers and senior officials, but they also had direct bearing on the planning of the UK's own nuclear warhead test programme. Test-ban imperatives led directly to the rescheduling of one test series (Grapple X) and shaped the firing programme of another (Antler). Had there been no anxiety about international and domestic pressures for a test ban, it is hard to see why it was so urgent for the UK to test in the Pacific in November 1957. Indeed, the possibility of a test ban had led directly to AWRE having to test devices earlier than originally planned or deemed desirable in relation to the weapon programme.[1] Until the UK had put itself in a position where it could make its own one megaton – one ton thermonuclear weapons, it could not contemplate a test ban or indefinite moratorium. Moreover, the UK needed to be able to demonstrate that it could make a two-stage radiation implosion thermonuclear weapon if it were to convince the Americans to share their nuclear secrets. Nevertheless, the government of the day felt that it had to respond to international and domestic anxieties over radioactive fall-out, hence the 1956 interim proposals to limit the world annual average to 15 megatons of fission for the three weapon states until such times as a better system of limitation was devised.

Given the success at Grapple Z in August/September 1958 and the resumed collaboration with the Americans, the way was cleared for the UK to become an enthusiastic supporter of a test-ban treaty. Macmillan and Lloyd were certainly in this category; whether those responsible for the UK's nuclear weapons programme in the MOD, Ministry of Supply (replaced by the Ministry of Aviation) or AWRE shared this view is another matter. Certainly if left to their own devices, they would have probably preferred to continue to test as the contingency planning for Grapple M and Lighthouse series indicates. However, for a period of about two to three years AWRE was busy accommodating itself to the massive flow of US design information. If international politics had not intruded it seems clear that, purely scientifically, MOD and AWRE would have wished to continue testing even with the flow of information from the US. Even taking into account the contingency planning on the basis that there would be no test limitations, it is clear that there were research and developmental reasons for further testing.

Once the prospect for testing opened up again after the Soviet resumption in September 1961, MOD's nuclear scientists and AWRE were very quick to come

---

1 TNA AVIA 65/1248, Nuclear Tests Policy Committee (Brief for meeting on 18 November), E.S. Jackson, DGAW, Appendix A, 15 November 1957.

forward with a requirement for a new UK design that they wished to test as a matter of priority: a new implosion system known as Super Octopus. Work on this had evidently been in hand since the moratorium had started back in 1958. Experience was already indicating that the key to small and effective thermonuclear weapons was efficient and compact fission triggers (primaries), therefore success of the Super Octopus implosion system was a fundamental factor in future UK designs. Continued testing was also considered to be an essential route to maintaining the flow of US information and was one of the main reasons for AWRE's proposed research tests programme in the mid 1960s. Without such a continuing programme, there was a very real fear that the flow of US data would dry up, and indeed, it did for a period of years after Zuckerman had told the US that there were to be no new UK nuclear weapons after Polaris and the WE177A and B. Testing was seen then and throughout the late sixties as the route back to closer collaboration with the US. AWRE and the MOD also feared that without testing and a significant challenging research programme, it would be impossible to retain the best scientific personnel needed to sustain the then current capability at Aldermaston and to provide the skills required if a future government were to decide upon a replacement for Polaris. Despite Treasury depredations and the efforts of Zuckerman, AWRE and MOD managed to sustain a basic capacity and once the Polaris Improvement Programme began following fears of Soviet ABM developments and was formally endorsed in 1973 to become the Chevaline programme, the need for future testing and effects tests was accepted and agreed to by Ministers. The UK would continue to test periodically right up until 1991. A CTBT was finally concluded in 1996 and the UK was amongst the first states to ratify it.

Polaris modernisation to counter potential Soviet ABM capabilities meant that the UK needed access to materials, components and effects and warhead tests in the late 1960s and early 1970s. In so doing, Britain was once again inhibited when it came time to promote a CTBT given its acute dependence on the US for provisions of testing facilities and support. So we end as we began: nuclear defence requirements taking precedence over disarmament proposals. Although the UK lobbied hard in Washington against a test moratorium in 1957 and 1958, at least until such times as UK requirements as a nuclear power were met, there were plans to do so again in the early 1970s, but these do not appear to have been necessary. International pressures for a CTBT had diminished by the early 1970s, at least in contrast to the 1950s and early 1960s largely because of the PTBT and NPT. This enabled the UK's defence requirements for a new series of tests for the Polaris Improvement Programme to proceed without undue international or domestic opposition. The political costs of the UK's new reticent position were consequently much diminished, but the costs of dependence upon the US were all too evident.

Throughout this period, the UK made a significant contribution to the science of seismology. Establishment of the Blacknest site and building from scratch a seismic research programme from the early 1960s onwards, provided the bedrock of UK thinking on CTBT verification. Even though the Wilson government, advised by

Zuckerman, decided privately that it would not insist on on-site inspections as the final deterrent to cheating under a CTBT, scientific understanding of the problems of detection and identification meant that for the rest of this period the official UK position remained that some on-site inspections would always be needed. There would always be unidentified seismic events behind which clandestine nuclear tests might have been lurking and for which only an on-site inspection could have a chance of investigating satisfactorily.

Atmospheric and underground testing were fundamental to the UK's thermonuclear programme, without testing there would have been no prospect of the UK developing thermonuclear weapons and mastering radiation implosion and small efficient and economical primaries. Indeed without this capability, the UK would not have unlocked the door to US assistance in the shape of design information, materials (fissile and non-fissile) and non-nuclear weapon components. In the absence of testing, or where only very limited testing has taken place, the UK programme suggests that other powers would struggle to master thermonuclear weapon design principles.

We should not overlook the fact that the minor trials with radioactive materials that took place in Australia – tamper movement, single-point safety, initiator and safety tests – were also critically important to the development of effective and reliable weapons that could be manufactured, handled, transported and stored safely. This is perhaps one of the key lessons to emerge from the UK experience and may have implications for more contemporary concerns over nuclear proliferation. Even though other states might be less fastidious than the British were in this regard, no one can afford to be completely cavalier when it comes to the safety of nuclear weapons. In addition, implosion trials with non-nuclear components, vibration, temperature, humidity and drop tests of casings and components for instance were also indispensable to the acquisition of a reliable, effective and safe weapon. Surveillance programmes of weapons in storage were an important and inevitable part of maintenance too. Without such programmes confidence in weapons would rapidly fall. In this respect, for example, it appears that the Air Ministry caused AWRE some surprise in early 1959. It mentioned that one of the environmental conditions to be satisfied in any aircraft carried weapon was the ability to stand up to quite a lot of flying during what would otherwise have been shelf life. This appears to have been news to AWRE. By the late 1960s Red Beard was not robust enough to survive carriage and delivery at maximum speeds and minimum altitudes that were required by the RAF to ensure that defended targets could be attacked successfully. This deficiency was one of the main reasons for the new WE177 weapon. No government or armed force would want an unreliable weapon that was prone to accidental detonation or to shake itself to bits when carried on the wing or in the bomb-bay of an aircraft. There is more to be said about non-nuclear aspects of testing and its place in a nuclear weapons programme judging by the quantity of National Archive files on this subject.

The high watermark of the UK's efforts to secure a CTBT was undoubtedly during Macmillan's premiership. Macmillan's very strong personal commitment to this cause was evident in his unstinting efforts to cajole, persuade, and badger US Presidents and Khrushchev. Although the 1958–62 Conference came close to completing a CTBT in 1960 it foundered on the on the rocks of US USAEC/DOD/JCCAE obstructionism, Soviet bloody mindedness and Cold War realities, but Macmillan was at least instrumental in securing a second prize in the PTBT. However, American concerns over ABMs and the need for further testing to counter such developments, Soviet withdrawal of its original acceptance of some on-site inspections and the effectiveness of underground testing quickly resulted in superpower loss of interest in a CTBT. A clear lesson for arms control and disarmament is that success requires the sustained and patient diplomacy of political leaders to overcome bureaucratic obstacles and the reluctance of other states to engage.

The priority in the mid 1960s was the NPT. Macmillan won three prizes: the 1958 renewal of much needed nuclear cooperation with the US; the PTBT and the Polaris Sales Agreement, both in 1963. His close relationship with Eisenhower and then with Kennedy helped enormously, and he was lucky to be supported by someone of the calibre of Sir William Penney both on weapons issues and disarmament matters. Unlike Macmillan, Wilson was much less driven by a personal commitment to a test-ban treaty. Paradoxically the ending of atmospheric testing by the three original weapon states with the consequent move underground had removed the initial international pressures for a comprehensive test ban and the improvements in East-West relations after the Cuban missile crisis reduced the immediacy of the test-ban issue. Britain's weapons programme was of course reaching maturity at the same time. Post-1958 atmospheric and underground tests along with the assessments trials at Maralinga and flow of US design and engineering information had laid a firm basis for Britain's deterrent future. No sooner had the UK achieved this, with the scheduled entry into service of the Polaris force in 1968, than doubts about its long-term effectiveness opened up new requirements for nuclear testing.

Looking back from the vantage point of 2008, 50 years since the 1958 Mutual Defence Agreement entered into force, three retired Aldermaston scientists commented that throughout its strategic deterrent history the UK had been ever conscious that a ban on nuclear testing might be just around the corner.[2] UK policy on test bans and weapons development throughout this period was thus always primarily driven by the perceived need to acquire and retain the appurtenances of a credible nuclear weapons state which in turn drove a necessary minimum programme of nuclear warhead and weapons development: no real surprise there. Arms control played an important but rarely decisive part. Hopefully this book has chartered the chronology of these intertwined issues in greater detail and laid the basis for further archival research into these topics.

---

2 Clive Marsh, Peter Roberts and Ken Johnston, 'Nuclear Testing: a UK Perspective' in Mackby and Cornish, *US-UK Nuclear Cooperation*, p. 228.

Bibliography

Primary Sources: The National Archives, Kew

AB 16/1444 Initiator Trials Australian T.I.M, 1955.

AB 16/1541 H Bomb: political repercussions, 1954.

AB 16/1888 Grapple and Green Bamboo (Finance), 1956–58.

AB 16/2228 Search for non-nuclear weapons test site and acquisition of Orfordness, 1956–59.

AB 16/2371 Grapple X, Y etc Committee Papers and Minutes, 1957–59.

AB 16/2778 Orpheus Operations, 1959–60.

AB 16/2971 UK Experimental Work on Underground Explosions, 1959–60.

AB 16/3139 Seismology: Underground Bomb Detection, 1960–65.

AB 16/3140 Seismology: Underground Bomb Detection, 1960–66.

AB 16/3447 Expanded Seismic Programme (AWRE) Operation Seagull Phase I, 1961.

AB 16/3707 Geneva negotiations for nuclear tests treaty: correspondence, 1960–61.

AB 16/3708 Geneva negotiations for nuclear tests treaty: correspondence, 1960.

AB 16/3141 Nuclear Test-ban control; scientific and other representatives, 1959–63.

AB 16/3446 Expanded Seismic Programme (AWRE) Operation Seagull Phase 1, 1960–62.

AB 16/3447 Expanded seismic programme, Atomic Weapons Research Establishment, Aldermaston: operation SEAGULL phase 1, 1961.

AB 16/3666 Weapons Group Management Board papers, 1958–59.

AB 16/3680 The effect of AWRE explosions at Porton, 1961.

AB 16/3841 AWRE Site Requirements for Seismic Testing Stations, 1960–62.

AB 16/40522 Weapons Group Management Board papers, 1960.

AB 16/4186 Seismic Research, 1960–61.

AB 16/4190 MOA Research and Development contracts, 1962–63.

AB 16/4550 Chairman's classified correspondence, 1960–64.

AB 43/196 Seismic Research (including Working Group on Disarmament), 1962–64.

AB 48/234 Expanded Seismic Programme (AWRE) Operation Seagull Phase 2, 1962–65.

AB 48/296 Defence planning Self initiated R and D, 1962–65.

AB 48/306 Tendrac, 1962.

AB 48/546 Seismic Research (Including Working Group on Disarmament), 1965–67.

AB 48/643 Case for resumption of nuclear tests, 1 January 1960–31 December 1961.
AB 49/13 NRDC General Correspondence, 1959–66.
AB 49/14 British underground nuclear test: Operation Pampas, 1 March 1962, 1 January 1961 to 31 December 1962.
AB 49/15 British underground nuclear test: Cormorant; Nevada, USA, September 1964, 1 January to 32 December 1964.
AB 49/17 British underground nuclear test: Charcoal; Nevada, USA, September 1965, 1 January to 31 December 1965.
ADM 1/31023 The Nuclear Deterrent, 1961–62.
ADM 205/202 British Nuclear Deterrent Blue Streak, Skybolt-Polaris, 1958–61.
AIR 2/166 Air Force Board Memoranda Annex nos. 13 and 20, 1964.
AIR 2/167 Air Force Board Memoranda Annex to No. 14, 1965.
AIR 2/13680 Megaton Bomb (OR 1136), 1954–56.
AIR 2/13681 Megaton Bomb (OR 1136), 1957.
AIR 2/13682 Megaton Bomb OR 1136, 1957–58
AIR 2/13683 Tactical Atomic Bomb, OR 1127, 1957.
AIR 2/13745 Warhead for the medium range ballistic missile (OR 1139 and 11420), 1955–56.
AIR 2/13746 Warhead for the medium range ballistic missile (OR 1139 and 11420), 1956–60.
AIR 2/13753 Atomic Trials Policy, 1957–60.
AIR 2/13774 Red Beard, 1958–60.
AIR 2/13775 Red Beard, 1960–61.
AIR 2/13705 Yellow Sun: Policy, 1957–66.
AIR 2/13718 Violet Club – Policy, 1957–58.
AIR 2/13728 Red Beard Operational Requirement, 1955–59.
AIR 2/13733 Progress Reports on the development on new weapons to the Air Council, 1957–58.
AIR 2/13747 Warhead for propelled Inertia controlled ASM – OR 1141, 1954–56.
AIR 2/13748 Warhead for propelled Inertia controlled ASM – OR 1141, 1956–58.
AIR 2/13749 Warhead for propelled Inertia controlled ASM – OR 1141, 1958–59.
AIR 2/13753 Atomic Trials Policy, 1957–61.
AIR 2/14999 Atomic Weapons Trials – Operation ‘Lighthouse’, 1959–61.
AIR 2/16372 Blue Steel: carriage in V-bombers, 1962–63.
AIR 2/16576 Fire hazards involving nuclear weapons, 1963–65.
AIR 2/16577 Blue Steel: carriage in V-Bombers, 1963–65.
AIR 2/17065 Blue Steel: Controller of Aircraft (CA) release documents, 1964–65.
AIR 2/17371 British nuclear deterrent study group, 1959.

AIR 2/17372 Air Ministry Nuclear Weapons Health and Safety Committee: minutes,
1958–63.
AIR 2/18184 Operation Grapple, 1955–57.
AIR 2/18209 Nuclear Weapons policy, 1968–69.
AIR 2/18210 Nuclear Weapons policy, 1969–70.
AIR 2/18270 Future deterrent policy, 1960–63.
AIR 2/18608 Nuclear Weapons Transportation, 1970–71.
AIR 2/18902 Operation Grapple, 1958.
AIR 6/115 Air Council Memoranda, 1957.
AIR 6/116 Air Council Memoranda, 1958.
AIR 6/117 Air Council Memoranda, 1959.
AIR 6/130 Air Council Memoranda/Minutes, 1960–61
AIR 6/139 Air Council Memoranda, 1963.
AIR 6/142 Air Force Board Memoranda, 1965.
AIR 6/152 Air Council Memoranda Annexes to Nos 12 and 36, 1962.
AIR 6/152 Air Council memoranda/Minutes, 1963.
AIR 6/153 Air Council Memoranda/Minutes, 1964.
AIR 6/167 Air Force Board Memoranda Annex to No.14, 1965.
AIR 6/169 Air Force Board Annexes to Nos. 3 and 35, 1966.
AIR 6/171 Air Force Board Annexes to Nos. 7, 1967.
AIR 8/2393 Nassau Agreement – shape and size of RAF deterrent up to 1970s, 1963–64.
AIR 8/2467 Operation Grapple, 1957–59.
AIR 8/2468 Operation Grapple, 1955–57.
AIR 8/2469 Operation Grapple, 1955–56.
AIR 8/2785 Stockpiling policy and planning: including safety and reliability of UK nuclear warheads under a CTBT, 1972–80.
AIR 19/844 Blue Steel Development, 1957–61.
AIR 19/1014 Blue Steel Development, 1961–64.
AIR 20/1080 Blue Steel: Nuclear Warheads, 1959–63.
AIR 20/10088 Nuclear Tests Overseas, 1959–63.
AIR 29/12080 Nuclear Weapons Policy, 1967–69.
AVIA 65/760 Blue Steel warhead trials: requirements, 1957–62.
AVIA 65/774 Yellow Sun: Stage 1 introduction into service; A ARM 7 relevant papers, 1957–59.
AVIA 65/775 Nuclear Weapons: policy and programme, 1957–60.
AVIA 65/824 Operation Grapple Policy, 1955.
AVIA 65/825 Grapple Policy, 1956–58.
AVIA 65/826 Grapple Executive papers, 1955–57.
AVIA 65/827 Grapple Minutes of executive meetings, 1955–57.
AVIA 65/854 Operation 'Antler' Policy, 1955–57.
AVIA 65/857 Operation Antler – use of Balloons at AW Trials, 1956–61.
AVIA 65/859 Operation Grapple X, Y, Z, 1957–58.

AVIA 65/869 Future Atomic Weapons Trials, 1954/55 Policy (Maralinga), 1954–55.
AVIA 65/870 Future Atomic Weapons Trials, 1956/1959 Policy, 1954–59.
AVIA 65/877 Procedure: service orders for nuclear weapons setting up of NRDC and AWPC, 1958–61.
AVIA 65/878 Procedure: service orders for nuclear weapons setting up of NRDC and AWPC, 1958–61.
AVIA 65/879 Functions of the Directorate General of Atomic Weapons, 1953–61.
AVIA 65/906 Development of new SAGW system: nuclear warheads, 1955–58.
AVIA 65/1041 Development of atomic weapons: Blue Water, 1958–62.
AVIA 65/1047 Atomic weapons policy: ADMS, 1957–63.
AVIA 65/1050 A fractional yield nuclear weapon, 1959–62.
AVIA 65/1064 Inspection of atomic weapons, 1958–60.
AVIA 65/1109 Miscellaneous tests in Australia: Kittens and Tim series 1 and 2, 1954–55.
AVIA 65/1110 Maralinga trials 1957, 1956–57.
AVIA 65/1111 Maralinga tests minor trials, 1957–59.
AVIA 65/1114 Ministry of Defence – Defence Research Policy Committee – Atomic Energy Sub-committee Supporting Documents, 1956.
AVIA 65/1116 Ministry of Defence – Defence Research Policy Committee – Atomic Energy Sub-committee – agendas and minutes, 1957–60.
AVIA 65/1160 10,000 lb HE MC bomb: acceptance trials and acceptance for service use, 1951–60
AVIA 65/1161 10,000 lb HE MC bomb: development policy, 1954–58.
AVIA 65/1166 Information on nuclear weapons programme, 1955–60.
AVIA 65/1167 Ordnance board: participation in nuclear weapon development trials, 1956–60.
AVIA 65/1184 Operation Grapple: executive agenda and minutes, 1955–57
AVIA 65/1185 Operation Grapple: agenda and minutes, 1955–57
AVIA 65/1189 Operation Grapple: associated tests, 1956–57.
AVIA 65/1193 Warhead for a medium range missile: Air Staff requirement OR 1142 Orange Herald (DAW plans action), 1955–58.
AVIA 65/1195 Operation Antler: policy, 1956–57.
AVIA 65/1198 Operation Antler: associated tests, 1957.
AVIA 65/1201 Operation ‘Antler’ Firing Programme, 1957.
AVIA 65/1211 Operation Grapple: correspondence, 1955–57.
AVIA 65/1217 Future trials from 1956–1959: tentative programme, 1956–59.
AVIA 65/1218 Violet Club correspondence, 1957–59.
AVIA 65/1233 Operation Lighthouse, Lightex Executive: minutes and agenda on Director Atomic Warfare (DAW) Trials, 1958–59.
AVIA 65/1234 Operation Lighthouse, Lightex Executive: papers, 1957–59.
AVIA 65/1239 Atomic Weapons Trials Executive Sub-Committee agenda and minutes, 1957–66.

AVIA 65/1240 Operation Lighthouse, Lighthouse Executive; agenda and minutes,
1958–59.
AVIA 65/1241 Operation Lighthouse, Lighthouse Executive: papers, 1957–59.
AVIA 65/1242 Operation Lighthouse, Lighthouse Executive: correspondence, 1958.
AVIA 65/1243 Operation Grapple X-Y-Z Correspondence DGAW, 1957–59.
AVIA 65/1244 Operation Grapple X-Y-Z Correspondence DGAW, 1957–59.
AVIA 65/1248 Cabinet Nuclear Test Policy Committee D.G.A.W Correspondence, 1957.
AVIA 65/1249 Cabinet Nuclear Test Policy Committee D.G.A.W Correspondence, 1958.
AVIA 65/1255 Red Beard: servicing policy, 1955.
AVIA 65/1259 Ministry of Aviation: nuclear weapon safety policy, 1958–61.
AVIA 65/1353 Operations Grapple –X-Y-Z Air Ministry Requirements, 1958–60.
AVIA 65/1431 Operation Grapple-X-Y-Z Policy, 1957.
AVIA 65/1432 Operation Grapple X-Y-Z Policy, 1957.
AVIA 65/1440 Operation Grapple X-Y-Z, 1957–59.
AVIA 65/1441 Operation Grapple X-Y-Z Policy, 1958.
AVIA 65/1696 Nuclear Weapons acceptance trials policy, 1956–65.
AVIA 65/1699 Nuclear Weapon testing: safety policy, 1958–66.
AVIA 65/1771 Defence Committee on Nuclear Requirements, 1959–63.
AVIA 65/1783 Fuzing System for Nuclear Warhead, 1958–62.
AVIA 65/1784 Nuclear Weapons Safety Board: meetings, 1958–61.
AVIA 65/1792 History Atomic Weapon Production DAWP, 1963.
AVIA 65/1820 Operation Lighthouse: target response, 1958–61.
AVIA 65/1834 Maintenance of nuclear deterrent 1965–71, 1963.
AVIA 65/1836 Future UK Atomic Weapons Trials, 1961–66.
AVIA 65/1964 Operation Grapple X-Y-Z War office requirements, 1957–64.
CAB 21/3909 The Lord President's Committee on Nuclear Disarmament, 1959.
CAB 21 /4535 Nuclear Tests (General) Geneva Nuclear Test Conference, 1960.
CAB 21/4536 Nuclear Tests, 1961.
CAB 21/4755 Nuclear Tests, 1961–62.
CAB 21/4977 Nuclear Co-ordinating Committee: Christmas Island, 1962.
CAB 21/4979 Nuclear Deterrent Policy: Skybolt and Blue Steel, 1960–62.
CAB 21/5015 Ministerial Committee on Disarmament, 1962–64.
CAB 21/5323 Official committee on disarmament, 1960–65.
CAB 21/5346 Nuclear Requirements for Defence 1959–64 Composition and terms of Reference, 1959–64.
CAB 21/6037 Nuclear Tests (General), 1962–65.
CAB 21/6038 Nuclear Requirements for Defence Committee General, 1960–64.
CAB 21/6039 Nuclear Requirements for Defence Committee, 1965.
CAB 21/6040 Nuclear Co-ordinating Committee A: Nuclear Tests on Christmas Island, 1962–64.

CAB 21/6041 Nuclear Test Ban Treaty negotiations, Moscow, 1963.
CAB 124/84 C.P. (56) 285, Nuclear Weapon Tests Memorandum by the Secretary of State for Foreign Affairs and the Minister of Defence, 19 December 1956.
CAB 124/1969 Moscow talks on Nuclear Test-ban Agreement, 1963.
CAB 128/30 Part 1 C.M. (56) 16th Conclusions Item 12, 22 February 1956.
CAB 128/30 Part I C.M. (56) 39th Conclusions Item 3, Nuclear Weapons Tests in 1957, 5 June 1956.
CAB 128/30 Part II C.M. (56) 48th Conclusions Item 2, Nuclear Weapons Tests, 11 July 1956.
CAB 128/30 Part II C.M. (56) 49th Conclusions Item 8, Nuclear Weapons Tests, 12 July 1956.
CAB 128/26 Cabinet Conclusions C.M. (56) 51st Conclusions Item 6, Nuclear Weapons Tests, 20 July 1956.
CAB 128/030 Part II C.M. (56) 74th Conclusions Item 6, Nuclear Weapon Tests, 25 October 1956.
CAB 128/31 Part I CC 17 (57) Item 4, Nuclear Weapons Tests, 12 March 1957.
CAB 128/31 Part I CC 34 (57) Item 1, Nuclear Weapons Tests, 15 April 1957.
CAB 128/31 Part I CC 37 (57) Item 1, Nuclear Weapons Tests, 2 May 1957.
CAB 128/31 Part I CC 46 (57) Disarmament, 24 June 1957.
CAB 128/31 Part II C.C. (57) 47th Conclusions Disarmament, 24 June 1957.
CAB 128/31 Part II C.C. (57) 51st Conclusions Item 4, Disarmament, 11 July 1957.
CAB 128/31 Part 2 C.C. (57) 57th Conclusions Item 1, Nuclear Tests, 25 July 1957.
CAB 128/32 C.C. 28 (58) Item 1, Disarmament Nuclear Tests, 1 April 1958.
CAB 128/32 Part 2 CC (58) 80 Item 4, Disarmament, 11 November 1958.
CAB 128/34 C.C. 20 (60) Item 6, Geneva Conference on Nuclear Tests, 24 March 1960.
CAB 128/34 C.C. 28 (60) Item 2, Nuclear Tests, 28 April 1960.
CAB 128/35 C.C. 49 (61) Item 2, Disarmament Nuclear Tests, 5 September 1961.
CAB 128/36 C.C. 2 (62) Item 1, Nuclear Tests, 3 January 1962.
CAB 128/36 C.C. 6 (62) Item 2, Nuclear Tests, 18 January 1962.
CAB 128/36 C.C. 10 (62) Item 2, Nuclear Tests, 1 February 1962.
CAB 128/36 C.C. 18 (62) Item 3, Nuclear Tests, 1 March 1962.
CAB 128/36 C.C. 25 (62) Item 6, Disarmament Nuclear Tests, 30 March 1962.
CAB 128/36 C.C. 26 (62) Item 7, Disarmament, 5 April 1962.
CAB 128/36 C.C. 27 (62) Item 1, Disarmament, 9 April 1962.
CAB 128/36 C.C. 28 (62) Item 1, Foreign Affair, 17 April 1962.
CAB 128/36 C.C. (62) 57 Item 1, Foreign Affairs Nuclear Tests, 20 September 1962.
CAB 128/37 C.C. (63) 34th meeting, Item 3, Nuclear Test, 23 May 1963.
CAB 129/83 C.P. (56) 243 Nuclear Weapons Tests, Memorandum by the Secretary of State for Foreign Affairs and the Minister of Defence, 24 October 1956.

CAB 129/87 C (57) 143, 28 June 1957.
CAB 129/87 C (57) 146 Disarmament Memorandum by the Secretary of State for Foreign Affairs, 21 June 1957.
CAB 129/88 C. (57) 151 Disarmament Note by the Minister of Defence, 23 June 1957.
CAB 129/101 C. (60) 74 Nuclear Tests Memorandum by the Prime Minister, 27 April 1960.
CAB 129/107 C. (61) 179 Use of Christmas Island for United States Nuclear Tests, 10 November 1961.
CAB 129/178/10 Nuclear Testing. Note by the Prime Minister, 31 July 1974.
CAB 130/125 Nuclear Test Committee: Meeting 1, 1957.
CAB 130/127 Nuclear Tests: Meeting 1, 1957.
CAB 130/130 Working Group on Disarmament: Meetings 1–18, 1957.
CAB 130/147 Prime Minister's visit to Washington and Ottawa, Papers 1–27, June 1958. 1958 CAB 130/164 Lord President's Committee on Nuclear Disarmament: meeting 1; Lord President's Committee on Nuclear Disarmament: papers 1–5, 1959.
CAB 130/178 Nuclear tests and fall-out: meeting 1, 1961.
CAB 130/189 Facilities at Christmas Island Meetings 1–5, 1962–63.
CAB 130/544 Anglo/United States Nuclear Relations: Meeting 1, 1971.
CAB 131/1–28 Cabinet: Defence Committee: Minutes and Papers (DO, D and DC Series) 1946–63 (D.C.) (56) 6th Meeting 10 July 1956; D.C.(56) 11th Meeting item 2 Nuclear Weapons Tests; D (57) 14 27 July 1957 Fissile material For Nuclear Weapons; D (58)13 18 March 1958 Disarmament Note by the Minister of Defence; D (58) 5th Meeting 20 March 1958 Item 2 Disarmament; D (58) 15th Meeting 25 July 1958 Nuclear Weapons Programme; D (58) 16th Meeting 1 August 1958 Nuclear Weapons Programme; D (58) 21st Meeting 15 October Suspension of Nuclear Weapons Tests; D (58) 28th Meeting 26 November 1958 Item 2 Fissile Material.
CAB 134/1328 Ministerial Committee on Atomic Energy Confidential Annex to A. (57) 1st meeting, Minute 1, 24 July 1957.
CAB 134/1596 Ministerial Committee on Disarmament: Meeting 1; Papers 1 to 3, 1962.
CAB 134/2229 Ministerial Committee on Disarmament: Paper 1 (1963); Paper 1 (1964), 1963–64.
CAB 134/2235 Nuclear Co-ordinating Committee 'A': Meetings 1 to 3; Papers 1 to 8, 1961–62.
CAB 134/2236 Nuclear Co-ordinating Committee 'A': Meetings 1 to 17; Papers 1 to 43, 1962.
CAB 134/2237 Nuclear Co-ordinating Committee 'A': Meetings 1 to 3; Papers 1 to 13, 1963.
CAB 134/2238 Nuclear Requirements for Defence Committee (meetings and papers), 1960–61.

CAB 134/2239 Nuclear Requirements for Defence Committee Meetings 1 to 3; Papers 1 to 16, 1962.
CAB 134/2240 Nuclear Requirements for Defence Committee, 1963.
CAB 134/ 2241 Nuclear Requirements for Defence Committee: Meeting 1 (1964); Papers 1 to 11 (1964); Meetings 1 to 2 (1965); Papers 1 to 8 (1965); Meetings 1 to 2 (1966); Papers 1–7 (1966); Paper 1 (1967) 1964 January–May 1967.
CAB 134/2274 Cabinet Nuclear Tests Policy Committee, 1957–58.
CAB 134/2275 Cabinet Nuclear Tests Policy Committee, 1959.
CAB 134/2289 Official Committee on Disarmament: Meeting 1; Papers 1 to 7 February–November, 1958.
CAB 134/2290 Official Committee on Disarmament: Meetings 1–2 (1959); Papers 1 to 9 (1959); Meetings 1 to 2 (1960); Papers 1 to 19 (1960), 1959–60.
CAB 134/2291 Official Committee on Disarmament: Papers 1 to 8, 1961.
CAB 134/2292 Official Committee on Disarmament: Indexes 1 to 2; Papers 1 to 44, 1962.
CAB 134/2293 Official Committee on Disarmament: Papers 45 to 73, 1962.
CAB 134/2294 Official Committee on Disarmament: Papers 74 to 92, 1962.
CAB 134/2295 Official Committee on Disarmament: Indexes 1 to 2; Papers 1 to 19, 1963.
CAB 134/2296 Official Committee on Disarmament: Papers 20 to 29, 1963.
CAB 134/2297 Official Committee on Disarmament: Papers 1 to 15, 1964.
CAB 134/3120 Ministerial Committee on Nuclear Policy Meetings, 1966–67.
CAB 134/3121 Ministerial Committee on Nuclear Policy Meetings, 1968–70.
CAB 134/3123 Official Committee on Nuclear Policy 1969.
CAB 148/1 to 153 Cabinet Office: Defence and Overseas Policy Committees and Sub-committees: Minutes and Papers (DO, DOP, OPD and OPDO Series), 1964–76.
CAB 148/50 Defence and Overseas Policy (O) Committee Sub-committee on Disarmament, 1964.
CAB 148/55 Defence and Overseas Policy (O) Committee Defence Review Working Party, 1967.
CAB 148/60 Defence and Overseas Policy (O) Committee Defence review Working Party, 1967.
CAB 148/71 Defence and Overseas Policy (O) Committee Sub-committee on Disarmament, 1966.
CAB 148/80 Defence and Oversea sPolicy (Official) Committee, 1967.
CAB 148/81 Defence and Overseas Policy (Official) Committee, 1967.
CAB 148/106 Defence and Overseas Policy (O) Committee Sub-committee on Disarmament, 1970.
CAB 164/631 Seismic detection of nuclear explosions, 1966–69.
CAB 165/204 Oversea Defence Committee, 1965–67.
CAB 165/402 NRDC Composition and terms of reference, 1965–69.
CAB 165/600 Ministerial Committee on Nuclear Policy-Business, 1966–70.
CAB 165/601 Ministerial Committee on Nuclear Policy, 1966–70.

CAB 168/15 Comprehensive Nuclear Tests Ban Treaty, 1966–69.
CAB 168/27 Improvement to Polaris Weaponry, 1970.
CAB 168/174 Comprehensive test ban on nuclear underground testing: proposed tripartite scientific Talks, 1966–67.
CAB 178/2 Seismic Research – Detection and Identification of underground explosions, 1958–66.
CAB 195/20, Cabinet Secretaries' Notebooks; Nuclear Tests, January 1962
CO 1036/280 UK Nuclear Tests in Pacific (Operation Grapple), 1956–57.
CO 1036/281 UK Nuclear Tests in Pacific (Operation Grapple), 1957.
CO 1036/282 UK Nuclear Tests in Pacific (Operation Grapple), 1957.
CO 1036/283 UK Nuclear Tests in Pacific (Operation Grapple), 1957–58.
CO 1036/284 UK Nuclear Tests in Pacific (Operation Grapple), 1958.
DEFE 5/154 COS Committee Memoranda 278/64, The British Strategic Nuclear Capability, Note by the Secretary, 14 October 1964.
DEFE 5/164 Chiefs of Staff Committee Memoranda COS 210/65 Chiefs of Staff Committee, Future Nuclear Defence Policy, Note by the Secretary, 10 December 1965.
DEFE 7/763 Defence Research Policy Committee: Scientific Working Party of the Sub-Committee on Atomic Energy; papers, 1956–57.
DEFE 7/901 UK hydrogen bomb tests code name 'Grapple', 1956.
DEFE 7/902 UK hydrogen bomb tests code name 'Grapple', 1956–57.
DEFE 7/903 UK hydrogen bomb tests code name 'Grapple', 1957.
DEFE 7/904 UK hydrogen bomb tests code name 'Grapple', 1957.
DEFE 7/905 UK hydrogen bomb tests code name 'Grapple', 1957–59.
DEFE 7/911 Operation ANTLER: preparation for UK nuclear weapon trials at Maralinga 1957, 1956–57.
DEFE 7/916 Nuclear test series, code name BUFFALO, 1955–56.
DEFE 7/919 UK megaton weapon tests at Christmas Island, code name GRAPPLE X, 1956–57.
DEFE 7/920 UK megaton weapon tests at Christmas Island, code name GRAPPLE X, 1957.
DEFE 7/921 UK megaton weapon tests at Christmas Island, code name Grapple X, 1957–60.
DEFE 7/922 Nuclear Test Policy Committee, 1957–58.
DEFE 7/923 United Kingdom megaton weapon tests at Christmas Island: extension of 'X' series, code name Grapple Y, 1957–71.
DEFE 7/930 Disarmament: Suspension of Nuclear Tests, 1962.
DEFE 7/931 Disarmament: Suspension of Nuclear Tests, 1962.
DEFE 7/932 Minutes of meetings on disarmament: suspension of nuclear tests, 1962.
DEFE 7/933 Minutes of meetings on disarmament: suspension of nuclear tests, 1962.
DEFE 7/1519 Operation BUFFALO: nuclear test series, 1956–65.
DEFE 7 /1887 Sub-Committee on Strategic Aspects of Atomic Energy, 1950–62.

DEFE 7 /1888 Sub-Committee on Strategic Aspects of Atomic Energy, 1958–62.
DEFE 7/2217 Possible Future Nuclear Deterrent Weapon System, 1960–64.
DEFE 7/2289 Atomic Energy: limitations of tests, 1955–57.
DEFE 7/2294 MoD Nuclear Weapons Safety Committee: formation, 1958–60.
DEFE 7/2296 Underground and High Explosive Tests in the United Kingdom, 1959–62.
DEFE 7/2363 Nuclear weapon safety experiments at Maralinga (VIXEN A and B), 1961–67.
DEFE 7/2379 Nuclear Tests Policy Committee, 1958.
DEFE 7/2380 Nuclear Test Policy Committee, 1958–59.
DEFE 11/437 POLARIS: Policy, 1967.
DEFE 11/791 Nuclear – General 1974, 1977.
DEFE 13/150 Future megaton test programme, 1957–59.
DEFE 13/353 Regulation and limitation of tests of nuclear weapons, 1955–58.
DEFE 13/544 Nuclear weapons policy, 1967–68.
DEFE 13/547 POLARIS, 1967–68.
DEFE 13/548 POLARIS, 1968–69.
DEFE 13/700 POLARIS, 1966–67.
DEFE 13/748 Resumption of nuclear tests, 1970–74.
DEFE 13/770 POLARIS improvement, 1970.
DEFE 13/1012 Atomic Weapons Trials, 1955–57.
DEFE 13/1038 Nuclear Weapons Policy: POLARIS improvement policy, 1973.
DEFE 13/1050 POLARIS programme: background, management, discussions and annual reports, 1970–73.
DEFE 16/127 Operation Antler Theoretical predictions, 1958.
DEFE 16/201 Permanent atomic weapons proving ground, 1959.
DEFE 16/206 Permanent atomic weapons proving ground, 1957.
DEFE 16/257 Operation Antler trials: signals out, 1956–57.
DEFE 16/258 Operation Antler trials: signals in, 1957.
DEFE 16/259 Operation Antler trials: Antex Committee meetings; minutes, 1957.
DEFE 16/271 Operation Antler: technical aspects, 1957.
DEFE 16/272 Operation Antler: technical aspects, 1957
DEFE 16/297 Operation Antler: weapon assemblies, 1957.
DEFE 16/334 Atomic weapons trial and proving ground policy, 1957–66.
DEFE 16/337 Atomic trials policy, 1957.
DEFE 16/338 Atomic trials policy, 1957.
DEFE 16/339 Atomic trials policy, 1959.
DEFE 16/342 Operation Antler: policy, 1957.
DEFE 16/353 Operation Antler: policy, 1955–57.
DEFE 16/343 Maralinga project policy: report on alternative sites for weapon tests, 1957.
DEFE 16/361 Assessment trials, 1958–60.
DEFE 16/375 Operation Antler: policy, 1956–57.

DEFE 16/385 Operation Antler: Antex (57) minutes of meetings, 1957.
DEFE 16/391 United Kingdom atomic energy trials: Operation Antler, 1957.
DEFE 16/432 Proposal for survey of possible site for an underground atomic weapons test: allocation of plutonium for assessment tests, 1958–59.
DEFE 16/481 Atomic Weapons Test Committee: extracts from minutes and papers etc, 1955–57.
DEFE 16/488 Atomic Tests Maralinga: deferment of Op. Lighthouse, 1958–59.
DEFE 16/532 Atomic weapons tests: Board of Management correspondence, 1957.
DEFE 16/533 Atomic weapons tests: Board of Management correspondence, 1957.
DEFE 16/534 Atomic weapons tests: Board of Management correspondence, 1959.
DEFE 16/545 Atomic Weapons test: board of management: UK proposals for trials 1960 onwards: reduction in establishment, 1959.
DEFE 16/546 Operation Antler: proposal for trials and further minor trials, 1957.
DEFE 16/599 Overseas atomic weapon test programme 1956: details of weapon trials etc, 1955–59.
DEFE 16/603 Note on limitation of nuclear tests, megaton and kiloton trials and weapon trials in 1957.
DEFE 16/605 Maralinga Experimental Programme, nuclear warhead safety experiments and related papers, 1959–61.
DEFE 16/832 Minutes of 7th meeting of Board of Management Atomic Weapons Tests and Maralinga service establishment, 1956–60.
DEFE 16/863 letter from HJB Lintott to Pritchard about underground nuclear tests in Australia, 1958.
DEFE 16/864 Testing of missiles with nuclear warheads incorporating American design, 1959.
DEFE 16/893 Covering letter for 'General Report on Operation Antler' and DGAW comments, 1958.
DEFE 16/920 Board of Management Atomic Weapon Tests: minutes of 4th and 7th meetings and Maralinga staffing tables, 1957.
DEFE 19/1 Scientific aspects of detection of nuclear weapons explosion, 1959–61.
DEFE 19/5 General Nuclear Test Conference: papers on Nuclear Test Control systems (NTCS), 1958.
DEFE 19/12 UK Seismic Research Programme, 1960–65.
DEFE 19/18 Technical History of the Geneva Nuclear Tests Conference, 1957–62.
DEFE 19/45 Future Atomic Weapons programme, 1952–55.
DEFE 19/65 Nuclear tests: on site inspection of seismic events, 1959–62.
DEFE 19/87 Report by the BND Technical Sub-committee, 1960–62.
DEFE 19/92 Resumption of Nuclear Tests, 1961–62.

DEFE 19/96 Miscellaneous papers of Dr Robert Press on the 1966, Test Ban Treaty Talks, 1966.
DEFE 19/98 Setting up and correspondence on Enquiry into AWRE, 1968.
DEFE 19/103 Weapons Development Committee: Nuclear Sub-committee, 1964–66.
DEFE 19/115 Nuclear tests and ballistic missile defence systems: notes and correspondence, 1962.
DEFE 19/126 Comprehensive Test Ban Treaty, 1970–71.
DEFE 19/130 Visits of Chief Scientific Adviser to the USA to discuss topics relating to nuclear deterrents, 1971.
DEFE 19/167 Nuclear test ban monitoring: use of black boxes (data recorders) as automatic recording stations, 1962–64.
DEFE 19/170 Chevaline: Polaris re-entry body improvement programme, 1974–75.
DEFE 19/181 UK Underground nuclear tests 1978–79.
DEFE 19/182 Polaris improvement programme 1973–74.
DEFE 19/191 Ad Hoc Committee on Polaris Improvements, 1975.
DEFE 19/197 Correspondence between Sir William Cook (Deputy Chief Scientific Adviser (Projects)), Sir William Penney (Chairman of UK Atomic Energy Authority) and Sir Solly Zuckerman (Chief Scientist) on the future nuclear programme, 1966.
DEFE 19/207 Ad hoc committees on Polaris improvements, 1972–73.
DEFE 19/208 Ad hoc committees on Polaris improvements, 1974–75.
DEFE 19/240 Comprehensive Test-Ban Treaty negotiations, 1978.
DEFE 24/78 UK independent nuclear force, 1964–67.
DEFE 24/291 British Nuclear Test Programme, 1963–65.
DEFE 24/783 Strategic Arms Limitation Talks (SALT), 1973.
DEFE 24/895 Equipment – Ships Strategic Nuclear Deterrent, 1973–77.
DEFE 24/896 POLARIS Improvement Programme: policy options 1972.
DEFE 25/334 Strategic nuclear deterrent: Polaris Improvement Programme (PIP); Options, Phase 4, 1973.
DEFE 69/449 Size of POLARIS force, 1964.
DEFE 69/466 Way Ahead for POLARIS, 1971–72.
DEFE 69/501 Communications requirement for monitoring of the Nuclear Test Ban Treaty, 1965.
ES 1/388 Permanent Proving Ground Trials, 1957.
ES 1/389 Permanent Proving Ground Trials, 1958.
ES 1/390 Permanent Proving Ground Trials, 1959.
ES 1/492 Mosaic – Results Trails, 1957.
ES 1/578 Grapple X, Y, Z Trials Atomic Weapons Trials Executive, 1957–59.
ES 1/583 Operation Grapple Y Planning, 1957.
ES 1/584 Trial Y Papers Planning, 1958.
ES 1/585 Trial Y Papers Planning, 1958.
ES 1/586 Operation Grapple Y Planning, 1958.

ES 1/587 Operation Grapple Y Planning, 1958.
ES 1/588 Operation Grapple Y Planning, 1958.
ES 1/589 Operation Grapple Y Planning, 1958.
ES 1/590 Operation Grapple Y Planning, 1958.
ES 1/591 Operation Grapple Y Summary Statement of Numerical Results, 1958.
ES 1/592 Operation Grapple Scientific Plan, 1958.
ES 1/593 Operation Grapple Y – Interim Report, 1958.
ES 1/598 Grapple Z Correspondence, papers and telegrams, 1958.
ES 1/599 Grapple Z Correspondence, papers and telegrams, 1958.
ES 1/605 Minutes of AWRE meetings to discuss Ops Grapple Y and Z, 1958.
ES 1/606 Ops Grapple X and Z: planning meeting 1957–58.
ES 1/758 Antler – General Arrangements, 1957.
ES 1/897 Operation Grapple M, 1958–59.
ES 1/609 Signals to AWRE, 1958.
ES 1/612 Signals out from AWRE, 1958.
ES 1/621 Operation GRAPPLE X: signals to and from AWRE, 1957.
ES 1/642 GRAPPLE X, Y and Z: trials planning meetings, 1957–58.
ES 1/747 Atomic Weapons Trials Executive: minutes of meetings, 1957.
ES 1/756 Operation Antler: general arrangements, 1957.
ES 1/757 Operation Antler: general arrangements, 1957.
ES 1/758 Operation Antler: general arrangements, 1957.
ES 1/759 Operation Antler: general arrangements, 1957.
ES 1/760 Operation Antler: general arrangements, 1957.
ES 1/761 Operation Antler: general arrangements, 1957.
ES 1/762 Operation Antler: general arrangements, 1957.
ES 1/956 Minor Trials Maralinga Papers etc January–June, 1958.
ES 1/957 Minor Trials Maralinga Papers etc July–December, 1958.
ES 1/958 Minor Trials Maralinga Papers etc, 1959.
ES 1/959 Minor Trials Maralinga Papers etc, 1959.
ES 1/960 Minor Trials Maralinga Papers etc, 1959.
ES 1/961 Minor Trials Maralinga Papers etc, 1960.
ES 1/962 Minor Trials Maralinga Papers etc, 1961.
ES 1/1098 Operation Lighthouse, 1957–59.
ES 1/1137 Future Trials Investigation Committee, 1957.
ES 1/1138 Future Trial Investigation Committee, 1958.
ES 1/1139 Future Trial Investigation Committee, 1959.
ES 1/1140 Future Trial Investigation Committee, 1960.
ES 1/1349 Underground tests papers etc, 1956–59.
ES 1/1350 Underground tests papers etc, 1960.
ES 1/1351 Underground tests papers etc, 1961.
ES 1/1350 Project Orpheus – Trials Papers Rounds for Seismic Shots, 1960.
ES 1/1352 Project Orpheus Phase C: search for a suitable site, 1959.
ES 1/1353 Project Orpheus Possible sites in UK, 1959.
ES 1/1354 Project Orpheus Reports: phases 1 to 4, 1960.

ES 1/1355 Project Orpheus Reports: phases 5 to 8, 1961.
ES 1/1356 Project Orpheus Discussions in USA, 1960.
ES 1/1357 Provision of weapons for trials, 1954 to 1957.
ES 1/1503 UKAEA Weapons Group: Secretary's Correspondence, 1957–58.
ES 1/1655 Summary of UK nuclear weapons trials, 1965–66.
ES 4/399 Detection of Underground Nuclear Explosions, 1960.
ES 4/405 AWRE Report No. O.30/60 Control and Detection of Nuclear Explosions, 1960.
ES 4/879 Analysis of the amplitudes of teleseismic signals by Russians and French nuclear explosions, 1966.
ES 4/908 Some seismic results in the Long Shot explosion, 1966.
ES 4/1089 Comparison of the short period seismic noise at the four UKAEA type arrays and an estimate of their detection capabilities, 1969.
ES 4/1167 Verification of a comprehensive test ban treaty, 1969.
ES 5/285 AWRE Report No. T 11/60 Operation Orpheus Phase C, 1961.
ES 5/293 Operation Seagull: Phase 1 Operational Report, 1961.
ES 5/295 Operation MacSeagull, Operational Report, 1961.
ES 5/296 AWRE Report No. T/61 Operation Grapple Yield from Fireball Measurements Part 1 J.R. Swain, December, 1961.
ES 5/300 AWRE Report T 11/61 Project Orpheus Phase A, 1962.
ES 5/301 AWRE Report No. T 12/61 Project Orpheus Phase A, 1962.
ES 5/302 AWRE Report No. T1/62 Project Orpheus, January, 1962.
ES 5/303 AWRE Report No. T 2/62 Project Orpheus, February, 1962.
ES 5/304 AWRE Report No. T3/62 Operation Orpheus Phase A, March, 1962.
ES 5/305 AWRE Report No. T4/62 Operation Orpheus Phase A, January, 1962.
ES 5/306 AWRE Report No. T5/62 Project Orpheus, Phases A and C, Operational Report, February, 1962.
ES 5/313 AWRE Report No. T12/62 Operation Orpheus Phase A January, 1963.
ES 5/320 AWRE Report No. T3/63 Operation Seagull, Phase 2 March, 1963.
ES 5/329 AWRE report no. T 20/63 Operation Pampas January, 1964.
ES 12/271 Assessment Tests, Maralinga, 1959.
ES 12/272 Maralinga Experimental Programme (MEP), 1960.
ES 12/273 Maralinga Experimental Programme (MEP) Vixen B trial, 1960.
ES 12/276 Maralinga Experimental Programme (MEP) Vixen B3; facilities plan, 1962.
ES 12/521 Maralinga Experimental Programme (MEP) 1961: facilities plan for VIXEN A, VIXEN B1, TIMS and KITTENS, 1961.
ES 13/1 Report of the Working Party on Atomic Weapons Establishments, chaired by Lord Kings Norton: volumes 1 and 2 including correspondence about the possibility of an interview with Lord Kings Norton, 1968.
FCO 10/136 Nuclear Test Ban CTBT (General), 1967–68.
FCO 10/137 CTBT Seismology and Scientific Aspects, 1967–68.
FCO 10/138 CTBT Seismology and Scientific Aspects, 1967–68.

FCO 46/824 Strategic Arms Limitations Talks (SALT) between Soviet Union and USA, 1972.
FCO 46/829 Strategic Arms Limitations Talks (SALT) between Soviet Union and USA, 1972.
FCO 46/1002 Strategic Arms Limitation Talks (SALT) between the Soviet Union and USA, 1973.
FCO 66/84 United Kingdom comprehensive test-ban treaty policy, 1968.
FCO 66/85 United Kingdom comprehensive test-ban treaty policy, 1969.
FCO 66/86 Negotiations for a comprehensive nuclear test-ban treaty, 1971.
FCO 66/279 Negotiations for a comprehensive nuclear test-ban treaty, 1971.
FCO 66/366 Negotiations for a comprehensive nuclear test-ban treaty, 1972.
FCO 66/367 Negotiations for a comprehensive nuclear test-ban treaty, 1972.
FCO 66/368 Negotiations for a comprehensive nuclear test-ban treaty, 1972.
FCO 66/369 Negotiations for a comprehensive nuclear test-ban treaty, 1972.
FCO 66/371 Partial Test-ban Treaty: firing of assemblies containing uranium at Foulness, Essex, 1972.
FCO 66/475 Negotiations for a Comprehensive Test-ban (CTB) Treaty prohibiting testing of nuclear weapons in all environments, 1973.
FCO 66/476 Negotiations for a Comprehensive Test Ban (CTB) Treaty prohibiting testing of nuclear weapons in all environments, 1973.
FCO 66/477 Negotiations for a Comprehensive Test-Ban (CTB) Treaty prohibiting testing of nuclear weapons in all environments, 1973.
FCO 66/478 Negotiations for a Comprehensive Test-Ban (CTB) Treaty prohibiting testing of nuclear weapons in all environments, 1973.
FO 371/123121 UN Committee on Radiation Hazards 1956, Banning of nuclear test explosions: draft brief for Soviet leaders' visit, 1956.
FO 371/123125 UN Committee on Radiation Hazards 1956, Nuclear tests: draft statement for UK Foreign Secretary, 1956.
FO 371/123130 UN Committee on Radiation Hazards 1956, Nuclear weapon tests: limitation of nuclear tests to be studied and results reported to cabinet; UK prime minister's wish to make such a proposal to US government, 1956.
FO 371/123131 UN Committee on Radiation Hazards 1956, UK and US nuclear test explosions, September 1955–August 1956.
FO 371/123137 UK nuclear tests in Australia, 1956.
FO 371/123138 UK nuclear tests in Australia, 1956.
FO 371/123139 UK nuclear tests in Australia, 1956.
FO 371/123141 UK nuclear tests in Australia, 1956.
FO 371/123702 Limitations on nuclear tests: Anglo-US exchanges, 1956.
FO 371/123703 Limitations on nuclear tests: tabling of five-power resolution in UN, 1956.
FO 371/123705 Limitation of nuclear test explosions: UK policy, 1956.
FO 371/129244 UN Committee on Radiation Hazards 1956 Opinion of UK government that cessation of nuclear tests should follow prohibition of production of fissionable material for weapons purposes, 1957.

FO 371/123702 Limitations on nuclear tests: Anglo-US exchanges, 1956.
FO 371/123703 Limitations on nuclear tests: replies to UK parliamentary questions, 1956.
FO 371/123705 Limitation of nuclear test explosions: UK policy, 1956.
FO 371/129244 Nuclear Tests and Radiation Hazards, 1957.
FO 371/129266 Testing of nuclear weapons by UK, 1957.
FO 371/129279 Position on nuclear tests, 1957.
FO 371/129802 Limitation of nuclear tests in the near future: possible change in UK policy, 1957.
FO 371/129806 Nuclear tests: UK-US agreement, 1957.
FO 371/129806 Cessation or control of nuclear tests: comments on difficulties for UK parliament and UN disarmament sub-committee, 1957.
FO 371/129808 UN disarmament sub-committee: draft working paper on registration of nuclear tests, 1957.
FO 371/129811 Discussions on disarmament at UN, 1957.
FO 371/129812 Limitation of nuclear tests: US request confirmation of further consultation between UK and US governments before any public statement on new proposals, 1957.
FO 371/129813 Nuclear tests explosions paper produced by Foreign Office: request for comments from UK ministers, 1957.
FO 371/129815 UN disarmament sub-committee: UK memorandum on nuclear test explosions, 1957.
FO 371/129815 UN disarmament sub-committee: UK nuclear test explosions memorandum, 1957.
FO 371/129816 Discussions on disarmament at UN, 1957.
FO 371/129820 Limitation of nuclear test explosions: UK proposal, 1957.
FO 371/129820 Nuclear tests and production of weapons: record of discussion with US delegation, 1957.
FO 371/129825 UK policy on nuclear tests: Foreign Office suggestion, 1957.
FO 371/129826 Disarmament: memorandum by foreign secretary on US draft working paper, and suggestion for four power draft paper on nuclear tests, 1957.
FO 371/129827 UK redraft of US working paper on temporary suspension of nuclear testing, 1957.
FO 371/129833 Discussions on disarmament at UN, 1957.
FO 371/129836 US proposals for temporary suspension of nuclear tests: UK comments and suggested redrafts, 1957.
FO 371/129839 Revised US nuclear policy: letter from Harold Stassen to David Ormsby Gore, UK minister for foreign affairs, commenting on 'cut-off' arrangements and resumption of nuclear testing, 1957.
FO 371/129840 Discussions on disarmament at UN, 1957.
FO 371/132684 Control of nuclear tests, 1958.
FO 371/132685 Geneva conference: UN report on possibility of detecting violations of a possible agreement on the suspension of nuclear tests, 1958.

FO 371/132686 Control of suspension of nuclear tests (Extension of suspension beyond first year), 1958.
FO 371/132686 Control of suspension of nuclear tests (tactics and policy at forthcoming Geneva talks: differences in approach between UK and US), 1958.
FO 371/132687 Control of suspension of nuclear tests, (Brief for UK delegation to Geneva Conference), 1958.
FO 371/132689 Control of suspension of nuclear tests, 1958.
FO 371/132690 Control of suspension of nuclear tests, 1958.
FO 371/132691 Control of suspension of nuclear tests, 1958.
FO 371/132692 Control of suspension of nuclear tests (Geneva Conference; difficulties and policy concerning links under Article I), 1958.
FO 371/132692 Control of suspension of nuclear tests (Meeting between UK Secretary of State, Minster of Supply and Prime Minister), 1958.
FO 371/132693 Control of suspension of nuclear tests, 1958.
FO 371/132694 Control of suspension of nuclear tests, 1958.
FO 371/132695 Control of suspension of nuclear tests, 1958.
FO 371/132696 Control of suspension of nuclear tests, 1958.
FO 371/132697 Control of suspension of nuclear tests, 1958.
FO 371/132698 UK policy on nuclear weapons and test suspension, 1958.
FO 371/132699 UK policy on nuclear weapons and test suspension, 1958.
FO 371/135549 Unilateral Suspension of nuclear tests, 1958.
FO 371/135550 Unilateral Suspension of nuclear tests, 1958.
FO 371/135551 Unilateral Suspension of nuclear tests, 1958.
FO 371/135552 Unilateral Suspension of nuclear tests, 1958.
FO 371/135553 Unilateral Suspension of nuclear tests, 1958.
FO 371/135554 Unilateral Suspension of nuclear tests, 1958.
FO 371/135555 Technical conference on detection of nuclear tests, 1958.
FO 371/135556 Technical conference on detection of nuclear tests, 1958.
FO 371/135557 Technical conference on detection of nuclear tests, 1958.
FO 371/135558 Technical conference on detection of nuclear tests, 1958.
FO 371/135559 Technical conference on detection of nuclear tests, 1958.
FO 371/135560 Technical conference on detection of nuclear tests, 1958.
FO 371/135561 Technical conference on detection of nuclear tests, 1958.
FO 371/135562 Technical conference on detection of nuclear tests, 1958.
FO 371/135563 Technical conference on detection of nuclear tests, 1958.
FO 371/135564 Technical conference on detection of nuclear tests, 1958.
FO 371/135565 Technical conference on detection of nuclear tests, 1958.
FO 371/135566 Technical conference on detection of nuclear tests, 1958.
FO 371/135567 Technical conference on detection of nuclear tests, 1958.
FO 371/135568 Technical conference on detection of nuclear tests, 1958.
FO 371/135569 Technical conference on detection of nuclear tests, 1958.
FO 371/135570 Technical conference on detection of nuclear tests, 1958.
FO 371/135571 Technical conference on detection of nuclear tests, 1958.

FO 371/135572 UK Nuclear Weapons Tests, 1958.
FO 371/135573 UK Nuclear Weapons Tests, 1958.
FO 371/135574 UK Nuclear Weapons Tests, 1958.
FO 371/135575 UK Nuclear Weapons Tests, 1958.
FO 371/135576 UK Nuclear Weapons Tests 1958.
FO 371/140433 Suspension and control of nuclear tests, 1959.
FO 371/140434 Suspension and control of nuclear tests, 1959.
FO 371/140435 Suspension and control of nuclear tests, 1959.
FO 371/140436 Suspension and control of nuclear tests, 1959.
FO 371/140437 Suspension and control of nuclear tests, 1959.
FO 371/140438 Suspension and control of nuclear tests, 1959.
FO 371/140439 Suspension and control of nuclear tests, 1959.
FO 371/140440 Suspension and control of nuclear tests, 1959.
FO 371/140441 Suspension and control of nuclear tests, 1959.
FO 371/140442 Suspension and control of nuclear tests, 1959.
FO 371/140444 Conference on nuclear tests held in Geneva, 1959.
FO 371/140445 Conference on nuclear tests held in Geneva, 1959.
FO 371/140449 Conference on nuclear tests held in Geneva, 1959.
FO 371/140450 Conference on nuclear tests held in Geneva, 1959.
FO 371/140453 Conference on nuclear tests held in Geneva, 1959.
FO 371/140454 Conference on nuclear tests held in Geneva, 1959.
FO 371/140457 Conference on nuclear tests held in Geneva, 1959.
FO 371/140456 Conference on nuclear tests held in Geneva, 1959.
FO 371/140516 Briefs and records of high level discussions on nuclear weapons and tests, 1959.
FO 371/140517 Briefs and records of high level discussions on nuclear weapons and tests, 1959.
FO 371/140518 Briefs and records of high level discussions on nuclear weapons and tests, 1959.
FO 371/149269 Suspension and control of nuclear tests, 1960.
FO 371/149270 Suspension and control of nuclear tests, 1960.
FO 371/149271 Suspension and control of nuclear tests, 1960.
FO 317/149272 Geneva Conference on NTS (US discussion in Washington during Christmas Recess, threshold systems, possibility of US threshold proposal, threshold agreements and moratorium), 1960.
FO 371/149273 Geneva Conference on NTS (Correspondence between Mr Dillon, US representative and Secretary of State), 1960.
FO 371/149274 Geneva Conference on NTS (Moratorium: US and UK future tactics), 1960.
FO 371/149275 Geneva Conference on NTS (Correspondence between Mr Herter and Secretary of State), 1960.
FO 371/149276 Geneva Conference on NTS (Correspondence between Secretary of State and Prime Minister regarding Mr Herter's message), 1960.

FO 371/149277 Geneva Conference on NTS (Mr Herter's message of 2 February), 1960.
FO 371/149278 Geneva Conference on NTS (US consideration of a possible test ban), 1960.
FO 371/149282 Geneva Conference on NTS (request for instructions by Sir M. Wright regarding a moratorium), 1960.
FO 371/149283 Geneva Conference on NTS, 1960.
FO 371/149285 Geneva Conference on NTS (present problems and prospects: annex to a brief for UK delegation), 1960.
FO 371/149287 Geneva Conference on NTS (Anglo-American discussions on future progress of conference), 1960.
FO 371/149289 Geneva Conference on NTS (Conversation between Sir M. Wright and Wadsworth concerning possible winding up of conference), 1960.
FO 371/149291 Geneva Conference on NTS (US/UK information: Working Group 1960: UK brief), 1960.
FO 371/149292 Geneva Conference on NTS (Concern about future of conference), 1960.
FO 371/149299 Geneva Conference on NTS (PM request for statement on present position at Conference), 1960.
FO 371/149302 Geneva Conference on NTS (Correspondence between Makins and McCone, discussion with technical experts, Washington September 1960; definitions article), 1960.
FO 371/149303 Geneva Conference on NTS (brief for bilateral talks), 1960.
FO 371/149306 Geneva Conference on NTS (history of the Conference from its opening on 31 October), 1960.
FO 371/149309 Geneva Conference on NTS (Report Conference on Discontinuance of nuclear weapon tests; analysis of progress), 1960.
FO 371/149311 Geneva Conference on NTS (scientific definitions, definitions of a nuclear explosion), 1960.
FO 371/149312 Geneva Conference on NTS (Threshold and moratorium proposals: possibilities for including in the treaty), 1960.
FO 371/149313 Geneva Conference on NTS (Prime Minister's visit to Moscow in 1959: negotiations on quotas), 1960.
FO 371/149314 Geneva Conference on NTS (papers on thresholds), 1960.
FO 371/149317 Nuclear Tests Conference: possibility of a moratorium on small underground tests, 1960.
FO 371/149318 Nuclear Tests Conference and moratorium proposal: prime minister's visit to Washington 26 March, 1960.
FO 371/149319 Soviet moratorium proposal: Anglo-American agreement of text on nuclear testing, 1960.
FO 371/149323 Nuclear tests: memo by the Prime Minister, 1960.
FO 371/149325 Nuclear Test Conference, 1960.
FO 371/149326 Nuclear Tests Treaty: definitions article, 1960.
FO 371/149327 Nuclear Test Conference, 1960.

FO 371/149328 Nuclear Test Conference, 1960.
FO 371/149329 Nuclear Test Conference, 1960.
FO 371/149330 Nuclear Test Conference: difficulties over the research programme, 1960.
FO 371/149331 Nuclear Test Conference, 1960.
FO 371/149332 Nuclear Test Conference: co-ordinated programme of seismic research, 1960.
FO 371/149333 Nuclear Tests Treaty: US proposal on safeguards, 1960.
FO 371/149334 Nuclear Tests Conference: use of UK nuclear devices in the US as a result of the co-ordinated scientific research programme proposed in the Camp David Declaration 29 March, 1960.
FO 371/149335 Control Organisation for Discontinuance of Nuclear Tests, 1960.
FO 371/149336 Control Organisation for Discontinuance of Nuclear Tests, 1960.
FO 371/149337 Control Organisation for Discontinuance of Nuclear Tests, 1960.
FO 371/149338 Control Organisation for Discontinuance of Nuclear Tests, 1960.
FO 371/149339 Control Organisation for Discontinuance of Nuclear Tests, 1960.
FO 371/149340 Control Organisation for Discontinuance of Nuclear Tests, 1960.
FO 371/149341 Control Organisation for Discontinuance of Nuclear Tests, 1960.
FO 371/149342 Control Organisation for Discontinuance of Nuclear Tests, 1960.
FO 371/149343 Control Organisation for Discontinuance of Nuclear Tests, 1960.
FO 371/149344 Control Organisation for Discontinuance of Nuclear Tests, 1960.
FO 371/149345 Control Organisation for Discontinuance of Nuclear Tests, 1960.
FO 371/149346 Control Organisation for Discontinuance of Nuclear Tests, 1960.
FO 371/149347 Control Organisation for Discontinuance of Nuclear Tests, 1960.
FO 371/149348 Control Organisation for Discontinuance of Nuclear Tests, 1960.
FO 371/149349 Control Organisation for Discontinuance of Nuclear Tests, 1960.
FO 371/149472 Nuclear test control and safeguards: discussion between Sir Roger Makins of the UKAEA and Mr John McCone of the US Government, 1960.
FO 371/157076 Suspension and control of nuclear tests, 1961.
FO 371/157077 –157110 Conference on nuclear tests in Geneva, 1961.
FO 371/157111–157115 Establishment of control, monitoring and inspection procedures for nuclear tests under NTBT, 1961.
FO 371/157117–157127 Resumption of nuclear testing by Soviet Union, 1961.
FO 371/163091–163108 Geneva Disarmament Conference on Nuclear test ban treaty, 1962.
FO 371/163109–163112 Geneva Conference on nuclear tests: technical history of the Conference, 1958–62.
FO 371/163113–163119 Resumption of nuclear tests, 1962.
FO 371/163120 Controls inspection under NTBT, 1962.
FO 371/171117 – 171119 UK nuclear tests and disarmament policy in Geneva Disarmament Conference on NTBT, 1963.
FO 371 171214–171217 Negotiations between US and Soviet Union on NTBT: UK participation, 1963.

FO 371 171218–171228 Negotiations between US and Soviet Union on NTBT: UK participation, 1963.
FO 371 171229 Negotiations between US and Soviet Union on NTBT: UK participation, 1963.
FO 371 171230–171233 Verification of nuclear tests: seismic detection, 1963.
FO 371/171234 Technical Talks on verification between US, Soviet Union and UK, 1963.
FO 371/171235 – 171237 UK policy towards NTB negotiations; discussions with US and tripartite talks in Moscow between USSR, UK and US, 1963.
FO 371/171243 UK-US negotiations on nuclear tests, 1963.
FO 371/176435 Nuclear Test Ban, 1964.
FO 371/176436 Monitoring Threshold Test-ban treaty, 1964.
FO 371/176436/1 Monitoring Threshold Test-ban treaty, 1964.
FO 371/176437 CTBT, 1964.
FO 371/181013 Nuclear testing, 1965.
FO 371/181436 Nuclear testing and UK-US nuclear collaboration, 1965.
FO 371/181453 Comprehensive nuclear test-ban treaty, 1965.
FO 371/187021 Nuclear testing, 1966.
FO 371/187543 Comprehensive Test-ban Treaty, 1966.
FO 371/187544 Comprehensive Test-ban Treaty, 1966.
PREM 11/1676 Testing atomic weapons, 1954–56.
PREM 11/1677 Testing of H-Bombs, 1955–56.
PREM 11/2163 nuclear testing: protest from Japanese Prime Minister, 1957.
PREM 11/2166 Discussions on nuclear disarmament and limiting nuclear testing, 1957.
PREM 11/2168 Lord Cherwell's views on proceedings at UN Disarmament Committee on nuclear testing, 1957.
PREM 11/2554 UK-US Cooperation in Field of Atomic Energy, 1957–58.
PREM 11/2565 Proposed Conference on the Suspension of nuclear weapons tests, 1958.
PREM 11/2566 Disarmament and possible suspension of Nuclear Tests, 1958.
PREM 11/2567 Preliminary discussions on disarmament and suspension of nuclear tests, 1958.
PREM 11/2778 Efforts to counteract unfavourable public opinion towards nuclear tests, 1958–59.
PREM 11/2839 Approval for underground tests of conventional explosives in Cornwall sought from Prime Minister by Lord Plowden, 1959.
PREM 11/2844 Atomic weapons testing in Australia: Operation ANTLER, 1957–59.
PREM 11/2859 Further testing of nuclear weapons, November, 1957.
PREM 11/2860 Negotiations in Geneva on suspension of nuclear tests: Part 1, 1958–59.
PREM 11/2861 Negotiations in Geneva on suspension of nuclear tests: Part 2, 1959.

PREM 11/2862 Negotiations in Geneva on suspension of nuclear tests: Part 3, 1959.
PREM 11/2864 Meeting between Lord Plowden and President Eisenhower, 12 November 1959.
PREM 11/2865 Visit to UK by Special Adviser to US President on Atomic Matters, October and November 1958 and October 1959.
PREM 11/3157 Nuclear Weapons Safety Experiments, 1959–60.
PREM 11/3161 Continuation of negotiations in Geneva on suspension of nuclear tests, 1959–60.
PREM 11/3162 Negotiations in Geneva on disarmament and suspension of nuclear tests: part 4, 1960.
PREM 11/3218 Commonwealth Prime Ministers' Conference, March 1961: Discussions on disarmament and suspension of nuclear tests, 1960–61.
PREM 11/3246 Testing of Nuclear Weapons: US sought UK agreement to use of Christmas Island, 1961.
PREM 11/3582 Note by S. Zuckerman on military significance of Soviet nuclear tests, 1961.
PREM 11/3583 Testing of nuclear weapons: Operation Grapple Part 5, 1961.
PREM 11/3584 Seismic Research Programme, 1959–61.
PREM 11/3588 Negotiations on disarmament and suspension of nuclear tests: part 14, 1960.
PREM 11/3589 Negotiations on disarmament and suspension of nuclear tests: part 15, 1960.
PREM 11/3590 Negotiations on disarmament and suspension of nuclear tests: part 16, 1961.
PREM 11/3591 Negotiations on disarmament and suspension of nuclear tests: part 17, 1961.
PREM 11/3592 Negotiations on disarmament and suspension of nuclear tests: part 18, 1961.
PREM 11/3706 Underground testing of UK weapons in US, 1961–62.
PREM 11/3718 Use of Christmas Island Parts 1 and 2, 1962.
PREM 11/3719 Use of Christmas Island Part 3, 1962.
PREM 11/3720 Use of Christmas Island Part 4, 1962.
PREM 11/4041 Negotiations on disarmament part 19, 1961–62.
PREM 11/4042 Negotiations on disarmament part 20, 1962.
PREM 11/4043 Negotiations on disarmament part 21, 1962.
PREM 11/4044 Negotiations on disarmament part 22, 1962.
PREM 11/4045 Negotiations on disarmament part 23, 1962.
PREM 11/4046 Negotiations on disarmament part 24 A, 1962.
PREM 11/4047 Negotiations on disarmament part 24 B, 1962.
PREM 11/4554 Negotiations on disarmament Part 25, 1962–63.
PREM 11/4555 Negotiations on disarmament Part 26, 1963.
PREM 11/4556 Negotiations on disarmament Part 27, 1963.
PREM 11/4557 Negotiations on disarmament Part 28, 1963.

PREM 11/4558 Negotiations on disarmament Part 29, 1963.
PREM 11/4559 Negotiations on disarmament Part 30, 1963.
PREM 11/4560 Negotiations on disarmament Part 31, 1963.
PREM 11/4561 Negotiations on disarmament Part 32, 1963.
PREM 11/4717 Christmas Island: cessation of use as base for testing nuclear weapons; part 5, 1962–64.
PREM 11/4719 Development Trials of lay-down nuclear bomb for RN and RAF, 1964.
PREM 11/4720 Flight trials of atomic weapons, 1961–64.
PREM 11/4737 Discussion on intermediate range ballistic missiles (IRBMs) cancellation of Skybolt; Polaris sales, 1963–64.
PREM 11/4747 Christmas Island: cessation of use as base, Part 5, 1962–64.
PREM 11/5172 Nuclear weapons programme: requirements for fissile material, 1963–64.
PREM 11/5173 Underground testing of nuclear devices, 1964.
PREM 11/5181 Continuation of disarmament negotiations: test ban treaty; part 33, 1963–64.
PREM 11/5182 Negotiations on disarmament and suspension of nuclear tests: part 18, 1961.
PREM 13/123 Polaris Economy Test January–November, 1965.
PREM 13/228 Development of Poseidon missile in US: possible UK purchase under Polaris sales agreement January–March, 1965.
PREM 13/666 UK-US nuclear arrangements: record of meeting between Prime Minister, Foreign Secretary and Defence Secretary, 2 June 1965.
PREM 13/722 Future of atomic weapons proving ground at Maralinga, Australia January–September, 1966.
PREM 13/1316 UK nuclear policy January 1965–March 1967.
PREM 13/1317 UK nuclear policy: part 2 March–September 1967.
PREM 13/1388 Manufacture of WE177 and Assembly of Polaris Warheads, 1965–67.
PREM 13/2004 Supplies of tritium from US after 1968: production of lithium, tritium and uranium in UK; memorandum by Lord Penney, 1965–68.
PREM 13/2440 Disarmament Negotiations Possible International Agreement on Test Ban, 1966–67.
PREM 13/2493 UK nuclear policy in defence field: work of AWRE Aldermaston in connection with Polaris programme; part 3, 1967.
PREM 13/3053 UK nuclear policy: work of Atomic Weapons Research Establishment (AWRE) Aldermaston; part 4, 1969.
PREM 13/3126 Deployment and Movement of Nuclear Weapons, 1964–70.
PREM 13/3129 Understanding between UK and US on use of UK bases; exchange of nuclear information and cooperation on defence, 1964–68.
PREM 15/1359 Polaris improvement: consideration of various options; future of strategic nuclear deterrent, 1970–73.

PREM 15/1360 Polaris improvement: consideration of Poseidon/Super Antelope/ Stag options; future of strategic nuclear deterrent; part 2, 1973.

PREM 15/1361 Relations between UK and USA on nuclear matters, 1971–73.

PREM 15/1365 Strategic Arms Limitation Talks (SALT): Operation Hullabaloo; part 2, 1972–73.

PREM 15/2038 Polaris improvement: consideration of Super Antelope/Option M; future of UK strategic nuclear deterrent; part 3, 1973.

PREM 16/93 Soviet/US threshold treaty on underground nuclear weapons tests, 1974.

PREM 16/1181 The British Strategic Nuclear Deterrent: part 1, 1977.

T 225/1079 Nuclear Weapons Test Range, Christmas Island (Operation Grapple), 1958–59

T 225/1080 Nuclear Weapons Test Range, Christmas Island (Operation Grapple), 1959.

T 225/1200 Nuclear Weapons test Range Christmas Island (Operation Grapple), 1957–58.

T 225/1201 Nuclear weapons test range: Christmas Island; Operation 'Grapple', 1958.

T 225/2922 Future U.K. Nuclear weapons policy, 1967.

T 225/2923 Future U.K. Nuclear weapons policy, 1967.

T 225/2924 Future U.K. Nuclear weapons policy, 1967–68.

T 225/3217 Future nuclear weapon tests: policy, 1962.

T 225/3218 Future nuclear weapon tests: policy, 1963.

WO 32/17069 Research and Development of Nuclear Warheads, 1956–63.

WO 286/39 Violet Mist, 1957–61.

WO 291/1526 The Value of Live Indoctrination at a Nuclear Weapon Trial (Operation Buffalo), 1957.

Conference, UK, UN, US Government and UKAEA Publications

Conference on the Discontinuance of Nuclear Weapon Tests, Official Verbatim Records, Geneva, Meetings 1–340, 1958–61.

Cmd. 9770, *Report of the Proceedings of the Sub-committee of the United Nations Disarmament Committee held at Lancaster House, London 19 March–4 May 1956* (London: HMSO, 1956).

Cmnd. 333, *Report of the Proceedings of the Sub-committee of the United Nations Disarmament Commission held at Lancaster House, London, 18 March–6 September 1957* (London: HMSO, 1957).

Cmnd. 551, *Report of the Conference of Experts to Study the Methods of Detecting Violations of a possible Agreement on the Suspension of Nuclear Tests, Geneva, 1 July to 21 August* (London: HMSO, 1958).

Cmnd. 1937, Text of the Draft Treaty on the discontinuance of Nuclear Weapon tests tabled by the Delegations of the United Kingdom and the United States at the 292nd meeting of the Conference on the Discontinuance of Nuclear Weapon tests at Geneva on 18 April 1961 (London, HMSO, 1961).

Cmnd. 4141 *Further Documents on Disarmament: The Disarmament Negotiations July–December 1968* (London: HMSO, 1969).

Cmnd. 4725 *Further Documents on Disarmament: The Disarmament Negotiations 1970, August 1971* (London: HMSO, 1972).

Cmnd. 5344 *Further Documents on Disarmament The Disarmament Negotiations 1972, June 1973* (London: HMSO, 1973).

Department of State, *Documents on Disarmament 1945–1959 Volume II 1957–1959* (Washington, 1960).

Treasury Solicitor, *The Submission of the Government of the United Kingdom to the Royal Commission into British Nuclear Tests in Australia 1952* (London: Treasury Solicitor, 1985).

United Kingdom Atomic Energy Authority Sixth Annual Report 1959–60 (London: HMSO, 1960).

United Kingdom Atomic Energy Authority Seventh Annual Report 1960–61 (London: HMSO, 1960).

United Kingdom Atomic Energy Authority, Eighth Annual Report 1961–62 (London: HMSO, 1962).

United Kingdom Atomic Energy Authority, Ninth Annual Report 1962–63 (London: HMSO, 1962).

United Nations, *The United Nations and Disarmament 1945–1965* (New York, 1967).

United States Arms Control and Disarmament Agency, *Documents on Disarmament 1962 Volume I January–June* (Washington, 1963).

United States Arms Control and Disarmament Agency, *International Negotiations on Ending Nuclear Weapon Tests September 1961–September 1962* (Washington, 1962).

Secondary Sources

Arnold, Lorna, *A Very Special Relationship, British Atomic Weapon Trials in Australia* (London, HMSO, 1987).

—, *Britain and the H-Bomb* (Basingstoke: Palgrave, 2001).

—, and Smith, Mark, *Britain, Australia and the Bomb, The Nuclear Tests and Their Aftermath* (Basingstoke: Palgrave, 2006).

Atomic Weapons Research Establishment, *The Detection and Recognition of Underground Explosions, A Special Report of the United Kingdom Atomic Energy Authority Appendices* (London: United Kingdom Atomic Energy Authority, 1965).

Bates, Charles, Gaskell, Thomas and Rice Robert, *Geophysics in the Affairs of Man, A Personalized History of Exploration Geophysics and Its Allied Sciences of Seismology and Oceanography* (Oxford: Pergamon Press, 1982).

Baylis, John, Ambiguity *and Deterrence: British Nuclear Strategy 1945–1964 (Nuclear History Program)* (Oxford; Clarendon, 1995).

Carter, April, *Success and Failure in Arms Control Negotiations* (Oxford: OUP, 1989).

Clark, Ian, *Nuclear Diplomacy and the Special Relationship Britain's Deterrent and America 1957–1962* (Oxford: Clarendon Press, 1994).

—, Wheeler, Nicholas S., *The British Origins of Nuclear Strategy 1945–1955* (Oxford: Clarendon, 1989).

Cochran, Thomas B., Arkin, William M., Norris, Robert S., Hoenig, Milton M., *Nuclear Weapons Databook Volume II, US Nuclear Warhead Production* (Cambridge, Mass.: Ballinger Publishing Company, 1987).

—, —, — and Sands, Jeffrey I., *Nuclear Weapons Databook Volume IV, Soviet Nuclear Weapons* (New York: Harper and Row, 1989).

Dokos, Thanos P., *Negotiations for a CTBT 1958–1994* (Lanham University Press of America, 1995).

Douglas, Alan, Recognising 'UGTs', *Discovery. The Science Journal of AWE*, Issue 1 March 2000.

Fetter, Steve, *Toward a Comprehensive Test Ban* (Cambridge Mass.: Ballinger Publishing Company, 1988).

Freeman, John, *Britain's Nuclear Arms Control Policy in the Context of Anglo-American Relations, 1957–68* (London: Macmillan, 1986).

Goldblat, Jozef and Cox, David eds, *Nuclear Weapon Tests, Prohibition of Limitation* (Oxford: Oxford University Press, 1988).

Goodman, Michael, *Spying on the Nuclear Bear, Anglo-American Intelligence and the Soviet Bomb* (Stanford: Stanford University Press, 2007).

Haggas, John, *Christmas Island, The Wrong Place at the Wrong Time* (Upfront Publishing, Leicestershire, 2002).

Hansen, Chuck, *U.S. Nuclear Weapons, The Secret History* (New York: Aerofax, Orion Books, 1988).

Hawkings, David J., *Keeping the Peace, The Aldermaston Story* (London: Leo Cooper AWE, 2000).

Hennesy, Peter, *Cabinets and The Bomb* (Oxford: OUP, 2007).

Holloway, David, *Stalin and the Bomb* (Yale: Yale University Press, 1994).

Hubbard, Kenneth and Simmons, Michael, *Operation Grapple Testing Britain's First H-Bomb* (London: Ian Allan, 1985).

Mackby, Jenifer and Cornish, Paul, *US-UK Nuclear Cooperation After 50 Years* (London: CSIS, Chatham House, 2008).

Macmillan, Harold, *Riding the Storm 1956–1959* (London: Macmillan, 1971).

—, At *the End of the Day 1961–1963* (London: Macmillan, 1973).

Mark, David, *The Test Ban Treaty: Neither Success nor Failure, Six years of Negotiations for a Ban on Nuclear Weapon Tests as a Case Study in East-West Relations* (Harvard: Center for International Affairs, Harvard University, 1966).

Marshall, Peter, 'UK Contribution to Regional Seismic Discrimination', in AWE Report 632/01, David Bowers, *Test-ban treaty verification research: Seismic signals recorded at regional distances* (Aldermaston: AWE, 2001).

Melissen, Jan, *The Struggle for Nuclear Partnership. Britain, the United States and the Making of an Ambiguous Alliance 1952–1959* (Groningen: Styx Publications, 1993).

Moore, Richard, *The Royal Navy and Nuclear Weapons* (London: Routledge, 2001).

—, *Nuclear Illusion, Nuclear Reality Britain, the United States and Nuclear Weapons, 1958–64* (Basingstoke: Palgrave Macmillan, 2010).

Navias, Martin S., *Nuclear Weapons and British Strategic Planning* (Oxford: Clarendon Press, 1991).

Norris, Robert, Burrows, Andrew S., Fieldhouse, Richard W., *Nuclear Weapons Databook Volume V, British, French and Chinese Nuclear Weapons* (Oxford: Westview Press, 1994).

Oliver, Kendrick, *Kennedy, Macmillan and the Nuclear Test-Ban Debate*, 1961–63 Basingstoke: Macmillan Press Ltd, 1997).

Oulton, Air Vice Marshall Wilfrid E., *Christmas Island Cracker* (London: Thomas Harmsworth Publishing, 1987).

Pierre, Andrew, *Nuclear Politics, The British Experience with an Independent Strategic Force 1939–1970* (London: Oxford University Press, 1972).

Romney, Carl, *Detecting the Bomb The Role of Seismology in the Cold War* (Washington: New Academia Publishing, 2009).

Schrafstetter, Susanna and Twigge, Stephen, *Avoiding Armageddon. Europe, the United States, and the Struggle for Nuclear Nonproliferation, 1945–1970* (London: Praeger, 2004).

Seaborg, Glenn T. with Loeb, Benjamin S., *Stemming the Tide, Arms Control in the Johnson Years* (Lexington: Lexington Books, 1987).

Simpson, John, *The Independent Nuclear State, Britain, the United States and the Military Atom* (Basingstoke: Macmillan, 1984 and 1986).

Stocker, Jeremy, *Britain and Ballistic Missile Defence 1942–2002* (London: Frank Cass, 2004).

Wheeler, Nick, *The British Origins of Nuclear Strategy 1945–1955* (Oxford: OUP, 1989).

Wright, Sir Michael, *Disarm and Verify, An Explanation of the Central Difficulties and of National Policies* (London: Chatto and Windus, 1964).

Wynn, Humphrey, *RAF Nuclear Deterrent Forces* (London: HMSO, 1996).

Zuckerman, Solly, *Monkeys, Men and Missiles: An Autobiography 1946–1988* (London: Collins, 1988).

Index

For Product Safety Concerns and Information please contact our EU representative GPSR@taylorandfrancis.com
Taylor & Francis Verlag GmbH, Kaufingerstraße 24, 80331 München, Germany

www.ingramcontent.com/pod-product-compliance
Ingram Content Group UK Ltd.
Pitfield, Milton Keynes, MK11 3LW, UK
UKHW020023200926
475702UK00014B/2029

* 9 7 8 1 1 3 8 3 8 3 9 7 5 *